Henry Byron Spotton

The commonly occurring wild plants of Canada

A flora for the use of beginners

Henry Byron Spotton

The commonly occurring wild plants of Canada
A flora for the use of beginners

ISBN/EAN: 9783337272036

Printed in Europe, USA, Canada, Australia, Japan

Cover: Foto ©berggeist007 / pixelio.de

More available books at **www.hansebooks.com**

W. J. Gage & Co.'s Educational Series.

THE COMMONLY OCCURRING WILD PLANTS OF CANADA.

A FLORA FOR THE USE OF BEGINNERS

BY

H. B. SPOTTON, M.A., F.L.S.,

PRIN. HARBORD STREET COLLEGIATE INSTITUTE, TORONTO.

REVISED EDITION.

W. J. GAGE & COMPANY, Limited

TORONTO.

CONTENTS.

Preface to the First Edition	v
Preface to the Third Edition	vii
Abbreviations of Names of Authors	viii
How to Use the Key and the Flora	ix
Key to the Orders	xii
Flora:—	
Flowering or Phanerogamous Plants	1
Dicotyledons	1
Angiosperms	1
Polypetalous Division	1
Gamopetalous "	95
Apetalous "	82
Gymnosperms	214
Monocotyledons	217
Spadiceous Division	217
Petaloideous "	224
Glumaceous "	248
Flowerless or Cryptogamous Plants	252
Glossary	272
List of Common Cultivated Plants	280
Index	291

PREFACE TO THE FIRST EDITION.

A few words will not be out of place by way of preface to the List of Common Canadian Plants contained in the following pages. It will be observed that the List is confined to wild plants, the exclusion of cultivated Species having been determined on, partly because of the difficulty of knowing where to stop when an enumeration of them has once been entered upon, and partly because it was thought that, on the whole, more important results would be attained by directing attention exclusively to the denizens of our own woods and fields. What is really desired is, to create among our young people an interest in the Botany of Canada, and it seems not unreasonable to hope that this end may be better attained by placing within their reach some such handy volume as the present, dealing only with such plants as grow spontaneously within our limits.

The great majority of the plants described have been personally examined, and their occurrence verified, by the writer, his observations having been directed to what may fairly be regarded as representative districts of the older Provinces, but special acknowledgments are also due to Prof. Macoun, of the Geological Survey, for the free use of his valuable notes, and other friendly assistance.

Whilst diligence has been exercised that no commonly occurring plant should be omitted, yet it can hardly be that such omissions do not occur, and the writer will be most grateful to any observers into whose hands the List may come, if they will kindly draw his attention to any such defects, so that they may be remedied in subsequent editions.

The Classification and Nomenclature adopted are very nearly those of the Sixth Edition of Dr. Gray's Manual of the Botany of the Northern United States, and the writer most gratefully acknowledges the great assistance he has received from the admirable descriptions in that work.

Except in a very general way, no attempt has been made to define the limits of the range of the various Species, as observations tend to show that the range, in many cases, is undergoing constant alteration from various causes. When, however, a Species has appeared to be confined to a particular locality, mention has been made of that fact, but, as a rule Species known to be of rare occurrence have been excluded

Characters considered to be of special importance in the determination of the various Species have been emphasized by the use of italics, and where the Species of a Genus, or the Genera of an Order, are numerous, a system of grouping according to some prominent character has been adopted, so as to reduce the labour of determination as much as possible.

To assist the non-classical student, names which might be mispronounced have been divided and accentuated, the division having no reference whatever to the etymology of the words, but being simply based upon their sound when properly pronounced.

It need hardly be added that the writer's ELEMENTS OF STRUCTURAL BOTANY is designed to be the constant companion of the present Flora, in the hands of the young student, for the explanation of such technicalities as he may not have previously mastered.

BARRIE, November, 1883

PREFACE TO THE THIRD EDITION.

The greatly increased interest in Botany as a subject of study in the schools of the Dominion has necessitated a revision and enlargement of the List of Common Canadian Wild Plants. The following pages now include most of the wild plants of Manitoba and the prairie region generally, as well as those of the older Provinces. Plants known to be peculiar to the west are indicated by the addition of the letters "N.W." to the descriptions, while the note "Atl. Prov." sufficiently marks those peculiar to the east. No attempt has been made to enter the plants of the Pacific coast.

In accordance with suggestions from several teachers of Botany a short list of the commoner cultivated plants has been added, which it is hoped will be found useful.

In connection with this revision, grateful acknowledgments are due to many friends for valuable assistance, but more particularly to Mr. Wm. Scott, B.A., Vice-Principal of the Toronto Normal School, who has placed his splendid herbarium at the writer's disposal and has revised all the proofs; A. H. MacKay, LL.D., Chief Superintendent of Education for Nova Scotia, whose extensive acquaintance with the flora of the Atlantic Provinces has been of the greatest service; Professor Brittain of the Provincial Normal School at Fredericton, who has described most of the New Brunswick plants; Professor Bryce of Winnipeg; and last, but not least, Mr. Jno. Dearness of London, whose accuracy in determination and diligence as a collector in Western Ontario are well known.

Toronto, June, 1897.

PRINCIPAL ABBREVIATIONS OF NAMES OF AUTHORS CITED IN THE FLORA.

Adans	for	Adanson.	Less.	" Lessing.
Ait.	"	Aiton.	L'Her.	" L'Heritier.
Anders.	"	Anderson.	Lehm.	" Lehmann.
Bart.	"	Barton.	Lindl.	" Lindley.
Beauv.	"	Palisot de Beauvois.	Loisel.	" Loiseleur - Deslongchamps.
Benth.	"	Bentham.		
Bernh.	"	Bernhardi.	Marsh.	" Marshall.
Bigel.	"	Bigelow.	Medic.	" Medicus.
Boiss.	"	Boissier.	Michx.	" Michaux.
Borkh.	"	Borkhausen.	Mill.	" Miller.
Britt.	"	Britton.	Muhl.	" Muhlenberg.
Cass.	"	Cassini.	Murr.	" Murray.
Carr.	"	Carrière.	Nees.	" Nees von Esenbeck.
Cav.	"	Cavanilles.	Nutt.	" Nuttall.
Cham.	"	Chamisso.	Pall.	" Pallas.
Coult.	"	Coulter.	Pers.	" Persoon.
Darlingt.	"	Darlington.	Poir.	" Poiret.
DC.	"	De Candolle.	R. Br.	" Robert Brown.
A. DC.	"	Alphonse De Candolle.	Raf.	" Rafinesque.
Desf.	"	Desfontaines.	Rich.	" Richard.
Dill.	"	Dillenius.	Richards.	" Richardson.
Dougl.	"	Douglas.	Rostk.	" Rostkovius.
Eat.	"	Eaton.	Roem.	" Roemer.
Ehrh.	"	Ehrhart.	Rottb.	" Rottboell
Ell.	"	Elliott.	Salisb.	" Salisbury.
Endl.	"	Endlicher.	Sav.	" Savi.
Engelm.	"	Engelmann.	Schlecht.	" Schlechtendal.
Gært.	"	Gærtner.	Schrad.	" Schrader.
Ging.	"	Gingins de Lassaraz.	Schreb.	" Schreber.
Gr	"	Gray.	Schum.	" Schumacher.
Griseb.	"	Grisebach.	Schwein.	" Schweinitz.
Gronov.	"	Gronovius.	Scop.	" Scopoli.
Haussk.	"	Haussknecht.	Sibth.	" Sibthorp.
Haw.	"	Haworth.	Spreng.	" Sprengel.
Hoffm.	"	Hoffmann.	Torr.	" Torrey.
Holl.	"	Holleck.	Tourn.	" Tournefort.
Hook.	"	W. J. Hooker.	Tratt.	" Trattenick.
Hornem.	"	Horneman.	Tuckerm.	" Tuckerman.
H. B. K.	"	Humboldt, Bonpland, and Kunth.	Vaill.	" Vaillant.
			Vent.	" Ventenat.
Huds.	"	Hudson.	Vill.	" Villars.
Jacq.	"	Jacquin.	Wahl.	" Wahlenberg.
Juss.	"	Jussieu.	Walt.	" Walter.
L	"	Linnæus.	Wangh.	" Wangenheim.
Lam.	"	Lamarck.	Willd.	" Willdenow.

HOW TO USE THE KEY AND THE FLORA.

Assuming that the student has carefully read the Introductory part of this work, and is familar with the ordinary botanical terms, and the chief variations in plant structure as there set forth, it should, with the aid of the accompanying Key, be a very simple task to refer to its proper Family any Canadian wild plant of common occurrence. To illustrate the method of using this Key, let us suppose that specimens of the following plants have been gathered, and that it is desired to ascertain their botanical names, that is, the name of the Genus and the Species of each :—Red Clover, Strawberry, Blue Flag and Cat-tail Flag.

All of these produce flowers of some kind, and must therefore be looked for under the head of FLOWERING, OR PHANEROGAMOUS, PLANTS.

With the specimen of Red Clover in hand, and the book open at page xii., we find that we have first to determine whether our plant is Dicotyledonous or not. The veining of the leaves suggests that it is so; and this impression is confirmed by the fact that the parts of the flower are in fives. Then, is the plant an ANGIOSPERM? As the seed will be found enclosed in an ovary, we answer—Yes. Has the plant both calyx and corolla? Yes. Are the parts of the corolla separate? Here a little doubt may arise; but suppose we answer —Yes. Then our plant will be found somewhere in the POLYPETALOUS DIVISION. Proceeding with the enquiries suggested under this heading:—Are the stamens more than twice as many as the petals? We find that they are not.

Turn, then, to the heading marked B, page xv, "**Stamens not more than twice as many as the petals.**" Under this we find two subordinate headings, designated by asterisks * and **. The first of these is not applicable to our plant. Under the second, marked thus **, we find two minor headings, designated by daggers, + and + +. The first of these, "*Corolla irregular,*" is clearly the one we want. We have now, therefore, five families to select from. We cannot choose any one of the first four, because our plant has ten stamens, but the characters of the fifth are precisely the characters exhibited by Clover. Our Clover, therefore, belongs to the Order LEGUMINOSÆ. Turning to page 50, and running through the "Synopsis of the Genera" as there given, we observe that No. 2, TRIFOLIUM, is the only Genus in which the flowers are in *heads*. Clover answers the description in the other respects also—viz.: "leaves of three leaflets," and "stamens diadelphous." The only question then remaining is, which Species of TRIFOLIUM have we in hand? Turning to page 52, we find we have eight Species to choose from. No. 2, TRIFOLIUM *pratense*, is the only one of them with *purplish flowers*. TRIFOLIUM *pratense* must, consequently, be the botanical name we are looking for.

Possibly the observer may decide that the parts of the corolla are not separate from each other, because in some instances it is really a doubtful question. He must then turn to page xvii, and under II. GAMOPETALOUS DIVISION, he must pursue his enquiries as before. Is the calyx superior? Plainly not. Proceed then to the heading B, "**Calyx inferior.**" Are the stamens more than the lobes of the corolla? Yes. Then the choice of the six Orders in the section marked * is easily made as before, and the plant is referred to LEGUMINOSÆ.

Now let us take the Strawberry. As with Clover, we decide without difficulty that the plant is a DICOTYLEDON. The carpels are separate, and produce achenes in fruit; the plant must, therefore, be an ANGIOSPERM. And there is no doubt that it is Polypetalous. As the stamens are very numerous it must come under the section marked A. Under this section we **have** three subordinate headings, marked by one, two, and three asterisks, respectively. The stamens are clearly inserted on the calyx, and so our plant must be found under the heading marked **. Without hesitation, we refer it to the Order ROSACEÆ. Turning to page 62, we find seventeen Genera to select from. A very little consideration will show us that No. 11, FRAGARIA, is the Genus we must fix upon. Referring to page 69, we have to choose between two species, *Virginiana* and *vesca*, and the choice is found to depend upon such obvious characters as to furnish no difficulty.

The leaves of Blue Flag are straight-veined; the parts of the flower, also, are in threes. We therefore decide that the plant is Monocotyledonous, and on turning to page xxii, we find three Divisions of Monocotyledons. The Flag clearly belongs to the PETALOIDEOUS DIVISION. Then, is the perianth superior or inferior? Clearly the former. Next, are the flowers diœcious or perfect? Clearly perfect. And as the flower has three stamens, it must belong to the Order IRIDACEÆ, described on page 235. The Genus is at once seen to be IRIS, and the Species is determined without difficulty.

The Cat-tail Flag is also manifestly Monocotyledonous, from the veining of the leaves. But it is not Petaloideous. The flowers are collected on a more or less fleshy axis at the top of a scape. It therefore belongs to the SPADICEOUS DIVISION, in which there are four Orders. The only practical question is, whether our plant belongs to ARACEÆ or TYPHACEÆ. On the whole, we choose the latter, and find our decision confirmed on reading the fuller account of the two Orders on pages 217 and 219. The Genus is immediately seen to be TYPHA, and the Species *latifolia*.

These examples need not be extended here; but the beginner is recommended to run down, in the same manner, a few plants whose names he already knows. If successful in these attempts, he will naturally acquire confidence in his determinations of plants previously unknown to him.

KEY TO THE FAMILIES OR ORDERS

INCLUDED IN THIS WORK.

SERIES I. PHANEROGAMS.

Plants producing true flowers and seeds.

CLASS I. DICOTYLEDONS.

Distinguished ordinarily by having net-veined leaves, and the parts of the flowers in fours or fives, very rarely in sixes. Wood growing in rings, and surrounded by a true bark. Cotyledons of the embryo mostly two.

SUB-CLASS I. ANGIOSPERMS.

Seeds enclosed in an ovary.

I. POLYPETALOUS DIVISION.

Two distinct sets of Floral Envelopes. Parts of the corolla separate from each other.

    A. Stamens more than twice as many as the petals.

      * *Stamens hypogynous (inserted on the receptacle).*

      + *Pistil apocarpous (carpels separate from each other).*

RANUNCULACEÆ.—Herbs. Leaves generally decompound or much dissected............................	2
ANONACEÆ.—Small trees. Leaves entire. Petals 6, in 2 sets......................................	10
MAGNOLIACEÆ—Trees. Leaves truncate. Fruit resembling a cone...............................	9
MENISPERMACEÆ.—Woody twiners. Flowers diœcious. Leaves peltate near the edge.................	10

Brasenia, in

NYMPHÆACEÆ.—Aquatic. Leaves oval, peltate; the petiole attached to the centre................	12
MALVACEÆ.—Stamens monadelphous. Calyx persistent. Ovaries in a ring.........................	38

Podophyllum, in

BERBERIDACEÆ.—Calyx fugacious. Leaves large, peltate, deeply lobed. Fruit a large fleshy berry, 1-celled. 10

╬ ╬ *Pistil syncarpous. (Stigmas, styles, placentæ, or cells, more than one).*

Actæa, in

RANUNCULACEÆ, might be looked for here. Fruit a many-seeded berry. Leaves compound......... 2
NYMPHÆACEÆ.—Aquatics. Leaves floating, large, deeply cordate................................ 12
SARRACENIACEÆ.—Bog-plants. Leaves pitcher-shaped.. 13
PAPAVERACEÆ.—Juice red or yellow. Sepals 2, caducous. 14
CAPPARIDACEÆ.—Corolla cruciform, but pod 1-celled. Leaves of 3 leaflets...................... 25
HYPERICACEÆ.—Leaves transparent-dotted. Stamens usually in 3, but sometimes in 5, clusters..... 30
CISTACEÆ.—Sepals 5, very unequal, or only 3. Ovary 1-celled, with 3 parietal placentæ..... 29
MALVACEÆ.—Stamens monadelphous, connected with the bottom of the petals. Calyx persistent. Ovaries in a ring............ 38
TILIACEÆ.—Trees. Flowers yellowish, in small hanging cymes, the peduncle with a leaf-like bract attached...................................... 39

* * *Stamens perigynous (inserted on the calyx).*

Portulaca, in

PORTULACACEÆ.—Low herbs, with fleshy leaves. Sepals 2, adhering to the ovary beneath. Pod opening by a lid....................................... 37
ROSACEÆ.—Leaves alternate, with stipules. Fruit apocarpous, or a drupe, or a pome... 62
CACTACEÆ.—Very fleshy (commonly prickly) plants, of peculiar appearance; either globose or of flattened joints. Sepals and petals many. Stamens many, inserted on the tube of sepals and petals......... 86

* * * *Stamens epigynous (attached to the ovary).*

Nymphæa, in

NYMPHÆACEÆ.—Aquatic. Leaves floating. Flowers white, large, with numerous petals gradually passing into stamens........... 12

B. Stamens not more than twice as many as the petals.

* *Stamens just as many as the petals, and one stamen in front of each petal.*

BERBERIDACEÆ.—Herbs (with us). Anthers opening by uplifting valves................................. 10
PORTULACACEÆ.—Sepals 2. Styles 3-cleft. Leaves 2, fleshy.. 37
VITACEÆ.—Shrubs, climbing by tendrils. Calyx minute. 44
RHAMNACEÆ.—Shrubs, not climbing.................... 45
PLUMBAGINACEÆ.—Herbs. Calyx plaited. Styles 5. Ovary 1-celled and 1-seeded.................. 144

Lysimachia, in
PRIMULACEÆ, is occasionally polypetalous. Flowers yellow, in axillary spikes; the petals sprinkled with purplish dots................................... 145

* * *Stamens either just as many as the petals and alternate with them, or not of exactly the same number.*

+ *Corolla irregular.*

FUMARIACEÆ.—Corolla flattened and closed. Stamens 6. 15
VIOLACEÆ.—Corolla 1-spurred. Stamens 5. Pod with 3 rows of seeds on the walls.................... 26
BALSAMINACEÆ.—Corolla 1-spurred, the spur with a tail. Stamens 5. Pod bursting elastically.......... 42
POLYGALACEÆ.—Lower petal keel-shaped, usually fringed at the top. Anthers 6 or 8, 1-celled, opening at the top. Pod 2-celled........................ 48
LEGUMINOSÆ.—Corolla mostly papilionaceous. Filaments often united. Ovary simple, with one parietal placenta. Leaves compound.................. 49

+ + *Corolla regular, or nearly so.*

1. Calyx superior (*i.e.*, adherent to the ovary, wholly or partially).

   (*a*) *Stamens perigynous (inserted on the calyx).*

Cratægus, in
ROSACEÆ.—Shrubs. Stamens occasionally from 5 to 10 only. Leaves alternate, with stipules. Fruit drupe-like, containing 1-5 bony nutlets.................. 62
SAXIFRAGACEÆ.—Leaves opposite or alternate, without stipules. Styles or stigmas 2; in one instance 4. Ovary 1-celled, with 2 or 3 parietal placentæ.... 75

HAMAMELACEÆ.—Shrubs. Stamens 8; styles 2. Flowers yellow, in autumn.................................. 79
HALORAGEÆ.—Aquatics. Stamens 4 or 8. Styles or sessile stigmas 4...................................... 79
ONAGRACEÆ.—Flowers symmetrical. Stamens 2, 4, or 8. Stigmas 2 or 4, or capitate..................... 81
MELASTOMACEÆ.—Anthers 1-celled, opening by a pore at the apex. Stamens 8. Style and stigma 1. Flowers purple....................................... 84
LYTHRACEÆ.—Calyx apparently adherent to, but really free from, the ovary. Stamens mostly unequal. Leaves mostly whorled. Flowers varying as to relative lengths of stamens and style......... 84
CUCURBITACEÆ.—Tendril-bearing herbs. Flowers monœcious... 85

(b) *Stamens epigynous (on the ovary, or on a disk which covers the ovary).*

Euonymus, in
CELASTRACEÆ.—Shrub, with 4-sided branchlets, not climbing. Leaves simple. Pods crimson when ripe. Calyx not minute................................. 46
UMBELLIFERÆ.—Flowers chiefly in compound umbels. Calyx very minute. Stamens 5. Styles 2. Fruit dry, 2-seeded................................. 87
ARALIACEÆ.—Umbels not compound; but sometimes panicled. Stamens 5. Styles usually more than 2. Fruit berry-like............................... 92
CORNACEÆ.—Flowers in cymes or heads. Stamens 4. Style 1... 93

2. Calyx inferior (*i.e.*, free from the ovary).

(a) *Stamens hypogynous (on the receptacle).*

CRUCIFERÆ.—Petals 4. Stamens 6, tetradynamous. Pod 2-celled... 16
CISTACEÆ.—Petals 3. Sepals 5, very unequal; or only 3. Pod partly 3-celled............................ 29
DROSERACEÆ.—Leaves radical; beset with reddish glandular hairs. Flowers in a 1-sided raceme........ 30

Elodes, in
HYPERICACEÆ.—Leaves with transparent dots. Stamens 9, in 3 clusters..................................... 30

CARYOPHYLLACEÆ.—Styles 2–5. Ovules in the centre or bottom of the cell. Stem usually swollen at the joints. Leaves opposite...................... 32

LINACEÆ.—Stamens 5, united below. Pod 10-celled, 10-seeded 40

GERANIACEÆ.—Ovaries or lobes of the ovary 5 or 3, with a common central style or axis which remains after the ripe ovaries separate from it.......... 41

OXALIDACEÆ.—Stamens 10. Pod 5-celled. Styles 5, distinct. Leaflets 3, obcordate, drooping at night-fall. 42

ERICACEÆ.—Anthers opening by pores at the top, or across the top. Leaves mostly evergreen, sometimes brown beneath; but in some instances the plant is white or tawny........................... 137

RUTACEÆ.—Shrubs, with polygamous flowers and leaves of 3 leaflets, dotted with transparent glands. Stigmas 2. Fruit a 2-seeded samara winged all round......... 43

(b) *Stamens perigynous (plainly attached to the calyx).*

SAXIFRAGACEÆ.—Leaves opposite or alternate, without stipules. Styles or stigmas 2; in one instance 4. Carpels fewer than the petals................. 75

CRASSULACEÆ.—Flowers *symmetrical*. Stamens 10 or 8. Leaves sometimes fleshy..................... 78

LYTHRACEÆ.—Stamens 10, in two sets. Calyx enclosing, but really free from, the ovary. Leaves mostly whorled.. 84

(c) *Stamens attached to a fleshy disk in the bottom of the calyx-tube.*

ANACARDIACEÆ.—Trees, or shrubs, not prickly. Leaves compound. Stigmas 3. Fruit a 1-seeded drupelet. 43

CELASTRACEÆ.—Twining shrub. Leaves simple. Pods orange when ripe 46

SAPINDACEÆ.—Shrubs, or trees. Fruit 2-winged, and leaves palmately-veined. *Or*, Fruit an inflated 3-celled pod, and leaves of 3 leaflets. Styles 2 or 3...... 47

(d) *Stamens attached to the petals at their very base.*

Claytonia, in

PORTULACACEÆ.—Sepals 2. Leaves fleshy. Style 3-cleft. 37

AQUIFOLIACEÆ.—Shrubs, with small axillary flowers, having the parts in fours and sixes. Fruit a red berry-like drupe. Stigma sessile. Calyx minute.... 145

II. GAMOPETALOUS DIVISION.

Corolla with the petals united together, in however slight a degree.

A. Calyx superior (adherent to the ovary).

* Stamens united by their anthers.

CUCURBITACEÆ.—Tendril-bearing herbs..................	85
COMPOSITÆ:—Flowers in heads, surrounded by an involucre	102
LOBELIACEÆ.—Flowers not in heads. Corolla split down one side...	135

* * Stamens not united together in any way.
+- Stamens inserted on the corolla.

DIPSACEÆ.—Flowers in heads, surrounded by an involucre. Plant prickly......................................	101
VALERIANACEÆ.—Flowers white, in clustered cymes. Stamens fewer than the lobes of the corolla........	101
RUBIACEÆ.—Leaves, when opposite, with stipules; when whorled, without stipules. Flowers, if in heads, without an involucre..............................	98
CAPRIFOLIACEÆ.—Leaves opposite, without stipules; but, in one genus, with appendages resembling stipules.	95

+- +- Stamens not inserted on the corolla.

CAMPANULACEÆ.—Herbs with milky juice. Stamens as many as the lobes of the corolla..................	136
ERICACEÆ.—Chiefly shrubby plants or parasites. Stamens twice as many as the lobes of the corolla........	137

B. Calyx inferior (free from the ovary).

* Stamens more than the lobes of the corolla.

LEGUMINOSÆ.—Ovary 1-celled, with 1 parietal placenta. Stamens mostly diadelphous...................	49

Adlumia, in

FUMARIACEÆ.—Plant climbing. Corolla 2-spurred......	15
MALVACEÆ.—Filaments monadelphous. Carpels in a ring.	38
ERICACEÆ.—Chiefly shrubby plants, with simple entire leaves. Stamens twice as many as the lobes of the corolla...	137
POLYGALACÆ.—Anthers 6 or 8, 1-celled, opening at the top. Pod 2-celled. Flowers irregular; lower petal keel-shaped, and usually fringed at the top......	48

OXALIDACEÆ.—Stamens 10, 5 of them longer. Styles 5, distinct. Leaflets 3, obcordate, drooping at nightfall.. 42

* * *Stamens just as many as the lobes of the corolla, one in front of each lobe.*

PRIMULACEÆ.—Stamens on the corolla. Style 1. Ovary 1-celled, with a free central placenta rising from the base... 145

PLUMBAGINACEÆ.—Styles 5. Ovary 1-celled and 1-seeded. 144

* * * *Stamens just as many as the lobes of the corolla, inserted on its tube alternately with its lobes.*

+ *Ovaries 2, separate.*

APOCYNACEÆ.—Plants with milky juice. Anthers converging round the stigmas, but not adherent to them. Filaments distinct............................. 179

ASCLEPIADACEÆ.—Plants with milky juice. Anthers adhering to the stigmas. Filaments monadelphous. Flowers in umbels............................. 179

+ + *Ovary 4-lobed around the base of the style.*

Mentha, in

LABIATÆ.—Stamens 4. Leaves opposite, aromatic...... 160
BORRAGINACEÆ.—Stamens 5. Leaves alternate......... 167

+ + + *Ovary 1-celled; the seeds on the walls.*

HYDROPHYLLACEÆ.—Stamens 5, usually exserted. Style 2-cleft. Leaves lobed and sometimes cut-toothed. 171

GENTIANACEÆ.—Leaves entire and opposite; or (in Menyanthes) of 3 leaflets............................. 176

+ + + + *Ovary with 2 or more cells.*

AQUIFOLIACEÆ.—Shrubs. Corolla almost polypetalous. Calyx minute. Fruit a red berry-like drupe. Parts of the flower chiefly in fours or sixes....... 145

PLANTAGINACEÆ.—Stamens 4. Pod 2-celled. Flowers in a close spike..................................... 148

Verbascum, in

SCROPHULARIACEÆ.—Corolla nearly regular. Flowers in a long terminal spike. Stamens 5; the filaments, or some of them, woolly............................ 151

POLEMONIACEÆ.—Style 3-cleft. Corolla salver-shaped, with a long tube. Pod 3-celled, few-seeded; seeds small...................................... 172

KEY TO THE ORDERS. xix

CONVOLVULACEÆ.—Style 2-cleft. Pod 2-celled, generally 4-seeded; seeds large. Chiefly twining or trailing plants..... ... 173
SOLANACEÆ.—Style single. Pod or berry 2-celled, many-seeded 174

* * * * *Stamens fewer than the lobes of the corolla; the corolla mostly irregular or 2-lipped.*

LABIATÆ.—Ovary 4-lobed around the base of the style. Stamens 4 and didynamous, or occasionally only 2 with anthers. Stem square................ 160
VERBENACEÆ.—Ovary 4-celled, but not lobed; the style rising from the apex. Or, Ovary 1-celled and 1-seeded. Stamens didynamous................ 158
LENTIBULARIACEÆ.—Aquatics. Stamens 2. Ovary 1-celled, with a free central placenta............. 149
OROBANCHACEÆ —Parasitic herbs, without green foliage. Ovary 1-celled, with many seeds on the walls. Stamens didynamous........................ 150
SCROPHULARIACEÆ.—Ovary 2-celled, with many seeds. Stamens didynamous, or only 2 151
ACANTHACEÆ.—Stamens 2, the anther-cells separated. Ovary flat, 2-celled, 4-seeded. Aquatics........ 159

III. APETALOUS DIVISION.

Corolla (and sometimes calyx also) wanting.

A. Flowers not in catkins.

* *Calyx and corolla both wanting.*

PIPERACEÆ.—Flowers white, in a dense terminal spike, nodding at the end. Carpels 6 or 4, nearly separate 182
CERATOPHYLLACEÆ.—Immersed aquatics, with whorled finely dissected leaves. Flowers monœcious.... 213

* * *Calyx superior (i.e., adherent to the ovary).*

SAXIFRAGACEÆ.—Small, smooth herbs, with inconspicuous greenish-yellow flowers. Stamens twice as many as the calyx-lobes, on a conspicuous disk........ 75
HALORAGEÆ.—Aquatics. Leaves finely dissected or linear. Stamens 1-8. Ovary 3-4-lobed or (Hippuris) 1-celled. 79
ONAGRACEÆ.—Herbs, in ditches. Stamens 4. Ovary 4-celled, 4-sided.................................. 81

KEY TO THE ORDERS.

ARISTOLOCHIACEÆ.—Calyx 3-lobed, dull purple inside. Ovary 6-celled.................................... 182.

SANTALACEÆ.—Low plants with greenish-white flowers in terminal clusters. Calyx-tube prolonged, and forming a neck to the 1-celled nut-like fruit..... 195

ELÆAGNACEÆ.—Shrubs with scurfy leaves. Flowers perfect or diœcious. Calyx 4-parted, in the fertile flowers apparently adherent to the ovary, and becoming fleshy in fruit............................ 194

    * * * Calyx inferior (plainly free from the ovary).

    + Ovaries more than one and separate from each other.

RANUNCULACEÆ.—Calyx present, coloured and petal-like. Achenes containing several seeds, or only one.... 2

RUTACEÆ.—Prickly shrubs, with compound transparent-dotted leaves, and diœcious flowers............. 43

    + + Ovary only one, but with more than one cell (except in Glaux).

CRASSULACEÆ.—Herbs, in wet places. Pod 5-celled and 5-horned.................................... 78

PHYTOLACCACEÆ.—Herbs. Ovary 10-celled and 10-seeded 183

EUPHORBIACEÆ.—Herbs. Ovary 3-celled, 3-lobed, protruded on a long pedicel. Juice milky......... 195

SAPINDACEÆ.—Trees. Ovary 2-celled and 2-lobed. Fruit two 1-seeded samaras joined together. Flowers polygamous..................................... 47

RHAMNACEÆ.—Shrubs. Ovary 3-celled and 3-seeded; forming a berry................................ 45

FICOIDEÆ.—Prostrate herbs with whorled leaves. Ovary 3-celled, many-seeded...................... 86

URTICACEÆ.—Trees. Leaves simple. Ovary 2-celled, but fruit a 1-seeded samara winged all round. Stigmas 2.. 198

EMPETRACEÆ.—Low shrubby evergreens with heath-like leaves.. 213

Glaux, in

PRIMULACEÆ.—Herbs. Ovary 1-celled, the placenta central... 145

    + + + Ovary only one, 1-celled and 1-seeded.

POLYGONACEÆ.—Herbs. Stipules sheathing the stem at the nodes...................................... 183

KEY TO THE ORDERS.

URTICACEÆ.—Herbs. Stigma 1. Flowers monœcious or diœcious, in spikes or racemes. No chaff-like bracts among the flowers. Or, Stigmas 2; leaves palmately-compound.................... 198

ILLECEBRACEÆ.—Herbs, small, tufted. Flowers greenish or whitish, terminal. Stamens as many as the divisions of the 5-parted calyx and opposite them, or fewer, perigynous. Sepals hooded at the top and bristle-pointed. Style 2-cleft............ 183

AMARANTACEÆ.—Herbs. Flowers greenish or reddish, in spikes, *with chaff-like bracts interspersed*. Stigmas 2. 187

CHENOPODIACEÆ.—Herbs. Flowers greenish, in spikes. *No chaff-like bracts*. Stigmas 2................ 184

OLEACEÆ.—Trees. *Leaves pinnately compound*. Fruit a 1-seeded samara........................ 181

URTICACEÆ.—Trees. *Leaves simple*. Fruit a 1-seeded samara winged all round, or a drupe............ 198

LAURACEÆ—Trees or shrubs. Flowers diœcious. Sepals 6, petal-like. Stamens 9, opening by uplifting valves. 193

THYMELEACEÆ.—Shrubs with leather-like bark, and jointed branchlets. Flowers perfect, preceding the leaves. Style thread-like........................ 194

B. Flowers in catkins.

* *Sterile or staminate flowers only in catkins.*

JUGLANDACEÆ.—Trees with pinnate leaves. Fruit a nut with a husk............................ 202

CUPULIFERÆ.—Trees with simple leaves. Fruit one or more nuts surrounded by an involucre which forms a scaly cup or bur..................... 204

* * *Both sterile or fertile flowers in catkins, or catkin-like heads.*

SALICACEÆ.—Shrubs or low trees. Ovary 1-celled, many-seeded; seeds tufted with down at one end...... 209

PLATANACEÆ.—Large trees. *Stipules sheathing the branchlets*. The flowers in heads................. 201

MYRICACEÆ.—Shrubs with resinous-dotted, usually fragrant, leaves. Fertile flowers one under each scale. Nutlets usually coated with waxy grains.. 203

BETULACEÆ.—Trees or shrubs. Fertile flowers 2 or 3 under each scale of the catkin. Stigmas 2, long and slender............................... 207

SUB-CLASS II. GYMNOSPERMS.

Ovules and seeds naked, on the inner face of an open scale; or, in Taxus, without any scale, but surrounded by a ring-like disk which becomes red and berry-like in fruit.

CONIFERÆ.—Trees or shrubs, with resinous juice, and mostly awl-shaped or needle-shaped leaves. Fruit a cone, or occasionally berry-like.............. 214

CLASS II. MONOCOTYLEDONS.

Distinguished ordinarily by having straight-veined leaves (though occasionally net-veined ones), and the parts of the flowers in threes, never in fives. Wood never forming rings, but interspersed in separate bundles throughout the stem. Cotyledon only 1.

I. SPADICEOUS DIVISION.

Flowers collected on a spadix, with or without a spathe or sheathing bract. Leaves sometimes net-veined.

ARACEÆ.—Herbs (either flag-like marsh-plants, or terrestrial,) with pungent juice, and simple or compound leaves, these sometimes net-veined. Spadix usually (but not always) accompanied by a spathe. Flowers either without a perianth of any kind, or with 4–6 sepals........................... 217

TYPHACEÆ.—Aquatic or marsh plants, with linear straight-veined leaves erect or floating, and monœcious flowers. Heads of flowers cylindrical or globular, no spathe, and no floral envelopes........ 219

LEMNACEÆ.—Small aquatics, freely floating about 218

NAIADACEÆ.—Immersed aquatics. Stems branching and leafy. Flowers perfect, in spikes, generally on the surface 221

II. PETALOIDEOUS DIVISION.

Flowers not collected on a spadix, furnished with a corolla-like. or occasionally herbaceous, perianth.

A. Perianth superior (adherent to the ovary).

* Flowers diœcious or polygamous, regular.

HYDROCHARIDACEÆ.—Aquatics. Pistillate flowers only above water; perianth of 6 pieces 226
DIOSCOREACEÆ.—Twiners, from knotted rootstocks. Leaves heart-shaped, net-veined. Pod with 3 large wings. 236

* * Flowers perfect.

ORCHIDACEÆ.—Stamens 1 or 2, gynandrous. Flowers irregular 226
IRIDACEÆ.—Stamens 3......... 235
AMARYLLIDACEÆ.—Stamens 6. Flowers on a scape from a bulb.......... 236

B. Perianth inferior (free from the ovary).

ALISMACEÆ.—Pistil apocarpous: carpels in a ring or head, leaves with distinct petiole and blade........... 224
SMILACEÆ.—Climbing plants, with alternate ribbed and net-veined petioled leaves. Flowers diœcious... 237

Triglochin, in

ALISMACEÆ.—Rush-like marsh herbs. Flowers in a spike or raceme. Carpels when ripe splitting away from a persistent axis............... 224
LILIACEÆ.—Perianth of similar divisions or lobes, mostly 6, but in one case 4. One stamen in front of each division, the stamens similar.................. 237

Trillium, in

LILIACEÆ.—Perianth of 3 green sepals and three coloured petals..................... 237
PONTEDERIACEÆ.—Stamens 6, 3 long and 3 short. Perianth (blue or yellow) tubular, of 6 lobes. Aquatics.. 247
JUNCACEÆ.—Perianth wholly glumaceous, of similar pieces. 243
XYRIDACEÆ.—Perianth partly glumaceous; inner set of 3 yellow petals. Flowers in dense heads, perfect. 247
ERIOCAULONACEÆ.—In shallow water. Flowers monœcious or diœcious, in a small woolly head, at the summit of a 7-angled scape. Leaves in a tuft at the base....... 247

III. GLUMACEOUS DIVISION.

Flowers without a true perianth, but subtended by thin scales called glumes.

Cyperaceæ.—Sheaths of the leaves not split............ 248
Gramineæ.—Sheaths of the leaves split on the side away from the blade.................................. 251

SERIES II. CRYPTOGAMS.

Plants without stamens and pistils, reproducing themselves by spores instead of seeds.

CLASS III. PTERIDOPHYTES.

Stem containing vascular as well as cellular tissue.

Filices.—Spores produced on the fronds............... 257
Equisetaceæ—Spores produced on the under side of the shield-shaped scales of a terminal spike or cone. 267
Lycopodiaceæ.—Low, long-stemmed, moss-like evergreens. Spores-cases produced in the axils of the simple leaves or bracts. Spores of one kind only.. 269
Selaginellaceæ.—Spores of two kinds, large and small. Spore-cases solitary in the axils of small 4-ranked leaves, or at the bases of linear radical leaves.... 270
Salviniaceæ.—Sporocarps beneath the stem. Floating aquatics, pinnately branched, with minute imbricated leaves.................................... 271

THE COMMONLY OCCURRING

WILD PLANTS OF CANADA.

SERIES I.

FLOWERING OR PHANEROG'AMOUS PLANTS.

Plants producing flowers (that is to say, stamens and pistils, and usually floral envelopes of some kind), and seeds containing an embryo.

CLASS I. DICOTYLE'DONS.

Embryo having at least two cotyledons. Parts of the flower usually in fours or fives. Leaves net-veined. Wood in circular layers between the pith and the bark.

SUB-CLASS I. AN'GIOSPERMS.

Seeds enclosed in a seed-vessel or pericarp.

I. POLYPET'ALOUS DIVISION.

Plants with flowers having both calyx and corolla, the latter consisting of petals entirely separate from each other. (In some genera and species, however, petals are absent.)

ORDER I. **RANUNCULA'CEÆ.** (CROWFOOT FAMILY.)

Herbs or woody climbers, with an acrid colourless juice. Parts of the flower separate from each other. Corolla sometimes wanting. Stamens numerous. Pistil (with one or two exceptions) apocarpous. Fruit an achene, follicle, or berry. Leaves exstipulate, with the blades usually dissected, and petioles spreading at the base.

Synopsis of the Genera.

1. **Clem'atis.** Real petals none or stamen-like. Coloured sepals 4 or more, valvate in the bud. Fruit an achene, with the long and feathery style attached. Leaves all opposite. Plant climbing by the bending of the petioles.

2. **Anemo'ne.** Petals none or stamen-like. Coloured sepals imbricated in the bud. Achenes many, in a head, pointed or tailed, not ribbed. Stem-leaves opposite or whorled, *forming an involucre remote from the flower.*

3. **Hepat'ica.** Petals none. Coloured sepals 6–12, whitish or bluish. Achenes many, not ribbed. Leaves all radical. *An involucre of 3 leaves close to the flower*, and liable to be mistaken for a calyx.

4. **Thalic'trum.** Petals none. Coloured sepals 4 or more, greenish. Achenes several, angled or grooved. No involucre. Stem-leaves alternate, decompound. Flowers in panicles or corymbs, mostly diœcious.

5. **Myosu'rus.** Petals 5, white, with slender claws. Sepals 5, *spurred at the base. Achenes in a long slender spike.* Small plants with tufted narrow root-leaves, and 1-flowered scapes.

6. **Ranun'culus.** Sepals 5, deciduous. Petals generally 5, each with a pit or little scale on the inside of the claw. Achenes many, in heads, short-pointed. Stem-leaves alternate. Flowers solitary or corymbed, mostly yellow, rarely white.

7. **Cal'tha.** Petals none. Sepals 5–9, yellow. Fruit a many-seeded follicle. Leaves large, glabrous, heart-shaped or kidney-shaped, mostly crenate. Stem hollow and furrowed.

8. **Cop'tis.** Sepals 5–7, white, deciduous. Petals 5–7, yellow with slender claws, and somewhat tubular at the apex. Carpels 3–7, on slender stalks. Fruit a follicle. Flowers on naked scapes. Leaves radical, shining, divided into three wedge-shaped leaflets, sharply toothed. Root fibrous, golden yellow.

9. **Aquile'gia.** Sepals 5, coloured. Petals 5, *each a long hollow spur.* Carpels 5. Follicles erect, many-seeded. Flowers very showy, terminating the branches. Leaves decompound.

10. **Delphin'ium.** Sepals 5, unequal, petal-like, *the upper one produced backward into a spur.* Petals 4, irregular, the upper pair spurred, and projecting backward into the spur of the upper sepal (rarely only 2). Flowers blue or whitish, in terminal racemes. Leaves palmately dissected.

11. **Actæ'a.** Sepals 4-5, caducous. Petals 4-10, with slender claws. Stamens many, with long filaments. *Fruit a many-seeded berry.* Flowers in a short thick raceme. Leaves decompound, leaflets sharply toothed.

12. **Cimicif'uga.** Sepals 4-5, caducous. Petals several, small, two-horned at the apex. Carpels 1-8, becoming pods. Flowers in long plume-like racemes.

13. **Hydras'tis.** Petals none. Flowers solitary. Sepals 3, petal-like, greenish-white. Carpels 12 or more, forming a head of crimson 1-2-seeded berries in fruit. Stem low, from a knotted yellow rootstock. Leaves simple, lobed.

1. CLEM'ATIS, L. VIRGIN'S BOWER.

1. **C. Virginia'na**, L. (COMMON VIRGIN'S BOWER.) A woody-stemmed climber. Flowers in panicled clusters, often diœcious, white. Leaves of 3 ovate leaflets, which are cut or lobed. Feathery tails of the achenes very conspicuous in the autumn.—Along streams and in swamps.

2. **C. ligusticifo'lia**, Nutt. Very much like the last, but the leaves have usually 5 leaflets.—N. W.

3. **C. verticilla'ris**, DC. Shrubby climber. Peduncles bearing single large flowers, with thin, wide-spreading, bluish-purple sepals. Tails of the achenes plumose. Leaves of three leaflets, which are entire, or sparingly toothed or lobed.—Rocky places.

2. ANEMO'NE, L. ANEM'ONE.

* *Styles long and hairy, forming feathery tails in fruit.*

1. **A. patens**, L., var. **Nuttallia'na**, Gray. (PRAIRIE A. OR PASQUE FLOWER.) Villous with long silky hairs. Flower erect, appearing before the leaves. Sepals 5-7, purplish or white, an inch or more in length, spreading at maturity. Involucre sessile, its lobes numerous, all united at the base to form a shallow cup. Leaves radical, the blades deeply cut into 3 well-marked divisions, the middle one stalked,

and all again dissected into narrow lobes. Feathery tails of the carpels 2 inches long. A low plant, flowering in early spring.—N. W.

** *Styles short, not plumose.*

2. **A. parviflo'ra,** Michx. (SMALL-FLOWERED A.) Stem 8–12 inches high, one-flowered. Sepals 5 or 6, white. Involucre 2-3-leaved far below the flower. Head of carpels *woolly*, globular. Root-leaves small, 3-parted, their divisions crenately lobed.—Rocky river-margins.

3. **A. multif'ida,** DC. (MANY-CLEFT A.) Silky-hairy. Principal involucre 2-3-leaved, bearing one naked and one or two 2-leaved peduncles. Leaves of the involucre short-petioled, twice or thrice 3-parted and cleft, their divisions linear. Sepals red, greenish-yellow, or whitish. Head of carpels spherical or oval, *woolly*.—Rocky river-margins, etc.

4. **A. cylin'drica,** Gray. (LONG-FRUITED A.) Carpels very numerous, in an oblong *woolly* head about an inch long. Peduncles 2-6, long, upright, leafless. Stem-leaves in a whorl, twice or thrice as many as the peduncles, *long-petioled*. Sepals 5, greenish-white. Plant about two feet high, clothed with silky hairs.—Dry woods.

5. **A. Virginia'na,** L. (VIRGINIAN A.) Very much like the last, but larger. Also, the central peduncle only is naked, the others having each a pair of leaves about the middle, from whose axils other peduncles occasionally spring. Sepals greenish. Head of carpels oval or oblong. —Dry rocky woods and river banks.

6. **A. Pennsylvan'ica,** L. (*A. dichotoma*, *L.*, in Macoun's Catalogue.) (PENNSYLVANIAN A.) Carpels fewer and the head *not woolly*, but pubescent and spherical. *Stem-leaves sessile*, primary ones 3 in a whorl, but only a pair of smaller ones on each side of the flowering branches. Radical leaves 5-7-parted. Sepals 5, obovate, large and white. Plant hairy, scarcely a foot high.—Low meadows.

7. **A. nemoro'sa,** L. (WOOD A. WIND-FLOWER.) Plant not more than six inches high, nearly smooth, one-flowered.

Stem-leaves in a whorl of 3, long-petioled, 3-5-parted. Sepals 4-7, oval, white, or often purplish on the back.—Moist places.

3. HEPAT'ICA, Dill. LIVER-LEAF. HEPATICA.

1. **H. acutil'oba,** DC. (SHARP-LOBED H.) Leaves with 3 (sometimes 5) acute lobes, appearing after the flowers. Petioles silky-hairy.—Woods in spring.

2. **H. tril'oba,** Chaix. (ROUND-LOBED H.) Leaves with 3 rounded lobes; those of the involucre also obtuse.—Dry rich woods in spring.

(The two species just described are included under ANEMONE in Macoun's Catalogue, the first named being *A. acutiloba, Lawson*, and the second *A. Hepatica, L.*)

4. THALIC'TRUM, Tourn. MEADOW-RUE.

1. **T. anemonoi'des,** Michx. (*Anemonella thalictroides,* Spach.) (RUE-ANEMONE.) Stem low. Stem-leaves all in a whorl at the top. *Roots tuberous. Flowers several in an umbel,* by which character this plant is easily distinguished from Wood Anemone, which it otherwise resembles.—Southwestward, in spring.

2. **T. dioi'cum,** L. (EARLY M.) Stem smooth, pale and glaucous, 1-2 feet high. *Flowers diœcious,* in ample panicles, purplish or greenish ; the yellow anthers drooping and very conspicuous. Leaves alternate, decompound ; leaflets with 5-7 rounded lobes.—Woods.

3. **T. polyg'amum,** Muhl. (*T. Cornuti,* L.) (TALL M.) Stem smooth or nearly so, 2-6 feet high. *Leaves sessile;* leaflets very much like No. 2. Flowers white, in compound panicles; *anthers not drooping;* filaments club-shaped.—Low wet meadows, and along streams.

4. **T. purpuras'cens,** L. (PURPLISH M.) Stem mostly purplish, 2-4 feet high. Stem-leaves sessile or nearly so ; leaflets veiny beneath, *often glandular-pubescent,* and so distinguished from No. 3. Flowers resembling those of No. 2.

5. MYOSU'RUS, Dill. MOUSETAIL.

1. **M. min'imus,** L. Scapes 2-6 inches high. Leaves linear-spathulate. Achenes blunt, in a spike 1-2 inches long when ripe.

2. **M. arista'tus,** Benth., is easily distinguished from the above by the *persistent style* nearly as long as the achene.—N.W.

6. RANUN'CULUS, L. CROWFOOT. BUTTERCUP.

1. **R. aquat'ilis,** L., var. **trichophyl'lus,** Gray. (WHITE WATER-CROWFOOT.) *Foliage under water*, filiform, usually collapsing when withdrawn from the water. *Flowers white*, floating, each petal with a little pit on the inside of the claw.—Stagnant pools and slow-flowing streams.

2. **R. circina'tus,** Sibth. Much like No. 1, but the immersed leaves are mostly *sessile,* and do not collapse when withdrawn from the water.—Toronto harbour; and abundant in N. W.

3. **R. Cymbala'ria,** Pursh. (SEA-SIDE CROWFOOT.) Low, smooth, spreading by runners which take root at the joints. Leaves long-petioled, roundish, crenate, rather fleshy. Petals 5-8, yellow. Carpels striate, in an oblong head.—Seashore, and beside brackish streams and springs.

4. **R. multif'idus,** Pursh. (YELLOW WATER-CROWFOOT.) Like No. 1, but larger, and with *yellow flowers*, sometimes creeping in the mud; the leaves round kidney-shaped, and more or less deeply lobed and toothed.—Ponds and ditches.

5. **R. Flam'mula,** L., var. **reptans,** Meyer. (CREEPING SPEARWORT.) Stem reclining, rooting at the joints, *only 3-6 inches long.* Leaves linear, entire, remote. Flowers yellow, ¼ of an inch broad.—Sandy and gravelly shores of ponds and rivers.

6. **R. rhomboi'deus,** Goldie. Stem erect, low (3-8 inches), *hairy;* root-leaves roundish or rhombic-ovate, mostly crenate; lowest stem-leaves similar or 3-5-lobed, the upper nearly sessile and deeply cut into linear lobes. Petals large, exceeding the calyx; *achenes orbicular, with a minute beak*, in a globular head.—Dry plains, in early summer.

7. **R. affi'nis,** R. Br. (Rough-fruited C.) Taller than No. 6, more or less pubescent. Root-leaves petioled, *usually pedately multifid;* stem-leaves nearly sessile, with broadly linear lobes. Petals light-yellow, about half an inch long. Achenes, with recurved beaks, forming an *oblong head.—* N. W.

8. **R. aborti'vus,** L. (Small-flowered C.) Petals shorter than the reflexed calyx. Stem erect, *very smooth,* slender. Radical leaves roundish, crenate, petiolate; stem-leaves 3–5-parted, sessile. Carpels in a globular head, each with a minute curved beak.—Shady hill-sides and wet pastures. Var. **micranthus,** Gray, is *pubescent,* with more slender peduncles and fewer achenes.

9. **R. scelera'tus,** L. (Cursed C.) Petals about the same length as the calyx. Stem thick, hollow, *smooth.* Radical leaves 3-lobed ; stem-leaves 3-parted, uppermost almost sessile. *Head of carpels oblong.—*Wet ditches.

10. **R. recurva'tus,** Poir. (Hooked C.) Petals shorter than the reflexed calyx. *Stem hirsute,* with stiff spreading hairs. Radical and cauline leaves about alike, long-petioled. Head of carpels globular, *each with a long recurved beak.—* Woods.

11. **R. Pennsylvan'icus,** L. (Bristly C.) Petals not longer than the reflexed calyx. *Stem hirsute.* Leaves ternately divided, divisions of the leaves stalked, unequally 3-cleft. *Head of carpels oblong, achenes with straight beaks,* and so easily distinguished from No. 10.—Wet places.

12. **R. his'pidus,** Michx. Resembling the last species, but with few-leaved ascending or reclining stems, not always hirsute. Root a cluster of stout fibres. Calyx hardly reflexed, soon deciduous, much shorter than the petals. Achenes strongly margined, with straight beaks; in a globular or oval head.—Wesley Park, Niagara.

13. **R. septentriona'lis,** Poir. Petals much longer than the calyx. Early-flowering stems ascending, *putting forth long runners during the summer.* Leaves ternate, divisions

generally stalked, deeply and sharply lobed, petioles appressed-pubescent. *Achenes large, compressed, strongly margined, in globular heads, and with long flat beaks.* Peduncles furrowed.—Wet places.

14. **R. repens,** L. Much resembling the last in habit, but smaller, and the leaves not so deeply and sharply cut. Flowering later. Leaves often blotched with white. The style shorter than in No. 13, and stigmatic along the whole inner side, persistent.—Low ground, chiefly eastward.

15. **R. bulbo'sus,** L. (BULBOUS C. or BUTTERCUP.) Petals much longer than the calyx. Stem erect, *from a bulb-like base.* Flowers an inch broad, on *furrowed peduncles.*—Pastures. Rather rare.

16. **R. a'cris,** L. (TALL C. or BUTTERCUP.) Much taller than No. 15. Petals much longer than the calyx. Stem upright, *no bulb at the base.* Peduncles *not furrowed.*

17. **R. fascicula'ris,** Muhl. (EARLY C.) Petals much longer than the calyx. Plant 5-9 inches high, erect, pubescent with silky hairs. Radical leaves appearing pinnate, the terminal division long-stalked, the lateral ones sessile. *Root a bundle of thickened fleshy fibres.*—Rocky woods and fields in spring.

7. CAL'THA, L. MARSH-MARIGOLD.

C. palustris, L. (MARSH-MARIGOLD.) Stem about a foot high, hollow, round, forking, very glabrous. Flowers golden yellow, 1-1½ inches broad.—Swamps and wet meadows. A very conspicuous plant in early spring.

8. COP'TIS, Salisb. GOLDTHREAD.

C. trifolia, Salisb. (THREE-LEAVED GOLDTHREAD.) Low and stemless. Scapes 1-flowered, with a single bract above the middle. Petals much smaller than the sepals.—On logs and about stumps in cedar swamps.

9. AQUILE'GIA, Tourn. COLUMBINE.

1. **A. Canadensis,** L. (WILD COLUMBINE.) Stem branching, a foot or more in height, smooth. Leaves decompound; leaflets in threes. Flowers nodding, scarlet outside, yellow within.—Rocky woods and thickets.

2. **A. vulga'ris**, L. (GARDEN COLUMBINE.) This species has escaped from cultivation in some places. *Spurs hooked.* Flowers blue, purple, or whitish.

10. DELPHIN'IUM, L. LARKSPUR.

1. **D. azu'reum,** Michx. (PRAIRIE LARKSPUR.) *Carpels 3,* the pods erect. Lobes of the leaves numerous, narrowly linear. Raceme strict, but not dense. Spur usually curved upwards.—N.W.

2. **D. Consol'ida,** L. (FIELD L.) Has escaped from gardens in a few places. *The pistil is single,* and the flowers are scattered on the spreading branches. Petals 2, united.

11. ACTÆ'A, L. BANEBERRY.

1. **A. spica'ta,** L., var. **rubra,** Ait. (RED B.) *Raceme short,* breadth and length being about the same. Pedicels slender. *Berries red.*—Rich woods.

2. **A. alba,** Bigel. (WHITE B.) *Raceme longer than broad.* Pedicels thickened in fruit, cherry-coloured. *Berries white.* —Same localities as No. 1.

12. CIMICIF'UGA, L. BUGBANE.

C. racemo'sa, Ell. (BLACK SNAKEROOT.) Stem 3–6 feet high. Resembling a tall Actæa, but easily distinguished by its plume-like raceme of white flowers.—South-western Ontario.

13. HYDRAS'TIS, L. ORANGEROOT. YELLOW PUCCOON.

H. Canadensis, L. A low plant, bearing a single radical leaf, and a pair of cauline ones near the summit of the simple stem. Leaves rounded, cordate, 5–7-lobed, very large when fully grown.—Wet meadows, in early summer, south-westward.

ORDER II. **MAGNOLIA'CEÆ.** (MAGNOLIA FAMILY.)

Trees or shrubs, with alternate entire or lobed (not serrate) leaves. Sepals 3, coloured, deciduous. Petals 6–9, deciduous. Stamens hypogynous, indefinite, separate; anthers adnate. Carpels numerous, in many rows on an elongated receptacle. Fruit resembling a cone.

1. LIRIODEN'DRON, L. TULIP-TREE.

L. Tulipif'era, L. A large and stately tree, growing to a great height in many parts of the western peninsula of Ontario. Leaves large, truncate, or with a shallow notch at the end. Flowers large, showy, solitary; petals greenish-yellow, marked with orange. Fruit a dry cone, which, at maturity, separates into dry winged indehiscent carpels.

ORDER III. **ANONA'CEÆ.** (CUSTARD-APPLE FAMILY.)

Trees or shrubs, with alternate and entire leaves, and solitary, axillary, perfect, hypogynous flowers. Sepals 3. Petals 6, in two sets, deciduous. Stamens numerous. Carpels few or many, fleshy in fruit.

1. ASIM'INA, Adans. NORTH AMERICAN PAPAW.

A. tril'oba, Dunal. (COMMON PAPAW.) Found only in the Niagara Peninsula. A small tree, not unlike a young beech in appearance, and forming thickets near Queenston Heights. Flowers purple, appearing before the leaves; the three outer petals much larger than the three inner ones. Fruit 2 to 3 inches long, edible.

ORDER IV. **MENISPERMA'CEÆ.** (MOONSEED FAMILY.)

Woody twiners, with peltate alternate leaves and small diœcious flowers. Sepals and petals yellowish-white, usually six of each, the petals in front of the sepals. Stamens numerous. Fruit a drupe, in appearance something like a small grape, with moon-shaped seeds.

1. MENISPER'MUM, L. MOONSEED.

M. Canadense, L. (CANADIAN MOONSEED.) A twining plant, found, though not abundantly, in low grounds in rich woods. It may be pretty easily recognized by its usually 7-angled thin leaves, which are *peltate near the edge*. Fruit bluish-black.

ORDER V. **BERBERIDA'CEÆ.** (BARBERRY FAMILY.)

Herbs (or shrubs), with alternate petiolate leaves. Sepals and petals in fours, sixes, or eights (except in the genus

Podophyllum), with the petals in front of the sepals. Stamens (except in Podophyllum) as many as the petals, one before each. Anthers usually opening by a valve at the top. Fruit berry-like, or a pod.

Synopsis of the Genera.
Petals and stamens 6.

1. **Ber'beris.** Shrub, with yellow wood and yellow flowers in drooping racemes. Stamens irritable. Petals 6, obovate, concave, each with two glandular spots inside at the base. Fruit a sour berry, oblong, scarlet.
2. **Caulophyl'lum.** A purplish herb, flowering in early spring. Petals thick, much shorter than the sepals. Leaves decompound.

*Petals 6-9. Stamens 8-18.

3. **Podophyl'lum.** Petals 6-9. Stamens 12-18. Anthers *not* opening by uplifting valves. Fruit a large berry. Leaves peltate.
4. **Jeffersu'nia.** Petals and stamens mostly 8. Anthers opening by uplifting valves. Pod opening by a lid. Leaves divided in two.

1. BER'BERIS, L. BARBERRY.

B. vulga'ris, L. (COMMON BARBERRY.) Shrub. Leaves on the fresh shoots of the season mostly reduced to branched spines, from whose axils proceed the next year close clusters of obovate-oblong, bristly-toothed leaves, with short, *jointed* petioles, and many-flowered racemes. Sepals, petals and stamens 6 each. Outside of sepals are 2-6 bractlets. Petals yellow. Fruit an oblong, sour, scarlet berry.—Cultivated grounds.

2. CAULOPHYL'LUM, Michx. BLUE COHOSH.

C. thalictroi'des, Michx. (BLUE COHOSH.) Plant 1-2 feet high, very glaucous and dull purple when young. Flowers yellowish-green, in a terminal small raceme, appearing in spring before the *decompound leaves* are developed. Sepals 6, with 3 little bractlets at their base. Petals 6, thick and somewhat kidney-shaped, much smaller than the sepals. Stamens 6, one before each petal. Ovary bursting soon after the flowering, and leaving the two drupe-like seeds naked on their rather thick stalks. Fruit bluish, ¼ of an inch across.—Rich woods.

2. PODOPHYL'LUM, L. May Apple. Mandrake.

P. pelta'tum, L. Stem about 1 foot high. Flowerless stems with one large 7-9-lobed umbrella-like leaf, peltate in the centre; the flowering ones with two leaves, peltate near the edge, the flower nodding from the fork. Sepals 6, caducous. Petals 6-9, large and white. Stamens 12-18. Fruit large, oval, yellowish, not poisonous.—Found in patches in rich woods. The leaves and roots are poisonous.

5. JEFFERSO'NIA, Barton. Twin-leaf.

J. diphyl'la, Pers. A low plant, flowering in early spring; the solitary white flowers on naked scapes. Sepals 4, fugacious. Petals 8. Stamens 8. Ovary pointed. Stigma 2-lobed. Pod pear-shaped, the top forming a lid. Leaves radical, long-petioled; the blades *divided into two leaflets with the outer margins lobed.*—Woods, chiefly in the western peninsula of Ontario.

Order VI. NYMPHÆA'CEÆ. (Water-Lily Family.)

Aquatic herbs with cordate or peltate, usually floating, leaves. Floating flowers on long immersed peduncles. Petals and stamens generally numerous.

Synopsis of the Genera.

1. **Brase'nia.** Sepals and petals each 3 (occasionally 4). Stamens 12-24. Leaves oval, peltate.
2. **Nymphæ'a.** Sepals 4-6. Petals numerous, *white*, imbricated in many rows, gradually passing into stamens, hypogynous, or epigynous. Stamens epigynous. Stigmas radiating as in a Poppy-head.
3. **Nu'phar.** Sepals 5-6, *yellow*. Petals many, small and stamen-like. Stamens under the ovary.

1. BRASE'NIA, Schreber. Water-Shield.

B. pelta'ta, Pursh. Stems and under surface of the leaves coated with jelly. Leaves oval, two inches across, peltate. Flowers small, purplish.—Ponds and slow-flowing streams.

2. NYMPHÆ'A, Tourn. WATER-LILY.

1. **N. odora'ta,** Ait. (SWEET-SCENTED WATER-LILY.) Leaves orbicular, cleft at the base to the petiole, 5-9 inches wide, often crimson underneath. *Flower very sweet-scented.* Ponds and slow streams.

Var. **minor,** Sims, has much smaller leaves and flowers, and the latter are often pink-tinted.

2. **N. tubero'sa,** Paine. (TUBER-BEARING W.) Leaves larger and more prominently ribbed than in No. 1, reniform-orbicular, green on both sides. Flower not at all, or only slightly, sweet-scented. Root-stocks producing tubers, which come off spontaneously.—Mostly in slow waters opening into Lake Ontario.

3. NUPHAR, Smith. YELLOW POND-LILY.

1. **N. ad'vena,** Ait. (COMMON Y. P.) Leaves floating, or emersed and erect, thickish, roundish or oblong, cordate. *Sepals 6.*—Stagnant water. Var. **minus** is more slender, and has smaller leaves and flowers. Probably a hybrid between this and the next.

2. **N. Kalmia'num,** Ait. (SMALL Y. P.) (*N. luteum, var. pumilum,* Gray.) Floating leaves usually not more than two inches across, the sinus very narrow or closed; the radical ones very thin and with a wide obtuse sinus. Flowers hardly an inch across. *Sepals 5.*—Northward, in slow waters.

ORDER VII. **SARRACENIA'CEÆ.** (PITCHER-PLANT F.)

Bog-plants, easily distinguished by their pitcher-shaped leaves, all radical.

1. SARRACE'NIA, Tourn. SIDE-SADDLE FLOWER.

S. purpu'rea, L. (PURPLE S. HUNTSMAN'S CUP.) Leaves hollow, with a wing on one side, purple-veined, curved, with the hood erect and open. Sepals 5, coloured, with 3 small bractlets at the base. Petals 5, fiddle-shaped, curved over the centre of the flower, deep purple. Ovary 5-celled, globose, the short style expanding above into a 5-angled umbrella, with a hooked stigma at each angle. Flowers on naked scapes, nodding.—Bogs.

ORDER VIII. **PAPAVERA'CEÆ.** (POPPY FAMILY.)

Herbs, with milky or coloured juice and alternate leaves without stipules. Flowers polyandrous, hypogynous. *Sepals 2, caducous.* Petals 4-12. Stamens numerous, anthers introrse. Fruit a 1-celled pod (in Poppy imperfectly many-celled), with numerous seeds.

Synopsis of the Genera.

1. **Chelido'nium.** Juice yellow. Petals 4, crumpled in the bud, yellow.
2. **Sanguina'ria.** Juice red. Petals 8-12, not crumpled in the bud, white.
3. **Papa'ver.** Juice milky. Petals mostly 4. Ovary imperfectly many-celled. Stigmas united into a radiate sessile crown.

1. CHELIDO'NIUM, L. CELANDINE.

C. majus, L. Petals 4, deciduous, crumpled in the bud. *Juice of the plant yellow.* Flower-buds nodding. Flowers small, yellow, in a kind of umbel. Fruit a smooth 1-celled slender pod, from which the two valves fall away, leaving the parietal placentas as a slender framework, with the seeds attached.—Waste places.

2. SANGUINA'RIA, Dill. BLOOD-ROOT.

S. Canadensis, L. Petals 8-12, not crumpled in the bud. Flower-buds not nodding. A stemless plant, with a thick rhizome which emits a *red juice* when cut, and sends up in early spring a single rounded, 5-7-lobed, thickish leaf, and a 1-flowered scape. *Flowers white.*—Rich woods.

3. PAPA'VER, Tourn. POPPY.

1. **P. somnif'erum,** L. (COMMON POPPY.) *Smooth and glaucous.* Leaves clasping, wavy, cut-toothed. *Pod globose.* Petals white or purple, the *buds nodding.*—Waste places and old gardens.

2. **P. Rhœ'as,** L. (CORN POPPY.) *Sparingly hispid.* Leaves deeply pinnatifid, the lobes cut-toothed. *Pod globular,* smooth. Corolla 2-4 inches in diameter, scarlet, often with a dark centre.—Waste-heaps, Atl. Prov.

3. **P. dubium**, L. Smaller than the last, and the lobes of the leaves narrower. Pod oblong, narrowed at the base. Corolla a paler red.—Ballast-heaps, Atl. Prov.

ORDER IX. **FUMARIA'CEÆ.** (FUMITORY FAMILY.)

Smooth herbs, with brittle stems, watery juice, dissected leaves, and irregular flowers. Sepals 2, very small. Corolla flattened and closed, of 4 petals, the two inner united by their tips over the anthers of the 6 stamens. Stamens in two sets of 3 each; filaments often united; the middle anther of each set 2-celled, the others 1-celled. Fruit a 1-celled pod.

Synopsis of the Genera.

1. **Adlu'mia.** Corolla 2-spurred. Petals all permanently united. *Plant climbing.*
2. **Dicen'tra.** Corolla 2-spurred. Petals slightly united, easily separated. Not climbing.
3. **Coryd'alis.** Corolla 1-spurred. Fruit a slender pod, many-seeded.
4. **Fuma'ria.** Corolla 1-spurred. Fruit a globular 1-seeded nutlet, indehiscent.

1. ADLU'MIA, Raf. CLIMBING FUMITORY.

A. cirrho'sa, Raf. A smooth vine, climbing by the petioles of its decompound leaves. Flowers in axillary pendulous clusters, pale pink.—Low and shady grounds, and rocky woods.

2. DICEN'TRA, Borkh. DUTCHMAN'S BREECHES.

1. **D. Cucullaria**, DC. (DUTCHMAN'S BREECHES.) Leaves all radical, multifid; these and the slender scapes rising from a bulb-like rhizome of coarse grains. Flowers several in a raceme, whitish, spurs *divergent, elongated, acute, straight.*—Rich woods.

2. **D. Canadensis**, DC. (SQUIRREL CORN.) Underground shoots bearing small yellow tubers, something like grains of corn. Leaves very much as in No. 1. Corolla merely *heart-shaped; spurs very short and rounded.* Flowers greenish-white, fragrant.—Rich woods.

3. CORYD'ALIS, Vent. CORYDALIS.

1. **C. au'rea,** Willd. (GOLDEN CORYDALIS.) *Stems low and spreading.* Leaves dissected. *Flowers in simple racemes, golden yellow,* the outer petals keeled, but *not crested* on the back. Pods pendulous.—Rocky river-margins and burnt woods.

2. **C. flav'ula,** DC. (YELLOW C.) Stems low and spreading. Flowers pale yellow, the outer petals *wing-crested* on the back; crest 3-4-toothed.—South-western Ontario.

3. **C. glaucà,** Pursh. (PALE CORYDALIS.) *Stems upright,* 1-4 feet high. *Flowers in compound racemes, purplish tipped with yellow.* Pods erect.—Rocky woods.

4. FUMA'RIA, Tourn. FUMITORY.

F. officina'lis, L. (COMMON FUMITORY.) Corolla flesh-coloured, tipped with crimson. Flowers small, in dense racemes or spikes.—Waste places near dwellings.

ORDER X. CRUCIF'ERÆ. (CRESS FAMILY.)

Herbs with a pungent watery juice, alternate leaves without stipules, and regular hypogynous flowers in racemes or corymbs. Pedicels without bractlets. Sepals 4, deciduous. Petals 4, forming a cross-shaped corolla. Stamens 6, two of them shorter. Fruit a silique or silicle. (See Chap. IV., Part I. for dissection of typical flower.) The genera are distinguished by the pods and seeds, the flowers in all cases being much alike. The seeds are exalbuminous, consisting entirely of the embryo, which is folded up in a variety of ways. The radicle may be bent so as to lie against the *edge* of the cotyledons, and the seed when cut through cross-wise shows this section: o⇒; the cotyledons are then said to be *accumbent*. Or the radicle may be folded against the *back* of the cotyledon, showing this cross-section: ⇔, in which case the cotyledons are said to be *incumbent*; and if, besides being incumbent, the cotyledons are doubled round the radicle, thus: ⋓, they are then *conduplicate*.

Synopsis of the Genera.

* *Pod dehiscent by 2 valves, usually much longer than broad.*

1. **Nastur'tium.** Flowers white or yellow. Pod terete, oblong-linear or ellipsoid. Seeds in *two rows* in each cell, globular, without a wing. Cotyledons accumbent.
2. **Barbare'a.** Flowers yellow. Pod somewhat 4-sided, *the valves strongly 1-nerved.* Seeds in *one row* in each cell. Cotyledons accumbent.
3. **Vesica'ria.** Flowers yellow. Plant densely hoary with stellate hairs. Pods ovate or globose. Seeds flattened, in *two rows* in each cell. Cotyledons accumbent.
4. **Denta'ria.** Flowers white or pale purple. Pod lanceolate, flat. Seeds wingless, on broad seed-stalks. *Stem-leaves 2 or 3 in a whorl;* stem naked below. *Root-stock toothed* or tuberous. Cotyledons accumbent.
5. **Cardam'ine.** Flowers white or rose-coloured. Pod linear or lanceolate, flat, the valves nerveless. Seeds wingless, on slender seed-stalks. Stem leafy below. Cotyledons accumbent.
6. **Ar'abis.** Flowers white or whitish. Pod linear or *elongated,* flattened, *the valves usually with a distinct mid-rib.* Stem leafy. Cotyledons accumbent.
7. **Erys'imum.** Flowers yellow (or creamy). Pod linear, distinctly 4-sided. Stigma broadly 2-lobed. *Pedicels of the pods diverging from the stem.* Leaves simple, not clasping. Pubescent with appressed hairs. Cotyledons incumbent.
8. **Sisym'brium.** Flowers yellow, small. Pods awl-shaped or 4-6-sided, the valves 1-3-nerved. Stigma small. Cotyledons incumbent.
9. **Hes'peris.** Flowers large, purple. Pod terete, elongated. Stigma-lobes narrow, erect. Cotyledons incumbent.
10. **Bras'sica.** Flowers yellow. Pod linear or oblong, nearly terete, or 4-sided, *with a distinct beak extending beyond the end of the valves.* Seeds usually in *one row.* Cotyledons conduplicate.
11. **Diplotax'is.** Sepals, petals, and stamens almost as in Brassica. *Pods slender, on slender spreading pedicels.* Beak conical, usually *short.* Seeds in *two rows.*

* * *Pod dehiscent by 2 valves, comparatively short.*

← *Silicle compressed parallel with the broad partition or globular.*

12. **Draba.** Flowers white. Pod flat; seeds several or many, 2 rows in each cell. Cotyledons accumbent. Low herbs.
13. **Alys'sum.** Flowers pale yellow or white. Pod orbicular, flat, 2-4-seeded.
14. **Camel'ina.** Flowers yellow. Pod pear-shaped, pointed; valves 1-nerved. Cotyledons incumbent.

15. **Nes'lia.** Flowers very much as in the last. Pod nearly globular, *hardly one-twelfth of an inch across*, usually only 1-celled and 1-seeded, wrinkled, tipped with the slender style.

← ←*Silicle compressed contrary to the narrow partition.*

16. **Capsel'la.** Flowers white. Pod obcordate-triangular; valves boat-shaped, *wingless. Seeds numerous.* Cotyledons incumbent.
17. **Thlas'pi.** Flowers white. Pod obovate or obcordate, *winged.* Seeds several. Cotyledons accumbent.
18. **Lepid'ium.** Flowers white or whitish. Pod roundish, very flat; the valves boat-shaped and *winged. Seeds solitary.*
19. **Senebie'ra.** Flowers greenish-white, very small. Pod 2-seeded, the valves wrinkled and separating at maturity as 2 *closed* one-seeded nutlets. Prostrate diffuse herbs, with a disagreeable characteristic odour.

* * * *Pod indehiscent, fleshy, jointed.*

20. **Cakl'le.** Flowers purplish. Pod 2-jointed, fleshy. Leaves fleshy. Cotyledons accumbent.
21. **Raph'anus.** Flowers yellow, turning whitish or purplish. Pod elongated, several-seeded, mostly constricted between the seeds. Cotyledons conduplicate.

1. NASTUR'TIUM, R. Br. WATER-CRESS.

1. **N. officina'le,** R. Br. (WATER-CRESS.) Flowers white. Stem spreading and rooting. Leaves pinnate; leaflets 3–11, roundish or oblong, nearly entire. Pods oblong-linear.—Ditches and streamlets.

2. **N. palus'tre,** DC. (MARSH CRESS.) Flowers yellow. Stem erect. Leaves pinnately parted, the lobes cut-toothed. Pods ovoid.—Wet places.

3. **N. lacus'tre,** Gray. (LAKE CRESS.) Flowers white. An aquatic plant, with the submerged leaves finely dissected; the leaves out of the water oblong, and either entire, serrate or pinnatifid. Pods ovoid, 1-celled.

4. **N. Armora'cia,** Fries. (HORSERADISH.) Has escaped from gardens in many places. Flowers white. Root-leaves very large, oblong, and generally crenate; stem-leaves lanceolate. Pods globular. Roots very large.

5. **N. trachycar'pum,** Gray. Flowers white. Stem nearly glabrous, erect, branching. Leaves lyrate-pinnatifid. Pods oblong-linear, *somewhat roughened,* soon

recurved, shorter than the awl-shaped style.—N. W., in beds of pools and streams.

6. **N. amphib'ium**, R. Br. Flowers white. Stem tall, erect, branching above, sparingly pubescent. Leaves lyrate-pinnatifid, rough-pubescent, the lobes coarsely toothed. Pods several times longer than the style, but much shorter than the pedicels.—Wet places.

2. BARBARE'A, R. Br. WINTER CRESS.

B. vulga'ris, R. Br. (YELLOW ROCKET.) Stem smooth, 1-3 feet high. Lower leaves lyrate, the terminal division round and generally large, and 1-5 pairs of lateral ones (the latter sometimes absent); upper leaves obovate, more or less pinnatifid at the base. Pods linear, erect, or slightly spreading. In var. **stricta**, a common form in the North-West, the pods are *appressed*.—Low grounds.

3. VESICA'RIA, Lam. BLADDER-POD.

V. Ludovicia'na, DC. (WESTERN BLADDER-POD.) (*Lesquerella Ludoviciana*, Watson, *var. arenosa*, Watson.) Stem low, simple, or somewhat branched above. Flowers large. Radical leaves spathulate, entire; stem-leaves linear. Pods hairy, globose, rather longer than the style.—N. W.

4. DENTA'RIA, L. TOOTHWORT. PEPPER-ROOT.

1. **D. diphyl'la**, L. (TWO-LEAVED T.) Flowers white. *Stem-leaves 2, nearly opposite*, ternately divided. Root-stock toothed, pleasantly pungent to the taste.—Rich woods.

2. **D. lacinia'ta**, Muhl. (LACINIATE T.) Flowers purplish. *Stem-leaves 3, in a whorl*. Root-stock jointed, scarcely toothed.—Rich woods.

5. CARDAM'INE, L. BITTER CRESS.

1. **C. rhomboi'dea**, DC. (SPRING CRESS.) Flowers white or (in var. **purpurea**) rose-purple. *Stem tuberous at the base*. Lower leaves round-cordate; upper nearly lanceolate; all somewhat angled or toothed.—Wet meadows.

2. **C. pratensis**, L. (CUCKOO-FLOWER. LADIES' SMOCK.) Flowers white or rose-colour, showy. Stem from a short

root-stock. Leaves pinnate, leaflets 7-15, those of the lower leaves rounded and stalked, entire or nearly so.—Bogs.

3. **C. hirsu'ta,** L. (SMALL BITTER CRESS.) Flowers white, small. Root fibrous. Leaves pinnate, leaflets 5-11, the terminal leaflets largest. Pods erect, *slender*.—Wet places.

6. AR'ABIS, L. ROCK CRESS.

* *Seeds in one row in each cell, nearly as broad as the partition.*

1. **A. lyra'ta,** L. (LOW R.) Flowers white, petals twice as long as the calyx. Stem branching from the base. Radical leaves clustered, pinnatifid, the terminal lobe largest; stem-leaves scattered, linear, with tapering base. Pods slender, erect, and spreading.—Rocky or sandy shores.

2. **A. hirsu'ta,** Scop. (HAIRY R.) Flowers greenish-white, small, petals slightly longer than the calyx. Stem-leaves many, rough, sagittate; often a dense rosette at the base of the stem. *Pods erect, straight.* Stems 1-2 feet high, 2 or 3 from the same root.—Rocky shores and dry plains.

. **A. lævíga'ta,** DC. (SMOOTH R.) Flowers white, rather small. Leaves linear or lanceolate, entire or slightly toothed, sagittate, clasping. Pods long and narrow, *recurved-spreading*. Stem glaucous, 1-2 feet high.—Dry hill-sides. Easily recognized by the pods.

4. **A. Canadensis,** L. (SICKLE-POD.) Flowers whitish, with linear petals, about twice the length of the calyx. Stem-leaves pointed at both ends, downy, the lower ones toothed. Pods 2-3 inches long, *scythe-shaped, hanging.* Stem 2-3 feet high. A striking plant when the pods are fully formed.—Dry woods and ravines.

5. **A. petræ'a,** Lam. Petals rose-colour or whitish. Pods shorter and less flat than A. lyrata. Leaves spathulate or oblong, entire or sparingly toothed.—Rocks; reported from New Brunswick.

* * *Seeds in two distinct rows in each cell, narrower than the partition.*

6. **A. perfolia'ta,** Lam. (TOWER MUSTARD). Flowers yellowish-white. Petals scarcely longer than the calyx.

Stem 2-4 feet high, *glaucous*. Cauline leaves ovate-lanceolate or oblong, clasping with sagittate base. Pods long and *very narrow, on erect pedicels.*—Meadows and old fields. Pretty easily recognized by its strict habit.

7. **A. confi'nis,** Watson. (*A. Drummondii*, Gray.) Flowers white or rose-colour. Petals twice as long as the calyx. Stem 1-2 feet high, *smooth above*. Cauline leaves lanceolate or oblong-linear, with sagittate base; lowest leaves spathulate and toothed. Pods long and flat; *the pedicels not so strictly erect* as in the last species.—Rocky banks of streams.

8. **A. Holbœl'lii,** Hornem. Petals white or rose-colour, occasionally purplish, *becoming reflexed*, twice as long as the sepals. Stem 6-24 inches high, usually simple, more or less hairy. Lowest leaves small, spathulate, somewhat toothed, with margined petioles; upper sessile, sagittate, linear-lanceolate. *Pods reflexed. Style none.*—N. W.

7. ERYS'IMUM, L. TREACLE MUSTARD.

1. **E. cheiranthoi'des,** L. (WORM-SEED MUSTARD.) Flowers yellow, *inconspicuous*. Stem slender, branching. Leaves lanceolate, scarcely toothed, roughish with appressed pubescence. Pods *small and short*, on slender diverging pedicels.—Waste wet places.

2. **E. as'perum,** DC. (PRAIRIE ROCKET.) Flowers *showy*, bright orange-yellow, rarely purple. Stem stout, 1-2 feet high, simple, hoary, with minute appressed hairs. Leaves oblanceolate. Pods long, *ascending* on *stout* spreading pedicels.—Dry soil, N. W.

3. **E. parviflo'rum,** Nutt. Stem low and simple, hoary. Leaves all linear, densely clustered at the base of the stem. Flowers *small*, sulphur-yellow. Pods long, narrow, ascending, on short pedicels.—N. W.

4. **E. orienta'le,** R.Br. Stem slender and branching, 1-2 feet high. Leaves grayish-green, oblong to oval, slightly clasping; radical ones spathulate. Flowers pale yellow or cream-coloured, *small.*—N. W. and Atl. Prov.

8. SISYM'BRIUM, L. HEDGE MUSTARD.

1. **S. officina'le,** Scop. (HEDGE MUSTARD.) Flowers yellow, small. Leaves runcinate. Stem 1-2 feet high, with spreading branches. *Pods awl-shaped, close pressed to the stem.*—A very common roadside weed.

2. **S. canes'cens,** Nutt. (TANSY-MUSTARD.) Flowers yellowish, very small. Leaves twice pinnatifid, *hoary with short branching hairs,* the divisions toothed. Pods acute at each end, shorter than the slender *spreading* pedicels.—N. W.

3. **S. Soph'ia,** L. Hoary. Leaves dissected into very numerous small leaflets. Pods very slender, about an inch long, ascending.—Scarce.

4. **S. sinapistrum,** Crantz. Tall, branching. Loosely pubescent below, smooth above. Leaves deeply pinnatifid, the lobes narrowly linear. *Pods very long* (3-4 inches) *and slender, spreading.*—An introduced weed which has now become a pest in the North-West.

5. **S. inci'sum,** Engelm., var. **Hartwegia'num,** Watson, has *pinnate* leaves, the leaflets narrowly oblong and toothed. Pods only one-eighth to one-fourth of an inch long, in a crowded raceme.—N.W.

9. HES'PERIS, Tourn. ROCKET.

H. **matrona'lis,** L. (GARDEN ROCKET.) Tall, leaves lanceolate, acuminate, serrate, usually petiolate. Flowers large, purple. Pods 2-4 inches long, spreading.—Waste places.

10. BRAS'SICA, Tourn. CABBAGE, MUSTARD, ETC.

1. **B. Sinapis'trum,** Bois. (CHARLOCK.) Flowers bright yellow. Stem 1-2 feet high, branching, *it and the leaves hairy.*—Too common in our grain fields.

2. **B. ni'gra.** (BLACK MUSTARD.) Flowers sulphur-yellow. Stem 3-6 feet high, round, *smooth* and branching. Lower leaves lyrate.—Fields and waste places.

2. **B. campes'tris,** L. Stem tall and erect, it and the leaves *smooth.* Lower leaves ovate or elliptical, coarsely

toothed, on long petioles; upper ones narrow and sessile. Pods an inch long or more, on ascending pedicels; beak about ⅛ the length of the pod.—A common weed in Manitoba.

11. DIPLOTAXIS, DC.

1. **D. mura'lis**, DC. Branching from near the base. Smooth or sparingly hispid. Leaves oblanceolate, shallowly and bluntly toothed or pinnatifid. Petals pale yellow. Pods over an inch long, on slender spreading pedicels half as long as the pods. Valves 1-nerved.—Ballast-heaps, Atl. Prov.

2. **D. tenuifo'lia**, DC. Like the last, but the leaves are *deeply sinuate-pinnatifid, with narrow lobes*. Pedicels of the pods over an inch long.—Ballast-heaps, Atl. Prov.

12. DRABA, DC. WHITLOW-GRASS.

1. **D. inca'na**, L. *Hoary-pubescent.* Flowers white. Leaves lanceolate or oblanceolate to ovate, entire or sparingly toothed. Pods oblong to lanceolate, usually *straight*, on short erect pedicels. *Style short or none.*—Dry rocks.

Var. **arab'isans**, Watson, has the pod glabrous, *twisted*, and tipped with a distinct style.—Rocks.

2. **D. nemoro'sa**, L. A small slender plant with leaves about ¼ of an inch long. Flowers white. Leaves oblong, or somewhat lanceolate, hairy, sessile, serrate. Racemes *elongated*. Pods elliptical-oblong, half as long as the widespreading pedicels, pubescent or smooth. Style none.—N. W.

3. **D. Carolinia'na**, Walt. Flowers white. Stem 1-5 inches high. Leaves obovate, entire. Raceme *very short*. Pods broadly linear, smooth, longer than the ascending pedicels. Style none.—Rocks; Southern Ontario.

13. ALYS'SUM, Tourn. ALYSSUM.

A. calyci'num, L. A dwarf hoary annual, with linear-spathulate leaves. Calyx persistent. Pod 4-seeded, sharp-edged.—Rather rare; abundant at Queenston Heights.

14. CAMEL'INA, Crantz. FALSE FLAX.

C. sati'va, Crantz. (COMMON F. FLAX.) Flowers yellowish. Stem 1-2 feet high, straight, erect, branching. Leaves lanceolate, sagittate. Pods pear-shaped, large, margined; style slender.—In flax fields.

15. NESLIA, L. BALL MUSTARD.

N. panicula'ta, Desv. Flowers yellow, in small panicles.—A troublesome weed in Manitoba.

16. CAPSEL'LA, Vent. SHEPHERD'S PURSE.

C. Bursa-pasto'ris, Mœnch. Flowers small, white. Root-leaves clustered, pinnatifid; stem-leaves clasping, sagittate. —A very common weed.

17. THLASPI, Tourn. PENNYCRESS.

T. arvense, L. (FIELD PENNYCRESS. FRENCH WEED.) A low smooth plant, with undivided radical leaves, and stem-leaves sagittate and clasping. Pods half an inch broad, deeply notched at the top.—Waste places: common N.W.

18. LEPID'IUM, L. PEPPERGRASS.

1. **L. Virgin'icum,** L. (WILD P.) Flowers small; *petals present*, white. Stem 1-2 feet high. Leaves lanceolate, the upper linear or lanceolate and entire, the lower toothed or pinnatifid, tapering towards the base. Pods marginless or nearly so, oval or orbicular.—Railways and roadsides.

2. **L. interme'dium,** Gray. Distinguished from No. 1 by having the cotyledons incumbent instead of accumbent, and the pods minutely winged at the top.—Dry sandy fields.

3. **L. rudera'le,** L. *Petals always absent.* More branched than the preceding.

4. **L. campestre,** L. Well distinguished from other species by its *sagittate, clasping* leaves. Pods ovate, winged. —Rather rare.

5. **L. sati'vum,** L., has leaves variously divided and cut, with numerous roundish winged pods, and flowers white or rose-coloured.—Escaped from cultivation in some places.

6. **L. Draba,** L. An obscurely hoary perennial. Upper leaves auricled, lower sessile, oblong. *Pods heart-shaped,* the style conspicuous.—Abundant about Clifton, Ont.

19. SENEBIE'RA, DC. WART-CRESS. SWINE-CRESS.

1. S. did'yma, Pers. Pods notched at the apex.—Atl. Prov.
2. S. corono'pus, DC. Pods not notched.—Atl. Prov.

20. CAKI'LE, Tourn. SEA-ROCKET.

C. America'na, Nutt. (AMERICAN S.) Flowers purplish. Leaves obovate, fleshy, wavy-toothed. Pod fleshy, 2-jointed.—Seashore, and borders of the Great Lakes.

21. RAPH'ANUS, Tourn. RADISH.

R. Raphanis'trum, L. (WILD RADISH.) Pods linear or oblong, tapering, indehiscent, long-beaked, necklace-form when ripe. Style long. Cotyledons conduplicate. Leaves rough, lyrate. Petals yellow, veiny, turning whitish or purplish.—An introduced weed; chiefly eastward.

ORDER XI. **CAPPARIDA'CEÆ.** CAPER FAMILY.

Herbs (in Canada), with an acrid watery juice, and alternate palmately compound leaves. Flowers cruciform. Stamens 8 or more. Pod like that of a crucifer, *but only 1-celled.*

Synopsis of the Genera.

1. **Polanis'ia.** Stamens 8 or more. Pod *sessile or nearly so.*
2. **Cleo'me.** Stamens 6. Pod *long-stalked.*

1. POLANIS'IA, Raf. POLANISIA.

1. P. grave'olens, Raf. A strong-scented herb, with a viscid hairy stem. Leaflets 3. Flowers in terminal racemes. Sepals 4. Petals 4, yellowish-white, narrowed below into long claws. Stamens 8–12, exserted. Pod glandular-pubescent, 2 inches long, linear, with a very short stalk.—Shore of Lake Ontario, Hamilton to Niagara.

2. P. trachysper'ma, Torr. and Gray, has larger flowers and *strongly* exserted stamens, and a *sessile* pod.—N.W.

2. CLEO'ME, L.

C. integrifo'lia, Torr. and Gray. Stem 2–3 feet high, glabrous. Flowers showy, rose-coloured or white, in leafy-bracted racemes. Leaves of 3 lanceolate or oblong leaflets. Stipe of the pod as long as the pedicel of the flower.—Dry clay soil; N.W.

ORDER XII. **VIOLA'CEÆ.** (VIOLET FAMILY.)

Herbs, with alternate stipulate leaves. Flowers irregular, the lower of the 5 petals being spurred. Sepals 5, persistent. Stamens 5, the anthers more or less coherent, and surrounding the pistil. Fruit a 1-celled pod, splitting into 3 valves. Seeds in three rows on the walls of the ovary.

Synopsis of the Genera.

1. **Vi'ola.** Sepals auricled at the base. Lower petal *distinctly spurred*. Stamens only slightly, if at all, coherent; the 2 lower ones spurred.
2. **So'lea.** Sepals *not* auricled. Lower petal *saccate* at the base. Stamens completely united and sheathing the ovary. Plant 1-2 feet high.

1. VI'OLA, L. VIOLET.

* *Stemless Violets ; leaves and scapes all from root-stocks.*
+ *Flowers white.*

1. **V. blanda,** Willd. (SWEET WHITE V.) Lower petal streaked with purple. Leaves round-heart-shaped or reniform. Petals beardless. Flower sweet-scented.—Swamps and wet meadows, in spring.

Var. **renifo'lia,** Gray, has leaves *much larger and more pubescent* than those of the preceding.—Dry cedar swamps, and ravines in rich woods.

2. **V. lanceola'ta,** L. (LANCE-LEAVED V.) Flowers white. Petals beardless. Leaves lanceolate, erect, tapering into a long, margined petiole, nearly entire.—Damp ground, mostly eastward.

3. **V. primulæfo'lia,** L. (PRIMROSE-LEAVED V.) Flowers white, lateral petals usually somewhat bearded. Leaves ovate or oblong, heart-shaped, or abrupt at the base.—Damp or dry ground; Atl. Prov.

+ + *Flowers blue or purple.*

4. **V. peda'ta,** L. (BIRD-FOOT V.) Nearly smooth. Rootstock *short, thick and erect.* Leaves all deeply cut into 3-5 segments, the lateral divisions 2-3-parted. Flower about an inch across; stigma large and not beaked.—N W.

5. **V. pedatif'ida,** G. Don. Very much like the last, but the flowers are smaller and of a deeper blue, and the stigma is beaked.—Prairies, N. W.

6. **V. Selkirk'ii,** Pursh. (GREAT-SPURRED V.) A small and delicate plant, distinguished by the *slender* root-stock, and the *very large spur, thickened at the end.* The pale violet petals also are *beardless.*—Damp, shady places.

7. **V. palustris,** L. (MARSH V.) Very similar in foliage, etc., to No. 1, but the flowers are pale lilac, and the root-stock is jointed.—Wet swamps amongst moss.

8. **V. palma'ta,** L. (COMMON BLUE V.) *Leaves on very long petioles,* cordate or reniform, the sides folded inwards when young, *the later ones variously lobed or parted.* Lateral petals bearded. Spur short and thick—Low grounds everywhere.

Var. **cuculla'ta,** Gray, has the later leaves merely crenate.

9. **V. odora'ta,** L. (ENGLISH SWEET V.) has escaped from gardens in some places. Flowers very fragrant.

10. **V. sagitta'ta,** Ait. (ARROW-LEAVED V.) Smoothish. Leaves cordate, halberd-shaped, or sagittate, slightly toothed, *the first ones on short and margined petioles.* Side-petals bearded.—Dry hill-sides and old pastures.

+ + + *Flowers yellow.*

11. **V. rotundifo'lia,** Michx. (ROUND-LEAVED V.) Leaves round-ovate, cordate, repand-crenulate, about an inch wide at flowering, increasing later to 3 or 4 inches, and then flat on the ground, shining above. Lateral petals bearded and marked with brown lines. Spur very short.—Cold woods, chiefly eastward.

* * *Leafy-stemmed Violets.*

+ *Flowers yellow.*

12. **V. pubes'cens,** Ait. (DOWNY YELLOW V.) Plant downy, 6-12 inches high. Leaves broadly cordate, coarsely serrate; stipules large, entire. Lower petals veined with purple. Spur very short.—Rich woods.

Var **scabriuscula,** Torr. and Gray, is smaller, and less pubescent, often nearly smooth.

13. **V. Nuttal'lii,** Pursh. Low, densely pubescent, or sometimes nearly glabrous. Leaves oblong-ovate or oblong, obtuse, entire or obscurely sinuate, *decurrent on the petiole;* stipules mostly narrow, entire.—Dry soil, N.W.

← ← *Flowers not yellow.*

14. **V. Canadensis,** L. (CANADA V.) Tall, often a foot high. Leaves large, cordate, serrate, pointed. Petals white inside, *purplish outside.* Spur very short.—Flowering all summer.

15. **V. cani'na,** L., var. **sylvestris,** Regel. (DOG V.) Low, spreading by runners. Leaves broadly cordate or reniform, *with fringed-toothed stipules.* Spur cylindrical, half as long as the petals, which are *pale purple.*—Wet places.

Var. **lon'gipes,** Watson, of the N. W. plains, has *ovate* leaves, obscurely crenate. Spur as long as the sepals, stout, obtuse, and nearly straight.

16. **V. stria'ta,** Ait. (PALE V.) Stem angular, 6–10 inches high. Leaves cordate, finely serrate; stipules fringed-toothed. Spur thickish, much shorter than the *cream-coloured or white petals.*—Low grounds.

17. **V. rostra'ta,** Pursh. (LONG-SPURRED V.) Distinguished at once by its extremely long straight spur. Petals violet-coloured.

18. **V. tri'color,** L., var. **arvensis,** DC. (PANSY.) *Stipules large, leaf-like and lyrate-pinnatifid.* Stem angled and branched. Leaves roundish. Petals variable in colour, about as long as the sepals.—Dry soil.

2. **SO'LEA,** Spreng., in part. GREEN VIOLET.

S. con'color, Ging. A homely herb with oblong entire leaves pointed at each end, and 1-3 small greenish-white flowers in the axils, on short recurved pedicels. Pod an inch long.—Rare; Niagara River and the banks of the Thames.

ORDER XIII. CISTA'CEÆ. (ROCK-ROSE FAMILY.)

Herbs or low shrubs, with simple entire leaves and regular polyandrous flowers. Calyx persistent, usually of 3 large and 2 smaller sepals. Petals 5 or 3, convolute in the bud. Stamens 3-20. Pod 1-celled, 3-valved. Seeds on 3 parietal projections.

Synopsis of the Genera.

1. **Helian'themum.** Petals 5, fugacious. Style none.
2. **Hudso'nia.** Petals 5, fugacious. Style long and slender.
3. **Lech'ea.** Petals 3, persistent. Style none.

1. HELIAN'THEMUM, Tourn. ROCK-ROSE.

H. Canadense, Michx. (FROST-WEED.) Flowers of two sorts, some solitary, with large yellow corolla and many stamens, the petals lasting but one day after the flower opens; others small, clustered in the axils of the leaves, and apetalous. Leaves lanceolate, downy beneath.—Sandy places.

2. HUDSO'NIA, L. HUDSONIA.

1. **H. tomento'sa,** Nutt. (DOWNY H.) *Hoary.* Leaves oval or narrowly oblong, short, close-pressed, or imbricated. Flowers small, *sessile, yellow, very numerous.*—A little heath-like shrub, on the shores of the Great Lakes and the River St. Lawrence.

2. **H. ericoi'des,** L. Downy, but *greenish.* Leaves slender, awl-shaped. Flowers *on slender naked stalks.*—Dry soil, Atl. coast.

3. LECH'EA, L. PINWEED.

1. **L. minor,** Lam. (SMALLER P.) Flowers inconspicuous, purplish, loosely racemose, on distinct pedicels. Stem slender, rough with appressed scattered hairs, producing radical shoots. Leaves scattered, linear. Pods the size of a pin's head.—Dry soil.

2. **L. major,** Michx. Stem 1-2 feet high, stout, very leafy, villous with *spreading* hairs, producing prostrate branches at the base. Leaves elliptical, mucronate. Flowers *densely crowded, on very short pedicels.*—Dry soil, S. W. Ontario.

ORDER XIV. **DROSERA'CEÆ.** (SUNDEW FAMILY.)

Low glandular-hairy marsh herbs, with circinate tufted radical leaves, and regular hypogynous flowers borne on a naked scape. Sepals, petals, and stamens, 5 each; anthers turned outwards. Styles 3-5, deeply 2-parted. Pod 1-celled, 3-valved. The only genus with us is

DROS'ERA, L. SUNDEW.

1. **D. rotundifo'lia,** L. (ROUND-LEAVED SUNDEW.) Flowers small, white, in a 1-sided raceme. Leaves orbicular, abruptly narrowed into the hairy petiole, clothed with reddish glandular hairs.—Bogs.

2. **D. longifo'lia,** L. (LONGER-LEAVED S.) has oblong-spathulate leaves gradually narrowed into erect naked petioles.—Bogs; not common.

ORDER XV. **HYPERICA'CEÆ.** (ST. JOHN'S-WORT F.)

Herbs or shrubs, with *opposite entire dotted leaves*, and no stipules. Flowers regular, hypogynous, mostly yellow. Sepals 5, persistent. Petals 5, deciduous. Stamens mostly numerous, and *usually in 3 or more clusters.* Styles 3-5, sometimes united. Pod 1-5-celled. Seeds numerous.

Synopsis of the Genera.

1. **Hyper'icum.** Petals 5, *unequal-sided*, convolute in the bud, *yellow.*
2. **Elo'des.** Petals 5, *equal-sided*, imbricated in the bud, *purplish.*

1. HYPER'ICUM, L. ST. JOHN'S-WORT.

* *Pod 3-celled. Styles 3, separate. Petals with black dots.*

1. **H. perfora'tum,** L. (COMMON ST. JOHN'S-WORT.) Stem much branched, *producing runners at the base,* slightly 2-edged. Leaves linear-oblong, *with transparent dots,* easily observed by holding the leaf up to the light. *Petals deep yellow.* Flowers in open leafy cymes.—Fields.

2. **H. corymbo'sum,** Muhl. (CORYMBED S.) (*H. maculatum,* Walt.) Stem rounded, not so branching as No. 1. *Leaves with both black and transparent dots,* oblong, somewhat

clasping. Flowers *small, pale yellow, crowded, corymbsa.—* Damp woods and wet places generally.

** *Pod 5-celled. Styles more or less united. Stamens very many, in 5 clusters, if clustered at all.*

3. **H. pyramida'tum,** Ait. (GREAT ST. JOHN'S-WORT.) Stem 3-5 feet high. Leaves 2-3 inches long, somewhat clasping. *Flowers very large, the petals about an inch long,* and narrowly obovate. Stamens showy. Pod conical, large.—Along streams; not common.

4. **H. Kalmia'num,** L. (KALM'S S.) *Shrubby,* a foot or more in height; leaves linear-lanceolate, crowded, revolute on the margins, thickly punctate, and sessile. Flowers about 1 inch across, in clusters.—Niagara Falls and westward.

*** *Pod 1-celled, purple.*

5. **H. ellip'ticum,** Hook. (ELLIPTICAL-LEAVED S.) Stem about 1 foot high, *not branched.* Leaves spreading, elliptical-oblong, obtuse, thin. Flowers rather few, showy, in a nearly naked cyme. Pod purple, ovoid, obtuse. Petals pale yellow.—Banks of streams, eastward.

6. **H. mu'tilum,** L. (SMALL S.) Stem slender, branching above, hardly a foot high. Leaves 5-nerved. Cymes leafy at the base. *Flowers small,* not ¼ of an inch across.—Low grounds.

7. **H. Canadense,** L. (CANADA S.) Stem upright, 6-15 inches high, with branches erect. Leaves linear or linear-lanceolate, 3-nerved at the base, the upper ones acute, sessile. Cymes naked. Pod much longer than the calyx. Flowers small, deep yellow.—Wet, sandy places.

2. **ELO'DES.** Adans. MARSH ST. JOHN'S-WORT.

E. Virgin'ica, Nutt. (*E. campanulata,* Pursh.) Stem smooth. Leaves oblong or oval, clasping, often purple-veined, obtuse, conspicuously dotted beneath. Flowers flesh-coloured, in the axils, and at the summit of the stem. The whole plant is of a purplish hue.—Marshes.

Order XVI. CARYOPHYLLA'CEÆ. (Pink Family.)

Herbs with opposite (occasionally whorled) and entire leaves, *the stems swollen at the joints.* Flowers regular, with the parts mostly in fives, occasionally in fours. Stamens not more than twice as many as the petals. Styles 2-5, stigmatic along the inner side. Pod usually 1-celled, with the seeds attached to the base, or to a column which rises from the centre of the cell. (Part I., Fig. 194.)

Synopsis of the Genera.

* *Sepals united into a tube or cup. Petals and stamens borne on the stalk of the ovary; petals with long narrow claws.*

1. **Sapona'ria.** Calyx cylindrical or 5-angled. Styles 2.
2. **Sile'ne.** Calyx 5-toothed. Styles 3.
3. **Lych'nis.** Calyx 5-toothed. Styles 5.

* * *Sepals separate to the base or nearly so. Petals without claws, they and the stamens inserted at the base of the sessile ovary. Low herbs.*

    + *Stipules none.*

4. **Arena'ria.** Petals not cleft at the apex. Styles usually 3. Pod splitting into 3 or 6 valves.
5. **Stella'ria.** Petals 2-cleft at the apex. Pod splitting to the base into twice as many valves as there are styles. Styles generally 3.
6. **Ceras'tium.** Petals 2-cleft, or notched. Styles 5. Pod opening at the apex by 10 teeth.
7. **Sagi'na.** Petals 4 or 5, entire. Stamens as many or twice as many. Styles 4 or 5. Pod 4-5-valved.

    + + *Stipules present.*

8. **Buda.** Styles 3. Pod 3-valved, short. Leaves filiform or linear, opposite.
9. **Sper'gula.** Styles 5. Valves of the pod opposite the sepals. Leaves thread-like, whorled.

1. SAPONARIA, L. SOAPWORT.

1. **S. officina'lis, L.** (Bouncing Bet.) A stout perennial, with rose-coloured or pinkish flowers clustered in corymbs. Leaves 3-5-ribbed, the lower ovate, upper lanceolate. Pod raised on a short stalk. Styles 2.—Old gardens and roadsides.

2. **S. vacca'ria, L.** (Common Cow-herb.) *Annual*, glabrous. Flowers pale red, in corymbed cymes. Calyx 5-angled, wing-angled in fruit.—Waste places.

2. SILE'NE, L. CATCHFLY. CAMPION.

1. **S. infla'ta,** Smith. (BLADDER CAMPION.) (*S. Cucubalus,* Wibel.) Pale or glaucous, very smooth. Stem erect, a foot high. Leaves ovate-lanceolate. *Calyx much inflated, purple-veined.* Stamens and styles exserted.—Not common westward.

2. **S. antirrhi'na,** L. (SLEEPY C.) Stem slender, simple or slightly branching above, a portion of the upper internodes sticky. Leaves linear or lanceolate. Flowers small, pink or purplish, opening only for a short time in sunshine. Calyx ovoid, shining.—Dry soil.

3. **S. noctiflo'ra,** L. (NIGHT-FLOWERING CATCHFLY.) *Stems very sticky, pubescent.* Lower leaves spathulate, upper lanceolate. Flowers few, *peduncled.* Calyx-tube with awl-shaped teeth. Petals white or whitish, 2-parted. Opening only at night or in cloudy weather.—A very common weed in cultivated grounds.

4. **S. Arme'ria,** L. (SWEET-WILLIAM CATCHFLY.) Glaucous. Leaves ovate-lanceolate. Flowers pink, in flat cymes. Calyx club-shaped.—Escaped from gardens in some places.

5. **S. Virgin'ica,** L. (FIRE PINK.) Occurs in southwestern Ontario, and may be recognized by its *crimson petals,* and bell-shaped calyx, nodding in fruit.

6. **S. acau'lis,** L. (MOSS CAMPION.) A very small tufted moss-like perennial, 1-2 inches high, with linear crowded leaves. Petals purple or rarely white.—Atl. Prov.

3. LYCH'NIS, Tourn. COCKLE.

1. **L. Githa'go,** Lam. (CORN COCKLE.) Plant clothed with long soft appressed hairs. *Calyx-lobes extremely long,* very much like the upper leaves, surpassing the *purple petals.* —Wheat-fields.

2. **L. vesperti'na,** Sibth. (EVENING L.) *Viscid-pubescent. Flowers commonly diœcious,* white or pinkish, opening at evening.—Waste grounds in a few localities.

3. **L. diurna,** Sibth., resembles the last, but has *red* flowers, opening in the morning.—Rare.

4. ARENA'RIA, L. SANDWORT.

1. **A. serpyllifo'lia,** L. (THYME-LEAVED S.) Much branched, 2-6 inches high, roughish-pubescent. Leaves small, ovate, acute. Petals white, hardly as long as the sepals. Sepals pointed, 3-5-nerved. Pod pointed, 6-toothed, —Sandy fields.

2. **A. stricta,** Michx. (*A. Michauxii*, Hook., in Macoun's Catalogue.) Stem erect, or diffusely spreading from a small root. Leaves awl-shaped or bristle-form, the upper ones reduced to 1-nerved bracts, crowded in the axils. Cyme diffuse, many-flowered. Sepals pointed, *3-ribbed*, half as long as the white petals.—Rocky fields.

3. **A. lateriflo'ra,** L. Stem erect, slender, minutely pubescent. Leaves oval or oblong, $\frac{1}{2}$-1 inch long. Peduncles usually three-flowered. Sepals obtuse. Petals white, large, twice as long as the sepals. Flower $\frac{1}{3}$ of an inch across when fully expanded.—Gravelly shores.

4. **A. Grœnlan'dica,** Spreng., is densely tufted, with soft filiform-linear leaves. Flowering stems erect, smooth, 2-4 inches high, few-flowered. Sepals oblong, obtuse, nerveless. —Atl. sea-coast.

5. **A. peploi'des,** L., with *very fleshy stems and leaves*, the latter somewhat clasping, occurs eastward towards the sea-coast.

5. STELLA'RIA, L. CHICKWEED. STARWORT.

1. **S. media,** Smith. (COMMON CHICKWEED.) Stems branching, decumbent, soft and brittle, *marked lengthwise with one or two pubescent lines.* Lower leaves on hairy petioles, *ovate*. Flowers small, white. *Petals shorter than the sepals.*—Extremely common in damp grounds and old gardens.

2. **S. longifo'lia,** Muhl. (LONG-LEAVED STITCHWORT.) Stems branching, very weak and brittle, supporting themselves on other plants. *Leaves linear.* Pedicels of the flowers long, slender, and spreading, reflexed. Petals white, longer than the 3-nerved sepals.—Low grassy banks of streams.

3. **S. lon'gipes,** Goldie. (LONG-STALKED STITCHWORT.) Leaves somewhat rigid, ascending, lanceolate, acute, broadest at the base. Cyme terminal, few-flowered, the long pedicels erect, scaly-bracted. Petals longer than the sepals. Seeds smooth.

4. **S. gramin'ea,** L. Like the last, but the leaves *broadest above the base, the pedicels widely spreading*, and the seeds strongly but finely rugose. (Int. from Eu.)

5. **S. uligino'sa,** Murr. (SWAMP S.) Stems weak, decumbent or diffuse. *Leaves lanceolate or oblong*, veiny. Petals and ripe pods *as long as the sepals.* Seeds roughened. *Cymes naked*, becoming lateral.—Swamps and rills, eastward.

6. **S. crassifo'lia,** Ehrh. Stems diffuse or erect, weak. *Leaves rather fleshy*, lanceolate to oblong, those of flowering branches smaller and thinner. Petals longer than sepals or wanting. *Seeds rugose-roughened.* Flowers terminal or in the forks of stem or branches.—Wet places.

7. **S. borea'lis,** Bigelow. (NORTHERN S.) Stem erect or spreading, weak, forking. Leaves broadly-lanceolate to ovate-oblong. Petals 2-5, shorter than the sepals or wanting. Cyme leafy. *Seeds smooth.*—Wet places.

8. **S. humifu'sa,** Rottb. Low, spreading or creeping. Leaves *fleshy*, ovate or oblong. Pedicels axillary or terminal, on leafy stems or branches. *Petals a little longer than the sepals.* Seeds smooth.—Atl. Prov.

6. CERAS'TIUM, L. MOUSE-EAR CHICKWEED.

1. **C. visco'sum,** L. (LARGER M.) Stem ascending, hairy and somewhat clammy. *Leaves ovate* or *obovate, obtuse.* Flowers in *close clusters.* *Pedicels not longer than the sepals.* Petals shorter than the calyx.—Not common, sometimes confounded with No. 2.

2. **C. vulga'tum,** L. (COMMON M.) Stems hairy, viscid, spreading. *Leaves lanceolate-oblong, rather acute.* Flowers in loose cymes. *Pedicels longer than the sepals.* Petals equalling the calyx.—Fields and copses; common.

3. **C. arven'se,** L. (FIELD CHICKWEED.) Stem decumbent at the base, pubescent, slender, 4-8 inches high. Leaves linear, or linear-lanceolate, *often fascicled in the axils*, longer than the lower internodes. Petals obcordate, more than twice as long as the calyx. Pod scarcely longer than the calyx. Cyme few-flowered.

Var. oblongifolium, Holl. and Britt. Taller, pubescent. Leaves oblong or oblong-lanceolate. Pod twice as long as the calyx.—S. W. Ontario.

4. **C. nu'tans,** Raf. Stems very clammy-pubescent and branching diffusely. The loose and open cymes many-flowered. Leaves lance-oblong. *Pods nodding on the stalks, curved upwards, thrice the length of the calyx.*—In places where water lies in spring.

7. SAGI'NA, L. PEARLWORT.

1. **S. procum'bens,** L. (PEARLWORT.) A low, matted herb with narrowly linear leaves. Flowers small, terminal, with their parts in fours, rarely in fives. *Petals shorter than the ovate, obtuse sepals*, or none. Pod many-seeded, 4-5-valved. *Top of peduncle often bent into a hook.*—Damp places, Atl. Prov.

2. **S. nodo'sa,** Fenzl. A low, tufted herb. Lower leaves thread-form; the upper short, awl-shaped, *with clusters of minute ones in their axils. Parts of flower in fives*, the stamens sometimes ten. *Petals much longer than sepals.* Flowers terminal. Pods as in S. procumbens.—Wet sandy shores, Atl. Prov.

8. BUDA, Adans. SAND-SPURREY.

1. **B. ru'bra,** Dumort. (*Spergula'ria ru'bra*, Presl.) Leaves linear, flat, hardly fleshy. Stipules lanceolate. Stems usually glandular-pubescent near the summit. Calyx rather longer than the pink-red corolla, and small pod. Seeds rough with projecting points.—Dry sandy ground, Atl. Prov.

2. **B. mari'na,** Dumort. (*Spergularia sali'na*, Presl.) More fleshy than B. rubra, usually pubescent. Stipules

ovate. Leaves terete. Sepals a little shorter than the pod. Petals pale. Seeds usually roughened with points.—Sea-coast, Atl. Prov.

3. **B. borea'lis,** Watson. (*Spergularia me'dia,* Presl.) Much branched, glabrous. Petals white. Pod about twice as long as the sepals, nearly or quite smooth.—Sea-coast, and N. W.

9. SPER'GULA, L. SPURREY.

S. arven'sis, L. (CORN SPURREY.) An annual herb resembling a Buda, with numerous thread-like leaves in whorls. Flowers white in panicled cymes. Pod 5-valved. —Grain fields, Atl. Prov. (Int. from Eu.)

ORDER XVII. **PORTULACA'CEÆ.** (PURSLANE F.)

Herbs with fleshy entire exstipulate leaves, and regular hypogynous or perigynous flowers. Sepals 2. Petals 5. Stamens 5-20. Styles 3-8, united below. Pod 1-celled, few or many-seeded.

Synopsis of the Genera.

1. **Portula'ca.** Stamens 8-20. Pod opening by a lid (Fig. 207, Part I.), many-seeded.
2. **Clayto'nia.** Stamens 5. Pod 3-valved, 3-6-seeded.
3. **Mon'tia.** Stamens usually 3. Petals 5, three of them somewhat smaller. Pod 3-valved and 3-seeded.

1. PORTU LA'CA, Tourn. PURSLANE.

P. olera'cea, L. (COMMON PURSLANE.) A low fleshy herb, very smooth, with obovate or wedge-shaped leaves. Calyx 2-cleft, the sepals keeled. Petals yellow, fugacious.—A common pest in gardens.

2. CLAYTO'NIA, L. SPRING-BEAUTY.

1. **C. Virgin'ica,** L. Leaves *linear-lanceolate,* 3-6 inches long.

2. **C. Carolinia'na,** Michx. Leaves *ovate-lanceolate* or oblong, tapering at the base. In both species the corolla is rose-coloured, with dark veins. The stem springs from a small tuber and bears two opposite leaves and a loose raceme of flowers.—Rich woods in early spring.

3. MON'TIA, L.

M. fonta'na, L. A small, spreading, somewhat fleshy herb, with opposite, spathulate leaves. Sepals 2. Petals 5, three of them somewhat smaller. Stamens usually 3, borne on the claws of the smaller petals. Pod 3-valved and 3-seeded.—Wet places, Atl. Prov.

Order XVIII. MALVA'CEÆ. (Mallow Family.)

Herbs, with palmately-veined alternate stipulate leaves. Flowers regular. Calyx valvate. Corolla convolute in the bud. Sepals 5, united at the base, Petals 5, hypogynous. Stamens numerous, monadelphous, hypogynous; anthers 1-celled. Carpels united in a ring, separating after ripening. Seeds kidney-shaped.

Synopsis of the Genera.

1. **Malva.** Carpels without beaks, 1-seeded. A circle of 3 bractlets at the base of the calyx. *Stigmas occupying the inner face of the style.*
2. **Malvas'trum.** Carpels without beaks, 1-seeded. Involucel of 3 bractlets, or none. *Stigmas terminal, capitate.* Low hoary herbs.
3. **Abu'tilon.** Carpels 2-beaked, 1-6-seeded. No circle of bractlets.
4. **Hibis'cus.** *Column of stamens naked and 5-toothed at the apex.* Pod 5-celled, many-seeded. Involucel of *many bractlets.*

1. MALVA, L. Mallow.

1. **M. rotundifo'lia,** L. (Round-leaved Mallow.) Stems several, procumbent, from a stout tap-root. Leaves long-petioled, round-heart-shaped, crenate, crenately-lobed. Petals obcordate, whitish, streaked with purple, twice as long as the sepals.—Waysides and cultivated fields.

2. **M. sylves'tris,** L. (High M.) *Stem erect,* 2 feet high. *Leaves sharply 5-7-lobed. Petals purple,* 3 times as long as the sepals.—Near dwellings.

3. **M. cris'pa,** L. (Curled M.) A tall erect annual, with round and angled toothed and crisped leaves, and small sessile flowers crowded in the axils.—Escaped from old gardens.

4. **M. moscha'ta,** L. (Musk M.) Stem erect, 1 foot high. *Stem leaves 5-parted, the divisions cleft.* Flowers large and

handsome, rose-coloured or white, on short peduncles, crowded on the stem and branches in the upper axils.—Roadsides near gardens.

2. MALVAS'TRUM, Gray. FALSE MALLOW.

M. coccin'eum, Gray. (PRAIRIE MALLOW.) A low hoary perennial, with rose-pink flowers in spikes or racemes. Leaves 5-parted or pedate. Petals much larger than the sepals.—N. W.

3. ABU'TILON, Tourn. INDIAN MALLOW.

A. Avicen'næ, Gærtn. (VELVET-LEAF.) Stem 2-5 feet high, branching. Leaves velvety, round-cordate, long-pointed. Corolla yellow.—Near gardens; not common.

4. HIBIS'CUS, L. ROSE-MALLOW.

1. **H. Moscheu'tos,** L. (SWAMP ROSE-MALLOW.) A tall perennial, with very large and showy pink or white flowers, in late summer. Calyx *not inflated*. Leaves ovate, pointed, toothed; the lower 3-lobed, all white-downy beneath.—S.W. Ontario, in marshes.

2. **H. Trio'num,** L. (BLADDER KETMIA.) A low hairy annual. Calyx *inflated in fruit, and 5-winged*. Corolla sulphur-yellow, with dark centre.—Escaped from gardens.

ORDER XIX. TILIA'CEÆ. (LINDEN FAMILY.)

Trees with fibrous bark, soft and white wood, and heart-shaped and serrate leaves, with deciduous stipules. Flowers in small cymes hanging on an axillary peduncle, to which is attached a leaf-like bract. Sepals deciduous. The only Canadian genus is

TILIA, L. BASSWOOD. WHITEWOOD.

1. **T. America'na,** L. (BASSWOOD.) A fine tree, in rich woods. Flowers yellow or cream-coloured, very fragrant. Leaves smooth and green on both sides, obliquely cordate or truncate at the base, sharply serrate. Sepals 5. Petals 5. Fruit a globular nut, 1-celled, 1-2-seeded.

2. **T. Europæ'a,** L. The European Linden is planted as a shade tree in some places. It is easily distinguished by the absence of *petal-like scales among the stamens*, which are present in the native Linden.

ORDER XX. **LINA'CEÆ.** (FLAX FAMILY.)

Herbs with entire exstipulate leaves (but sometimes with glands in place of stipules), and regular hypogynous flowers. Sepals, petals, stamens, and styles, 5 each. Filaments united at the base. Pod 10-celled, 10-seeded. Our only genus is

LINUM, L. FLAX.

1. **L. Virginia'num,** L. (VIRGINIA F.) *Flowers yellow*, small ($\frac{1}{4}$ of an inch long), scattered. Stem erect, it and the spreading branches terete. Leaves lanceolate and acute, the lower obtuse and opposite. No glands, styles distinct.—Dry soil.

2. **L. stria'tum,** Walt., has the branches *wing-angled*, broader leaves and more crowded flowers than No. 1. The whole plant is stouter.—Shores of the Great Lakes.

3. **L. sulca'tum,** Riddell, has yellow flowers twice as large as the preceding, wing-angled branches, and *a pair of dark glands in place of stipules*. Sepals strongly 3-nerved, with rough bristly margins. *Styles united to the middle.*—Dry soil, rare.

4. **L. rig'idum,** Pursh. Glaucous. Flowers yellow, large. Sepals lanceolate, glandular-serrulate. Branches rigid, angled. *Stipular glands usually present. Styles united.*—N. W.

5. **L. peren'ne,** L., var. **Lewisii,** Eat. and Wright. Flowers *blue*, on long peduncles. *Perennial*, glabrous and glaucous, leaves linear, acute. Styles distinct. Pod ovate. —N. W.

6. **L. usitatis'simum,** L. (COMMON F.) *Annual. Flowers blue*, the sepals ciliate. Leaves alternate, linear-lanceolate, acute, 3-veined.—Cultivated grounds.

ORDER XXI. **GERANIA'CEÆ.** (GERANIUM FAMILY.)

Herbs (often strong-scented) with symmetrical flowers, having the parts in fives or threes, the filaments usually united at the base, and glands on the receptacle alternate with the petals. Stigmas 5 or 3. Carpels 5 or 3, each 1-2-ovuled, they and the lower part of the styles *attached to an axis which rises from the receptacle.* In the fruit the styles split away from the axis, *carrying the carpels with them.*

Synopsis of the Genera.

1. **Geranium.** Stamens 10, all with anthers.
2. **Ero'dium.** Stamens with anthers, only 5.
3. **Floerk'en.** Stamens 6. Sepals, petals, and ovaries 3.

1. GERANIUM, L. CRANESBILL.

1. **G. macula'tum,** L. (WILD C.) *Perennial.* Stem erect, hairy, about a foot high. Leaves 5-7-parted, the wedge-shaped divisions lobed and cut. Flowers purple, an inch across. *Petals entire,* bearded on the claw, *much longer than the long-pointed sepals.*—Open woods and fields.

2. **G. inci'sum,** Nutt. *Perennial.* Villous and *glandular-pubescent.* Leaves 5-7-cleft nearly to the base, the lobes cut into narrow divisions. Petals deep purple.—N. W.

3. **G. Carolinia'num,** L. (CAROLINA C.) *Not perennial.* Stem usually decumbent, hairy. Sepals *awn-pointed,* as long as the *notched* rose-coloured petals.—Waste places.

4. **G. Robertia'num,** L. (HERB ROBERT.) *Not perennial.* Stems reddish, spreading, pubescent; branches weak. *Leaves 3-divided,* or *pedately 5-divided,* the divisions twice pinnatifid. Sepals awned, shorter than the reddish-purple petals. *Plant with a very strong odour.*—Shaded ravines and moist woods.

4. **G. pusil'lum,** L. (SMALL-FLOWERED C.) *Not perennial.* Stem procumbent, slender, minutely pubescent. Leaves rounded, kidney-shaped, deeply 5-7-cleft, the divisions wedge-shaped. *Sepals awnless,* about the same length as the purplish petals.—Waste places.

2. ERO'DIUM, L'Her. STORKSBILL.

E. cicuta'rium, L'Her. Stem low and spreading, hairy. *Leaves pinnate*, the leaflets sessile, pinnatifid. Peduncles several-flowered. Styles when they separate from the beak *bearded on the inside.*—Not common.

3. FLŒRK'EA, Willd. FALSE MERMAID.

F. proserpinacoi'des, Willd. A low tender annual with very small solitary flowers on axillary peduncles. Leaves alternate, pinnate, of 3-5 lanceolate leaflets.—Marshes and river-banks.

ORDER XXII. OXALIDA'CEÆ. (WOOD-SORREL F.)

Low herbs with an acid juice and alternate compound leaves, the 3 leaflets obcordate and drooping in the evening. Flowers very much the same in structure as in the preceding Order, but the fruit is a 5-celled pod, each cell opening in the middle of the back (loculicidal), and the valves persistent. Styles 5, separate. The only genus is

OX'ALIS, L. WOOD-SORREL.

1. **O. Acetosel'la,** L. (WHITE WOOD-SORREL.) Scape 1-flowered. Petals *white, with reddish veins.*—Cold woods.

2. **O. cornicula'ta,** L. (YELLOW W.) Annual, or *perennial by running underground shoots.* Stipules present. Peduncles 2-6-flowered, longer than the leaves. *Petals yellow.* Pod elongated, erect in fruit.—Rare.

Var. **stricta,** Sav. (*O. stricta,* L.), is less pubescent, has an erect stem, and is without stipules.—Common

ORDER XXIII. BALSAMINA'CEÆ. (BALSAM FAMILY.)

Smooth herbs, with succulent stems and simple exstipulate leaves. Flowers irregular, the sepals and petals coloured alike, *one of the coloured sepals spurred, the spur with a tail.* Stamens 5, coherent above. Pod bursting elastically, and discharging its seeds with considerable force. The only genus is

IMPA'TIENS, L. Touch-me-not. Jewel-weed.

1. **I. fulva,** Nutt. (Spotted Touch-me-not.) Flowers *orange-coloured, spotted with reddish brown.* Sac longer than broad, conical, tapering into a long recurved spur.—Cedar swamps and along streams.

2. **I. pal'lida,** Nutt. (Pale T.) Flowers *pale yellow, sparingly dotted with brown.* Sac dilated, broader than long, ending in a short spur.—Wet places.

Order XXIV. RUTA'CEÆ. (Rue Family.)

Shrubs, with compound *transparently-dotted* leaves, and an acrid taste. Flowers (with us) diœcious or polygamous, appearing before the leaves. Stamens hypogynous, as many as the petals.

Synopsis of the Genera.

1. **Xanthox'ylum.** Flowers diœcious. Ovaries separate, forming pods. Leaflets more than 3.
2. **Ptel'ea.** Flowers polygamous. Fruit a samara, winged all round. Leaflets 3.

1. XANTHOX'YLUM, L. Prickly Ash.

X. America'num, Mill. (Northern Prickly Ash. Toothache Tree.) A prickly shrub, with yellowish-green flowers in dense umbels in the axils. Sepals obsolete or none. Petals 5. Stamens in the sterile flowers 5. Carpels 3-5, forming fleshy 1-2-seeded pods. Fruit very pungent and aromatic. Leaves pinnate, 4-5 pairs, with an odd one at the end.—Forming thickets in low grounds along streams.

2. PTE'LEA, L. Shrubby Trefoil. Hop-tree.

P. trifolia'ta, L. A tall shrub with greenish-white small polygamous flowers in compound terminal cymes. Sepals 3-5. Stamens as many. Ovary 2-celled. Stigmas 2. Fruit a 2-celled and 2-seeded samara, winged all round. Leaves of 8 ovate pointed leaflets.—S. W. Ontario.

Order XXV. ANACARDIA'CEÆ. (Cashew Family.)

Trees or shrubs, with a milky or resinous juice, and alternate leaves without dots or stipules. Sepals, petals and stamens, each 5. Fruit a 1-seeded drupelet. The petals and stamens inserted under the edge of a disk which surrounds the base of the ovary. The only genus is

RHUS, L. SUMACH.

1. **R. typh'ina,** L. (STAGHORN SUMACH.) A small tree, 10-30 feet high, with *densely soft-hairy branches and stalks.* Flowers greenish-white, polygamous, forming a terminal thyrse. Fruit globular, *covered with crimson hairs.* Leaves pinnate, leaflets 11-31, oblong, lanceolate, serrate, pointed. —Dry hill-sides.

2. **R. glabra,** L., (SMOOTH S.) is *smooth,* and seldom exceeds 5 feet in height.

3. **R. Toxicoden'dron,** L. (POISON IVY. POISON OAK.) Shrub about a foot high, smooth, often climbing by rootlets. Leaves 3-foliolate, leaflets rhombic-ovate, notched irregularly. Flowers polygamous, in slender axillary panicles. Plant poisonous to the touch. Var. **radi'cans,** L., has the leaves *entire, and climbs high.*

4. **R. venena'ta,** DC. (POISON ELDER.) A tall shrub, smooth or nearly so. Leaves odd-pinnate; leaflets 7-13, obovate-oblong, entire. Greenish-white flowers as in No. 3. —Swamps.

5. **R. Canadensis,** Marsh. (FRAGRANT S.) (*R. aromatica,* Ait.) A shrub 2-3 feet high, or more, with 3-foliolate leaves, sweet-scented when crushed, and catkin-like spikes of pale yellow flowers appearing before the leaves.—Dry rocky banks. Var. **triloba'ta,** with small leaflets and an unpleasant odour, is rather common in the North-West.

ORDER XXVI. **VITA'CEÆ.** (VINE FAMILY.)

Shrubs climbing by tendrils, with small greenish flowers in panicled clusters opposite the leaves. Stamens as many as the petals and opposite them. Calyx minute. Petals 4 or 5, hypogynous or perigynous, very deciduous. Fruit a berry, 1-4-seeded. Leaves palmately-veined, or compound.

Synopsis of the Genera.

1. **Vitis.** Leaves simple, heart-shaped, and variously lobed.
2. **Ampelopsis.** Leaves compound-digitate, of 5 serrate leaflets.

1. VITIS, Tourn. GRAPE.

1. **C. æstiva'lis,** Michx. (NORTHERN FOX-GRAPE.) *Leaves and branches woolly.* Berries large, dark purple or amber-coloured.—Moist thickets.

2. **C. cordifo'lia,** Michx. (FROST GRAPE.) *Leaves smooth or nearly so,* bright green on both sides, heart-shaped, sharply serrate. Berries small, blue or black. Var. **ripa'ria,** Michx., has broader cut-lobed leaves.—Banks of streams.

2. AMPELOP'SIS, Michx. VIRGINIA CREEPER.

A. quinquefo'lia, Michx. A common woody vine in low grounds. Leaves digitate, of 5 oblong-lanceolate leaflets. Tendrils with sucker-like disks at the end, by which they cling to walls, trunks of trees, etc. Fruit a small black berry.

ORDER XXVII. **RHAMNA'CEÆ.** (BUCKTHORN FAMILY.)

Shrubs with simple stipulate leaves, and small regular perigynous greenish or whitish flowers. Stamens opposite the petals, and with them inserted on the margin of a fleshy disk which lines the calyx-tube. Fruit a berry-like drupe, or a pod.

Synopsis of the Genera.

1. **Rham'nus.** Petals minute, *or none.* Drupe berry-like. Calyx and disk free from the ovary.
2. **Ceano'thus.** Petals white, long-clawed, hooded. Fruit dry, dehiscent. Calyx and disk adherent to the base of the ovary.

1. RHAM'NUS, Tourn. BUCKTHORN.

R. alnifo'lia, L'Her. A low erect shrub, not thorny, with oval, acute, serrate leaves, and apetalous flowers. Fruit a 3-seeded berry.—Swamps.

2. CEANO'THUS, L. NEW JERSEY TEA.

1. **C. America'nus,** L. A shrubby plant with downy branches, and ovate, 3-ribbed, serrate leaves. Flowers in white clusters at the summit of the naked flower-branches.

Sepals and petals white, the latter hooded, and with slender claws. Pedicels also white.—Dry hill-sides.

2. **C. ova'tus**, Desf. (*C. ovalis*, Bigel.), has the leaves narrowly oval or elliptical-lanceolate, finely serrate, and glabrous or nearly so. The flowers, also, are larger than in No. 1.—South-western Ontario.

ORDER XXVIII. **CELASTRA'CEÆ.** (STAFF-TREE F.)

Shrubs with simple stipulate leaves, alternate or opposite, and small regular flowers, the sepals and petals both imbricated in the bud. Stamens 4-5, alternate with the petals, and inserted on a disk which fills the bottom of the calyx. Pods orange or crimson when ripe.

Synopsis of the Genera.

1. **Euon'ymus.** Flowers perfect. Sepals 4 or 5, united at the base, and forming a *flat calyx*. Branchlets 4-sided; *leaves opposite.* Flowers *axillary.*
2. **Celas'trus.** Flowers polygamous. Petals and stamens 5. Calyx cup-shaped. *Leaves alternate.* Flowers in *terminal racemes.*

1. EUON'YMUS, Tourn. SPINDLE-TREE.

1. **E. America'nus**, L., var. **obovatus**, Torr. and Gray. (STRAWBERRY BUSH.) A low, rather straggling shrub, with *short-petioled* or *sessile leaves*, the latter ovate or obovate, pointed. Flowers greenish, with the parts generally in fives. Pods *rough-warty, depressed*, crimson when ripe.—Wooded river-banks and low grounds.

2. **E. atropurpu'reus**, Jacq., (BURNING BUSH) occurs at Toronto and westward in Ontario, and may be distinguished from No. 1 by its greater size (4-8 feet high) its *long-petioled leaves, purplish flowers*, and smooth pods.

2. CELAS'TRUS, L. STAFF-TREE.

C. scandens, L. (WAX-WORK. CLIMBING BITTER-SWEET.) A twining smooth shrub, with oblong-ovate, serrate, pointed leaves. Flowers small, greenish, in terminal racemes. Pods orange-coloured. These burst in autumn and display a scarlet pulpy aril, presenting a highly ornamental appearance.—Twining over bushes on river-banks and in thickets.

Order XXIX. SAPINDA'CEÆ. (Soapberry Family.)

Trees or shrubs, with compound or lobed leaves, and usually unsymmetrical and often irregular flowers. Sepals and petals 4-5, both imbricated in the bud. Stamens 5-10, inserted on a fleshy disk which fills the bottom of the calyx-tube. Ovary 2-3-celled, with 1 or 2 ovules in each cell.

Synopsis of the Genera.

1. **Staphyle'a.** *Flowers perfect.* Lobes of the coloured calyx, the petals, and the stamens, each 5. *Fruit a 3-celled, 3-lobed, inflated pod.* Leaves pinnately compound.
2. **Acer.** *Flowers polygamous.* Leaves *simple*, variously lobed, opposite. Calyx coloured, usually 5-lobed. Petals none, or as many as the sepals. Stamens 3-12. *Fruit two 1-seeded samaras* joined together, at length separating.
3. **Negun'do.** Flowers diœcious. Leaves *pinnate*, of 3 or 5 leaflets. Fruit a double samara, as in Acer.

1. STAPHYLE'A, L. Bladder-Nut.

S. trifo'lia, L. (American Bladder-Nut.) Shrub, 4-6 feet high. Leaflets 3, ovate, pointed. Flowers white, in drooping racemes, at the ends of the branchlets.—Thickets and hill-sides.

2. ACER, Tourn. Maple.

1. **A. Pennsylvan'icum,** L. (Striped Maple.) A small tree, 10-20 feet high, with light-green bark striped with dark lines. Leaves 3-lobed at the apex, finely and sharply doubly-serrate, the lobes taper-pointed. Flowers greenish, in terminal racemes, appearing after the leaves. Samaras large, with divergent wings.—Rich woods.

2. **A. spica'tum,** Lam. (Mountain Maple.) A shrub or small tree, 4-8 feet high, growing in clumps in low grounds. Leaves 3-lobed, coarsely serrate, the lobes taper-pointed. Flowers greenish, appearing after the leaves, in dense *upright* racemes. Fruit with small widely-diverging wings.

3. **A. sacchari'num,** Wang. (*A. saccharum*, Marshall.) (Sugar Maple.) A fine tree, with 3-5-lobed leaves, breadth and length the same, dull above, a paler green underneath, *the rather narrow sinuses rounded, and the lobes sparingly*

sinuate-toothed, the petioles *without stipules*. Flowers greenish-yellow, *drooping on slender hairy pedicels*, appearing at the same time as the leaves. Calyx fringed on the margin. Var. **barbatum**, Michx., has rather glossy leaves, mostly broader than long, 3-lobed, with very open rounded sinuses, and no stipules. Var. **nigrum**, Torr. and Gray, may be distinguished from the ordinary form by its paler and more pubescent leaves, the petioles *with adnate foliaceous stipules*. —Rich woods.

4. **A. dasycar'pum**, Ehrhart. (*A. saccharinum* L.) (WHITE OR SILVER M.) Leaves deeply 5-lobed, the sinuses rather acute, silvery-white underneath, the divisions narrow, sharply-toothed. Flowers in erect clusters, greenish-yellow, *appearing much before the leaves; petals none*. Samara very large, *woolly when young*. River-banks and low grounds.

5. **A. ru'brum**, L. (RED M.) Leaves 3-5-lobed, the sinuses acute. *Flowers red*, appearing much before the leaves. *Petals linear-oblong. Samara small and smooth*, on drooping pedicels. A smaller tree than No. 4, with reddish twigs, and turning bright crimson in the autumn.—Swamps.

2. **NEGUN'DO**, Mœnch. ASH-LEAVED MAPLE. BOX-ELDER.

N. aceroi'des, Mœnch. Calyx minute, 4-5-cleft. Petals none. Staminate flowers in rather dense clusters on slender pedicels, the pistillate ones in drooping racemes. *Wings of the samaras incurved.*—N. W., common. Also, Don valley near Toronto, and planted as a shade-tree.

ORDER XXX. **POLYGALA'CEÆ.** (MILKWORT FAMILY.)

Herbs with entire exstipulate leaves, and irregular hypogynous flowers. Stamens 6 or 8, monadelphous or diadelphous, the anthers 1-celled, and opening at the top by a pore. Pod 2-celled and 2-seeded, flattened contrary to the partition. The only genus with us is

POLYG'ALA, Tourn. MILK-WORT.

Sepals 5, the upper one and the two lower ones small and

often greenish, the 2 lateral ones (called wings) larger and coloured like petals. Petals 3, connected with each other and with the tube of filaments, the lower one keel-shaped, and usually fringed or crested at the top. Style prolonged and curved.

1. **P. verticilla'ta,** L. Flowers small, greenish-white, in slender spikes. Stems 4-8 inches high, much branched. *Stem-leaves linear, 4-5 in a whorl,* the upper ones scattered. —Dry soil.

2. **P. Sen'ega,** L. (SENECA SNAKEROOT.) Flowers greenish white, in a solitary cylindrical close spike. Stems several, from a hard knotty rootstock, 6-12 inches high. Leaves lanceolate, with rough margins, alternate.—Dry hill-sides and thickets.

3. **P. polyg'ama,** Walt. Flowers rose-purple, showy, fringed, in a *many-flowered* raceme. Stems 5-8 inches high, tufted and very leafy, the leaves linear-oblong or oblanceolate. Whitish fertile flowers on underground runners.—Dry soil.

4. **P. paucifo'lia,** Willd. (FRINGED P.) Flowers rose-purple, very showy, fringed, *only 1-3 in number.* Stems 1-4 inches high, from long underground runners, which also bear concealed fertile flowers. Leaves ovate, crowded at the top of stem.—Dry woods.

5. **P. sanguin'ea,** L. Flowers usually bright red-purple, but sometimes pale. Corolla inconspicuously crested. Flowers in *dense globular heads,* at length oblong. True petals mostly shorter than the wings, the latter broadly ovate, closely sessile. Stem leafy to the top; leaves oblong-linear.—Sandy and moist places.

ORDER XXXI. **LEGUMINO'SÆ.** (PULSE FAMILY.)

Herbs, shrubs, or trees, mostly with compound alternate stipulate leaves, and papilionaceous corollas. (For description of a typical flower, see Part I., cap. v.) Stamens usually 10 (rarely 5), monadelphous, diadelphous, or distinct. Fruit a legume.

Synopsis of the Genera.

* *Flowers papilionaceous. Stamens monadelphous or diadelphous.*
  — *Not climbing or twining. Pod not jointed.*

1. **Lupi'nus.** Leaves palmately-compound, leaflets 7–9. Flowers in terminal racemes. *Stamens monadelphous.*

2. **Trifo'lium.** Leaves of 3 leaflets. *Flowers in heads.* Stamens diadelphous.

3. **Medica'go.** Leaves pinnate, of 3 leaflets. Flowers in axillary spikes or racemes. *Pod curved or coiled.* Stamens diadelphous.

4. **Melilo'tus.** Leaves pinnate, of 3 leaflets, the leaflets toothed. Flowers in slender axillary racemes. Pod wrinkled, 1-2-seeded. Stamens diadelphous.

5. **Hosack'ia.** Leaves pinnate, of 1–3 leaflets, *stipules gland-like, minute.* Flowers yellow, often turning reddish, *solitary* on axillary peduncles bearing each a leafy bract. Pod linear, several-seeded, partitioned between the seeds.—N. W. only.

6. **Robin'ia.** *Trees.* Leaves odd-pinnate, often with spines for stipules, and the leaflets with small stipules. Flowers in hanging axillary racemes. Pod margined on one edge. Stamens diadelphous.

7. **Astrag'alus.** Leaves odd-pinnate, leaflets numerous. Flowers in dense axillary spikes. Corolla long and narrow, the keel *not* tipped with a sharp point. Pod turgid, *one or both sutures* (see Part I., section 217) *projecting into the cell, thus partially or wholly dividing the cavity.* Stamens diadelphous.

8. **Oxyt'ropis.** Low perennials with very short tufted stems from a hard rootstock, covered with scaly stipules. Flowers nearly as in Astragalus, *but the keel is tipped with a sharp appendage.* Peduncles scape-like.

9. **Glycyrrhi'za.** Leaves odd-pinnate, glandular-viscid. Flowers whitish, in axillary spikes, like those of Astragalus. Root long, perennial, sweet. Pods few-seeded, clothed with rough glands or *short hooked prickles.*—N. W. chiefly.

10. **Tephro'sia.** Hoary perennials. Stem simple. Flowers yellowish-white marked with purple, in a terminal dense oblong raceme. Wings coherent with the keel. Pod linear, flat, several-seeded.

11. **Psora'lea.** Leaves palmate, of 3–5 leaflets. Stamens diadelphous (sometimes monadelphous). Flowers spiked or racemed, mostly bluish, sometimes white. Pod as long as the calyx, 1-seeded. Calyx, pods, &c., usually *glandular-dotted.*—N.W. only.

12. **Amor'pha.** Leaves odd-pinnate, of many leaflets, *glandular-dotted. Corolla of one petal only,* the standard, enfolding the stamens and style. Flowers violet or purple, in dense terminal spikes. Stamens 10, monadelphous *at the base only,* distinct above.—N. W. only.

LEGUMINOSÆ. 51

13. **Petaloste'mon.** Leaves odd-pinnate, crowded, *glandular-dotted*. Flowers small, in dense terminal heads or spikes. Corolla hardly papilionaceous. *Stamens only 5*. Petals on thread-like claws, 4 of them on the tube of filaments, the fifth on the bottom of the calyx. Pod 1-2-seeded.—N.W. only.

+ + *Climbing by tendrils at the ends of the leaves.*

14. **Vicia.** Leaves abruptly pinnate, *the leafstalk prolonged into a tendril.* Flowers axillary. *Style filiform, hairy at the apex.* Pod 2-several-seeded. Stamens diadelphous.

15. **Lath'yrus.** Leaves as in Vicia. *Style flattish*, flattened above, and *hairy down the side opposite the free stamen.* Stamens diadelphous.

+ + + *Twining herbs.*

16. **A'pios.** A twining herb, leaves pinnate, *of 5-7 leaflets*. Keel of the flower slender and *coiled inward*. Flowers brown-purple, in dense racemes. Stamens diadelphous.

17. **Strophosty'les.** *A twining herb.* Leaves pinnate, of 3 leaflets. Keel long, strongly incurved. Standard recurved-spreading. Style bearded along the upper side. Pod terete. Seeds oblong, pubescent. Stamens diadelphous.

18. **Amphicarpæ'a.** *A low and slender twiner*, the stem clothed with brownish hairs. *Leaves pinnate, of 3 leaflets.* Flowers polygamous, those of the upper racemes perfect, those near the base fertile, with the corolla inconspicuous or none. Stamens diadelphous.

+ + + + *Pods transversely jointed, the joints reticulated.*

19. **Desmo'dium.** Leaves pinnate, of 3 leaflets. *Calyx 2-lipped.* Flowers purple or purplish, in axillary or terminal racemes. Pod flat, *the lower margin deeply lobed*, thus making the pod jointed, roughened with hooked hairs, causing the pods to adhere to the clothing, etc. Stamens diadelphous.

20. **Hedys'arum.** Leaves pinnate, of numerous leaflets. Calyx 5-cleft. Pod flat, *several-jointed*, the joints *roundish, reticulated, not rough.* Flowers purple, deflexed, in racemes. Stamens diadelphous.

21. **Lespede'za.** Leaves pinnate, of 3 leaflets. *Calyx 5-cleft.* Pod flat, oval or roundish, *occasionally 2-jointed but only 1-seeded.* Flowers sometimes polygamous. Stamens diadelphous.

** *Flowers papilionaceous. Stamens all separate.*

22. **Baptis'ia.** Leaves palmate, of 3 leaflets. Flowers yellow. *Stamens all separate. The keel-petals nearly separate.* Racemes terminating the bushy branches. *Pods inflated.*

23. **Thermop'sis.** Leaves palmate, of 3 leaflets, and with broad leaf-like stipules. *Stamens all separate.* Flowers yellow. *Pod flat, linear, curved.*—N.W. only.

*** *Flowers not papilionaceous; polygamous. Trees.*

24. **Gledit'schia.** Thorny trees, with abruptly once- or twice-pinnate leaves. Flowers greenish, inconspicuous, in small spikes. Stamens separate, 3-10.

25. **Gymnoc'ladus.** Not thorny. Large trees with doubly pinnate leaves, the leaflets vertical. Pod very long and broad. Stamens 10. Petals on the summit of the long calyx-tube, whitish.

1. LUPI'NUS, Tourn. LUPINE.

1. **L. peren'nis,** L. (WILD LUPINE.) Stem erect, somewhat hairy. Leaflets 7-9, oblanceolate. Calyx deeply 2-lipped. Pods hairy.—Sandy soil.

2. **L. argen'teus,** Pursh., var. **argophyl'lus,** Watson, occurs in the far west. The petals are blue or cream-coloured, and the calyx is decidedly *spurred* at the base. Plant hoary-pubescent.

3. **L. King'ii,** Watson (**L. pusillus,** Pursh.), is also a N.W. species. *Seeds only 2 or 3.* Plant low, villous with white soft hairs. Racemes short, few-flowered, on long slender peduncles.

2. TRIFO'LIUM, L. CLOVER. TREFOIL.

1. **T. arvense,** L. (RABBIT-FOOT or STONE CLOVER.) Stem erect, 4-12 inches high, branching. Heads of whitish flowers oblong, *very silky and soft.* Calyx-teeth fringed with long silky hairs.—Dry fields.

2. **T. pratense,** L. (RED C.) Stems and leaves somewhat hairy, the latter marked with a pale spot on the upper side. Flowers purplish, in dense sessile heads.—Pastures.

3. **T. medium,** L. (ZIGZAG C.) Like the last, but smoother, and the leaves without the pale spot. Heads mostly stalked.—Atl. Prov.

4. **T. repens,** L. (WHITE C.) Smooth, *creeping.* Heads of white flowers rather loose.—Fields everywhere.

5. **T. hy'bridum,** L. (ALSIKE C.) Resembles the last, but the flowers are rose-tinted, and the stems *erect* or *ascending.*

6. **T. reflexum,** L. (BUFFALO C.) Only in south-western Ontario, in the neighbourhood of the Detroit river. Heads

large, on naked peduncles; standard rose-red, wings and keel whitish. Flowers reflexed when old.

7. **T. agrarium,** L. (YELLOW or HOP-C.) Flowers *yellow*, reflexed when old. Leaflets obovate-oblong, *all 3 from the same point.* Stem 6-12 inches high.—Sandy fields.

8. **T. procumbens,** L. (LOW HOP-C.) Flowers yellow, reflexed when old. Leaflets *wedge-obovate*, the lateral ones at a short distance from the terminal one. Stem smaller than in No. 6, spreading.—Sandy fields.

3. MEDICA'GO, L. MEDICK.

1. **M. lupuli'na,** L. (BLACK MEDICK.) Stem procumbent, downy. Leaflets obovate, toothed at the apex. Flowers yellow. Pods kidney-shaped.—Waste places.

2. **M. sati'va,** L., (LUCERNE) is *upright*, has *purple flowers* in a long raceme, and *spirally-twisted pods.*—Cultivated fields.

3. **M. macula'ta,** Willd. (SPOTTED M.) Spreading or procumbent. Leaflets obcordate, *with a purple spot.* Peduncles 3-5-flowered. *Pods compactly spiral*, furrowed on the thick edge, *bristly.*—Atl. sea-coast.

4. **M. denticula'ta,** Willd. *Pods loosely spiral*, deeply reticulated, with a thin keeled edge; otherwise like the last. —Atl. sea-coast.

4. MELILO'TUS, Tourn. SWEET CLOVER.

1. **M. officina'lis,** Willd. (YELLOW MELILOT.) Stem erect, 2-4 feet high. Leaflets obovate-oblong. Flowers yellow. Pod drooping, 1-2-seeded.—Waste places.

2. **M. alba,** Lam., (WHITE M.) is much like No. 1, but has white flowers.—Escaped from gardens.

5. HOSACK'IA, Douglas.

H. Purshia'na, Benth. A low annual, more or less silky-villous, sometimes glabrous. Leaves nearly sessile; leaflets varying from ovate to lanceolate. Calyx-teeth linear, much longer than the tube.—N.W.

6. ROBIN'IA, L. LOCUST-TREE.

1. R. Pseudaca'cia, L. (COMMON LOCUST.) *Racemes slender, loose.* Flowers white, fragrant. A large tree.

2. R. visco'sa, Vent. (CLAMMY L.) *Racemes crowded.* Flowers white, with a reddish tinge. *Branchlets and leafstalks clammy.* Smaller than No. 1.

7. ASTRAG'ALUS, Tourn. MILK-VETCH.

NOTE.—All the following species except Nos. 2, 7, 8 and 10 belong to the North-West.

* **Pod completely 2-celled, being divided lengthwise by the intrusion of one or both sutures.**

(a) Pod thick and fleshy, sessile.

1. A. caryocar'pus, Ker. (GROUND PLUM.) A pale perennial, with low leafy stems, and minutely pubescent. Flowers violet, large, in short spike-like racemes. Leaflets about 21, elliptical-oblong.

(b) Pod coriaceous or cartilaginous.

← *Pod sessile.*

2. A. Canadensis, L. (CANADIAN MILK-VETCH.) Stem erect, 1–4 feet high, somewhat pubescent (or glabrous). Leaflets 21–27, linear-oblong, comparatively large. Flowers greenish-yellow, in long dense spikes. Pod and ovary glabrous.—Common from Quebec to the Rocky Mountains.

3. A. adsur'gens, Pall. Stem 4–18 inches high, ascending or decumbent, grayish with appressed pubescence. Leaflets about 21, narrowly oblong. Flowers purplish, medium-sized, in dense short spikes. Pod finely pubescent, triangular-compressed, *with a dorsal furrow.*

← ← *Pod with a very short stalk.*

4. A. hypoglot'tis, L. Stem slender, 6–24 inches long. diffusely procumbent or ascending, sparingly pubescent. Leaflets 15–21. Flowers violet, in capitate clusters. *Calyx copiously pubescent.* Pod silky-villous, *with a dorsal furrow.*

** Pod incompletely 2-celled.
+ Pod sessile.

5. **A. Missourien'sis,** Nutt. *Stem very short*, prostrate or matted, hoary-silky-pubescent. Leaflets 5-15 or more. Flowers large, violet or purplish, in few-flowered heads *on scape-like peduncles.* Pod obcompressed (flattened *contrary to* the usual way), nearly straight, pubescent.

+ + *Pod with a very short stalk.*

6. **A. lotiflo'rus,** Hook. Stems very short, hoary with close pubescence. Leaflets 7-13. Flowers yellowish, in few-flowered heads. Pod oblong-ovate, acute at the base, *hoary.*—Rare.

7. **A. oroboi'des,** Hornem., var. **Americanus,** Gray. Stems 12-18 inches high, grayish-puberulent. Leaflets 11-15, oval to linear-oblong. Flowers in a long 1-sided raceme. Pod lens-shaped, gray-pubescent, grooved on the back.— Atl. Prov.

+ + + *Pod long-stalked.*

8. **A. alpi'nus,** L. Stems *diffuse*, 6-12 inches high, smooth or hairy-pubescent. Leaflets 13-25, oval or oblong. Racemes short, many-flowered, the flowers violet-purple. Pod narrowly oblong, *black-pubescent*, triangular-turgid, deeply grooved on the back, pendent.—Atl. Prov.

*** Pod strictly 1-celled, coriaceous or cartilaginous.
A. *Leaves pinnate, of many leaflets.*
+ *Pod sessile.*

9. **A. pectina'tus,** Dougl. Stem a foot high, grayish-pubescent, the branches striate and angled. Leaflets numerous, *nearly filiform.* Flowers large, *white.* Pod *short, thick-cartilaginous, pendulous, cuspidate*, smooth, with a very thick dorsal suture.

+ + *Pod with a very short stalk.*

10. **A. Coo'peri,** Gray. Stem erect, nearly smooth, 1-2 feet high. Leaflets 11-21, comparatively large, minutely hoary beneath. Flowers white, in short spikes. Pod *inflated*, smooth, slightly furrowed on both sides.—Ontario and Quebec.

11. **A. flexuo'sus**, Dougl. Stem 1-2 feet high, ascending, grayish-pubescent. Leaflets numerous, narrow. Flowers white or purplish, small, *in loose racemes*. Pod very finely pubescent, *not furrowed*.

→ → → *Pod exserted on its stalk.*

12. **A. bisulca'tus**, Gray. Stem stout, a foot or more high, with short rigid pubescence. Leaflets numerous. Flowers violet, in dense spike-like racemes. Pod straight, narrowly oblong, *the furrow of the ventral surface divided lengthwise into two by the projecting suture.*

13. **A. multiflo'rus**, Gray. Stem slender, somewhat glabrous. Leaflets 13-21, rather narrow. Flowers yellowish-white, tinged with purple, *in loosely few-flowered racemes*, the peduncles not exceeding the leaves. *Stipules mostly connate.* Pod 7-20-seeded, reflexed.

14. **A. aborig'inum**, Rich. Stems slender, rigid, clustered, hoary, pubescent or somewhat villous. Leaflets 7-13. Flowers white or bluish, the keel violet, in short spike-like racemes. Pod with a straight dorsal edge (*not furrowed*), and a rounded ventral edge.

B. Leaves apparently palmate, mostly of 3 leaflets.

15. **A. triphyl'lus**, Pursh. A low stemless glossy-silky perennial, with crowded long-petioled leaves, mostly having *3 lanceolate leaflets*, and whitish crowded sessile flowers. Pod conical-ovate, villous.

8. OXYTROPIS, DC.

* * *Leaves simply pinnate.*

1. **O. campes'tris**, DC., var. **cæru'lea**, Koch. Flowers white or yellowish, often tinged with purple or violet, in short spikes on naked scapes. Pods ovate or oblong, of a thin or papery texture. Leaves pinnate, of many leaflets. Gravelly river-margins.—N. W. and Atl. Prov.

** *Leaflets numerous, mostly in fascicles of 3 or 4 along the axis.*

2. **O. Lambert'i**, Pursh. Silvery-silky-pubescent. Flowers large, purple, violet, or white. Scapes often a foot high,

Pods silky-pubescent, cartilaginous, about half an inch long, imperfectly 2-celled.—N. W.

3. **O. montic'ola**, Gray, is reported as common in the prairie region. It is smaller than No. 2, and has denser spikes, usually smaller flowers, and less downy leaves.

4. **O. splendens**, Dougl. Silvery-silky-villous, 6-12 inches high. Flowers erect-spreading. *Pod erect, hardly surpassing the villous calyx.*—N. W.

9. GLYCYRRHI'ZA, Tourn. LIQUORICE.

G. lepido'ta, Nutt. (WILD LIQUORICE.) Two to three feet high. Leaflets 15-19, sprinkled with little scales when young.—Fort Erie and N. W.

10. TEPHRO'SIA, Pers. HOARY PEA.

T. Virginia'na, Pers. (GOAT'S RUE. CATGUT.) Stem 1-2 feet high, simple, leafy to the top. Leaflets 17-29, linear-oblong. Roots long and slender and tough.—Sandy soil, Lake Erie coast.

11. PSORA'LEA, L.

1. **P. argophyl'la**, Pursh. *Silvery-silky* all over. Leaves of 3-5 leaflets, the latter elliptical-lanceolate. Spikes of flowers *interrupted*. Roots *not* tuberous.—N. W. prairies.

2. **P. lanceola'ta**, Pursh. Smooth or nearly so. Leaflets 3. Flowers small, in short spikes. Roots not tuberous. —N. W.

3. **P. esculen'ta**, Pursh. *Rough-hairy* all over. Leaflets 5, obovate or lance-oblong. Spikes *dense*. *Roots tuberous.*—N. W.

12. AMOR'PHA, L. FALSE INDIGO.

1. **A. canes'cens**, Nutt. (LEAD-PLANT.) White-downy all over, 1-3 feet high. Leaflets less than half an inch long, crowded. Spikes generally clustered. *Pods 1-seeded.* —N. W.

2. **A. microphyl'la**, Pursh, differs from the last in being lower and nearly smooth. Spikes usually solitary. —N.W.

3. **A. frutico'sa,** L. A tall *shrub.* Leaflets larger and not so crowded as in the other species. *Pods 2-seeded.*—River-banks, N. W.

13. PETALOSTE'MON, Michx. PRAIRIE CLOVER.

1. **P. viola'ceus,** Michx. *Leaflets 5,* very narrow. Calyx silky-hoary. Corolla *rose-purple.*—Dry prairies, N. W.

2. **P. can'didus,** Michx. Leaflets 7-9, lanceolate or linear-oblong. Calyx smooth or nearly so. Corolla *white.* —Dry prairies, N. W.

14. VICIA, Tourn. VETCH. TARE.

1. **V. sati'va,** L. (COMMON VETCH or TARE.) Stem simple, somewhat pubescent. Leaflets 10-14, varying from obovate-oblong to linear. *Flowers purple, large, one or two together, sessile in the axils, or nearly so.*—Cultivated fields and waste grounds.

2. **V. tetrasper'ma,** L. Peduncles long and slender, 1-2-flowered. Flowers small. Calyx-teeth unequal. Corolla whitish. Pod 4-seeded, smooth. Leaflets about 4 pairs.— Atl. Prov.; also at Ottawa and opposite Montreal.

3. **V. Cracca,** L. (TUFTED V.) Downy-pubescent. *Leaflets 20-24,* oblong-lanceolate, strongly mucronate. *Peduncles long, bearing a dense one-sided raceme of blue flowers,* bent downward in the spike, and turning purple before withering. —Borders of thickets, and pastures. Chiefly eastward.

4. **V. Carolinia'na,** Walt. Smooth. Leaflets 8-12, oblong. Peduncles bearing a *rather loose raceme of whitish flowers, the keel tipped with blue.*—Low grounds and river-banks.

5. **V. America'na,** Muhl. Smooth. *Leaflets 10-14,* oval or ovate-oblong, very veiny. *Peduncles 4-8-flowered, flowers purple.*—Moist places.

Var. **linearis,** Watson, a low form, with *linear* leaflets, is common in the N.W. prairie region.

6. **V. hirsu'ta,** Koch. Stem weak. *Leaflets 12-16, linear.* Peduncles 3-6-flowered. *Pods hairy, 2-seeded.*—Chiefly eastward.

15. LATH'YRUS, L. EVERLASTING PEA.

1. **L. marit'imus,** Bigel. (BEACH PEA.) Stem stout, about a foot high. Leaflets 8–16, oval or obovate. *Stipules broadly halberd-shaped, about as large as the leaflets.* Flowers large, purple.—Sea-coast, and shores of the Great Lakes.

2. **L. veno'sus,** Muhl. (VEINY E.) Stem 2–3 feet high. Leaflets 10–14. *Stipules very small, slender, half arrow-shaped.* Flowers numerous.—Shady banks, chiefly westward and southward.

3. **L. ochroleu'cus,** Hook. (PALE E.) Stem slender. Leaflets 6–8, smooth and glaucous. *Stipules half heart-shaped, large. Corolla yellowish-white.*—Chiefly northward.

4. **L. palus'tris,** L. (MARSH E.) Stem slender, wing-margined. Leaflets 4–8, lanceolate, linear, or narrowly oblong, sharply mucronate. Stipules small, half arrow-shaped. Corolla blue-purple.—Moist places. Var. **myrtifolius** has oblong-lanceolate leaflets, and pale purple flowers. Upper stipules much larger than the lower ones.

16. A'PIOS, Boerhaave. GROUND-NUT. WILD BEAN.

A. tubero'sa, Mœnch. Flowers brown-purple.—A common twining plant in low grounds.

17. STROPHOSTY'LES, Ell.

S. angulo'sa, Ell. (*Phase'olus diversifolius,* Pers.) Annual. Stems 1–6 feet long. Leaflets ovate to oblong-ovate, often with a rounded lobe at the base. Flowers greenish-white and purplish, few, sessile, clustered on the long peduncles.—S. W. Ontario chiefly.

18. AMPHICARPÆ'A, Ell. HOG PEA-NUT.

A. mono'ica, Nutt. Flowers white or purplish.—Moist thickets and river-banks.

19. DESMO'DIUM, DC. TICK-TREFOIL.

' *Pod raised on a stalk much surpassing the calyx, the latter slightly toothed. Stipules bristle-form.*

1. **D. nudiflo'rum,** DC. Stem smooth, 4–8 inches high. Leaves crowded at the summit of sterile stems. Flowers in

a terminal raceme or panicle, *on a scape which rises from the root.* Leaflets broadly ovate.

2. **D. acumina'tum,** DC. Stem pubescent. Leaves all crowded at the summit of the stem, *from which the raceme or panicle arises.* Leaflets conspicuously pointed.—Rich woods.

3. **D. pauciflo'rum,** DC. Leaves *scattered* along the low ascending stems; leaflets rhombic-ovate, rather blunt. Racemes few-flowered, terminal.—Rich woods, western Ontario.

* * *Pod raised on a stalk hardly surpassing the calyx, the latter deeply cleft. Stipules ovate, taper-pointed.*

4. **D. rotundifo'lium,** DC. Stem *prostrate, soft-hairy.* Leaflets orbicular. Flowers purple. Pods indented on both edges.—Dry sandy woods, western Ontario. Also at Queenston Heights.

* * * *Pod hardly, if at all, stalked.*

5. **D. cuspida'tum,** Torr. and Gray. Stem tall, erect, very smooth. Leaflets ovate-lanceolate, taper-pointed, very large, green on both sides. Flowers and bracts large. Pod 1-6-jointed.—Thickets.

6. **D. Illinoen'se,** Gray, has been found in the Thames valley. Leaflets *obtuse* and *roughish*, grayish beneath, with prominent veins. Pod about an inch long, 3-5-jointed, indented on both margins. Otherwise nearly like the last.

7. **D. panicula'tum,** DC. Stem slender, nearly smooth. leaflets oblong-lanceolate, tapering to a blunt point. Flowers medium-sized. Pod 3-5-jointed, the joints triangular. Racemes panicled.—Rich woods.

8. **D. Dille'nii,** Darlingt. Distinguished from the last by the *pubescent* stem and finely pubescent leaflets, the latter *oblong* or *oblong-ovate*.—Dry and open thickets.

9. **D. Canadense,** DC. Stem erect, *hairy*, tall, furrowed. Leaflets oblong-lanceolate ($1\frac{1}{2}$-3 inches long), *with many straightish veins.* Flowers large, about $\frac{1}{2}$ inch long, in dense racemes. Joints of the pod roundish.—Dry woods.

10. **D. cilia're,** DC. Stem ascending, slender, hairy. Leaflets round-ovate (½-1 inch long). Flowers small, in *loose racemes.*—Dry thickets, south-western Ontario.

20. HEDYS'ARUM.

H. borea'le, Nutt. Leaflets 13-21, oblong or lanceolate. Stipules scaly, united opposite the petiole. Calyx 5-cleft, the awl-shaped lobes nearly equal. Pod of 3-4 flattened roundish joints, easily separated.—Rocky banks, Atl. Prov. and N. W. prairies.

21. LESPEDE'ZA. BUSH-CLOVER.

* *Flowers of two sorts; the larger perfect, the smaller pistillate and usually apetalous, mingled with the others.*

1. **L. viola'cea,** Pers. (*L. reticulata*, Pers., in Macoun's Catalogue.) Stems upright, branched. Leaflets varying from oblong to linear, downy underneath. *Flowers violet-purple.*—Dry borders of woods, western Ontario. Also at Queenston Heights.

* * *All the flowers perfect, in close spikes or heads.*

2. **L. hirta,** L. Stem erect, wand-like, tall, pubescent. Leaflets roundish or oval, pubescent. Spikes dense, on *peduncles longer than the leaves. Corolla yellowish-white*, with a purple spot on the standard.

3. **L. capita'ta,** Michx. *Peduncles and petioles short.* Leaflets varying from oblong to linear, silky underneath. Flowers in *dense heads; corolla as in No. 1.* Calyx much longer than the pod.—Both species are found in dry soil.

22. BAPTIS'IA, Vent. FALSE INDIGO.

B. tincto'ria, R. Br. (WILD INDIGO.) Smooth and slender, 2-3 feet high, branching. Leaves nearly sessile. Leaflets wedge-obovate, turning black on drying. Flowers yellow.—Dry soil. Lake Erie coast and Thames valley.

23. THERMOP'SIS, R. Br.

T. rhombifo'lia, Rich. A low perennial, with short, few-flowered racemes terminating the branches. Stipules as long as the petioles. Pods recurved, 10-14-seeded.—N. W. prairies.

24. GLEDIT'SCHIA, L. HONEY-LOCUST.

2. **G. triacan'thos,** L. Thorns stout, often triple or compound. Pods linear, often more than a foot long, with pulp between the flat seeds.—Common in cultivation, and established on Point Pelee.

25. GYMNOC'LADUS, Lam. KENTUCKY COFFEE-TREE.

G. Canadensis, Lam. Leaves 2–3 feet long, without stipules; leaflets ovate. Pod 6–10 inches long and 2 inches broad.—Established in a few localities.

ORDER XXXII. **ROSA'CEÆ.** (ROSE FAMILY.)

Herbs, shrubs, or trees, with alternate stipulate leaves, and regular flowers. The petals (mostly 5) and stamens (mostly more than 10) inserted on the edge of a disk which lines the calyx-tube. (See Part I., sections 48 to 57, for typical flowers.)

Synopsis of the Genera.

SUBORDER **AMYGDALEÆ.**

Carpel solitary, becoming a drupe, entirely free from the calyx, the latter deciduous. Ovules 2, but seed solitary as a rule. Trees or shrubs with simple leaves and deciduous stipules.

1. **Pru'nus.** Flowers perfect. Petals and calyx-lobes 5. Fruit a drupe.

SUBORDER **ROSACEÆ.**

Carpels few or many, free from the persistent calyx, becoming achenes, follicles, or drupe-like in fruit.

2. **Spiræ'a.** Carpels mostly 5, forming follicles in fruit. Calyx 5-cleft, short. Petals obovate, similar.
3. **Gille'nia.** Carpels and fruit as in Spiræa. Calyx elongated, 5-toothed. Petals slender, dissimilar.
4. **Agrimo'nia.** Carpels 2, forming achenes enclosed in the hardened calyx-tube. Calyx armed with hooked bristles. Flowers yellow, in slender spikes.
5. **Alchemil'la.** Carpels 1–4, forming achenes enclosed in the persistent calyx-tube. *Petals none.* Stamens 1–4. Calyx-tube inversely conical, the limb 4-parted, with 4 alternating bractlets. Low herbs, with palmately-lobed leaves.

ROSACEÆ. 63

6. **Pote'rium.** Carpels 1-3. Achene (mostly solitary) enclosed in the dry 4-angled closed calyx-tube. *Petals none.* Lobes of the top-shaped calyx 4, petal-like, spreading. Stigma tufted. Tall herbs, with pinnate leaves and a dense white spike of small flowers, often polygamous or diœcious.

7. **Geum.** Carpels numerous, one-ovuled, becoming dry achenes, the persistent styles becoming tails, plumose or naked, and straight or jointed. Calyx-lobes with 5 alternating bractlets.

8. **Waldstel'nia.** Carpels 2-6, forming achenes. Leaves radical, of 3 wedge-form leaflets. Bractlets of the calyx minute and deciduous. Flowers yellow, on bracted scapes.

9. **Potentil'la.** Carpels numerous, forming achenes heaped on a dry receptacle, the styles not forming tails. Lobes of the calyx with 5 alternating bractlets.

10. **Chamæ'rhodos.** Carpels 5-10, on a dry receptacle. Petals white, obovate. Stamens 5, opposite the petals. Calyx campanulate, 5-cleft. Small glandular-pubescent herbs with flowers in forked cymes, and many-cleft leaves.

11. **Fraga'ria.** Flowers as in Potentilla, but receptacle becoming fleshy or pulpy and scarlet in fruit. (See Part I., sec. 235.) Leaves all radical, of 3 leaflets. Low plants, producing runners.

12. **Dalibar'da.** Carpels 5-10, each 2-ovuled, forming nearly dry drupelets. Calyx 5-6-parted, 3 of the divisions larger than the others, and toothed. Calyx without bracts, persistent, enclosing the fruit. Leaves radical, round heart-shaped. Flowers white, on scapes.

13. **Rubus.** Carpels numerous, 2-ovuled, forming drupelets heaped on the receptacle. (See Part I., section 234.) Fruit edible. Calyx without bracts.

14. **Rosa.** Carpels numerous, 1-ovuled, forming achenes enclosed in the fleshy calyx-tube. (See Part I., section 49.)

SUBORDER **POMEÆ.**

Carpels enclosed in and adnate to the fleshy calyx-tube, forming a pome in fruit. Styles often united below. Trees or shrubs.

15. **Cratæ'gus.** Calyx-tube urn-shaped, becoming thick and fleshy in fruit, enclosing and combined with the 2-5 carpels. Fruit a pome, but drupe-like, containing 2-5 bony nutlets. *Thorny shrubs.* Flowers generally white.

16. **Pyrus.** Fruit a pome or berry-like, the 2-5 carpels or cells of a papery or cartilaginous texture (see Part I., sections 52 and 232), each 2-seeded. Shrubs or trees.

17. **Amelan'chier.** Pome berry-like, *10-celled, i.e.*, with twice as many cells as styles. Petals narrow. Otherwise as in Pyrus. Shrubs or small trees, *not thorny*.

1. PRUNUS, Tourn. PLUM. CHERRY.

1. **P. America'na,** Marshall. (WILD PLUM.) A thorny tree 8-10 feet high, with orange or red drupes half an inch or more in diameter; and *ovate, conspicuously pointed, coarsely or doubly serrate, veiny* leaves. Flowers white, appearing before the leaves, in umbel-like lateral clusters. —Woods and river-banks.

2. **P. pu'mila,** L. (DWARF CHERRY.) A small trailing shrub, 6-18 inches high. Leaves *obovate-lanceolate, tapering to the base*, toothed near the apex, *pale beneath*. Flowers in umbels of 2-4, appearing with the leaves. Fruit ovoid, dark red, as large as a good-sized pea.—Crevices of rocks, and sand-beaches and plains.

3. **P. Pennsylvan'ica,** L. (WILD RED CHERRY.) A tree 20-30 feet high, or shrubby. Leaves *oblong-lanceolate, sharply serrate*, green both sides. Flowers (appearing with the leaves) in large clusters, the pedicels elongated. Fruit globular, as large as a red currant, very sour.—Rocky thickets, and in old windfalls.

4. **P. Virginia'na,** L. (CHOKE-CHERRY.) A good-sized shrub, 3-10 feet high. Leaves *oval, oblong, or obovate, finely and sharply serrate, abruptly pointed*. Flowers in short erect racemes, appearing after the leaves. Fruit red, becoming darker, very astringent.—Woods and thickets.

5. **P. sero'tina,** Ehrhart. (WILD BLACK CHERRY.) A large tree, with reddish-brown branches. Leaves smooth, varying from *oval to ovate-lanceolate, taper-pointed, serrate, with short and blunt incurved teeth*, shining above. Flowers in long racemes. Fruit purplish-black, edible.—Woods and thickets.

2. SPIRÆ'A, L. MEADOW-SWEET.

1. **S. opulifo'lia,** L. (*Physocarpus opulifolius*, Maxim.) (NINE-BARK.) Shrub 3-7 feet high, the old bark separating

ROSACEÆ. 65

in thin layers. Leaves broadly ovate or cordate, 3-lobed, doubly crenate, smooth. Flowers white, in umbel-like corymbs terminating the branches. Folicles 2-5, *inflated*, purplish.—River-banks.

2. **S. salicifo'lia,** L. (COMMON MEADOW-SWEET.) Shrub 2-3 feet high, nearly smooth. Leaves wedge-lanceolate, doubly serrate. Flowers white or rose-coloured, in a dense terminal panicle.—Low grounds along streams.

3. **S. tomento'sa,** L. (DOWNY M.), with deep rose-coloured flowers, and the stems and under surface of the leaves densely woolly, occurs eastward toward the sea-coast, and in the northern counties of Ontario.

3. GILLE'NIA, Mœnch. INDIAN PHYSIC.

G. trifolia'ta, Mœnch. (BOWMAN'S ROOT.) Herb with 3-foliolate leaves; the leaflets ovate-oblong, pointed, rather coarsely serrate; stipules small, awl-shaped, entire. Flowers white or rose-coloured, in loose few-flowered corymbs.— Rich woods, chiefly south-westward.

4. AGRIMO'NIA, Tourn. AGRIMONY.

1. **A. Eupato'ria,** L. (COMMON AGRIMONY.) Stem herbaceous, hairy, 2-3 feet high. Leaves interruptedly pinnate, larger leaflets 5-7, oblong-ovate, coarsely serrate. Petals yellow, twice as long as the calyx.—Borders of woods.

2. **A. parviflo'ra,** Ait. (SMALL-FLOWERED A.), has crowded lanceolate leaflets, 11-19, with smaller ones intermixed, and small petals.—S. W. Ontario.

5. ALCHEMIL'LA, Tourn. LADY'S MANTLE.

A. vulga'ris, L. Leaves large, several-lobed, serrate. Flowers small, yellowish-green, in corymbose terminal clusters.—Atl. sea-coast.

6. POTE'RIUM, L. BURNET.

P. Canadense. Stamens 4, long-exserted, white. Stem 3-6 feet high. Leaflets numerous, heart-shaped, ovate or oblong-lanceolate, coarsely serrate.—Bogs and wet meadows, Atl. Prov.

7. GEUM, L. AVENS.

1. **G. album,** Gmelin. (WHITE AVENS.) Stem 2 feet high, slender, branching, smoothish or downy. Root-leaves pinnate, the cauline ones 3-divided, lobed, or only toothed. *Petals white,* as long as the calyx. Achenes bristly, tipped with the hooked lower joint of the style, the upper joint falling away. *Receptacle of the fruit bristly.*—Low rich woods and thickets.

2. **G. Virginia'num,** L. Stem stout, bristly-hairy. Leaves nearly as in No. 1. *Petals white, shorter than the calyx. Receptacle of the fruit nearly smooth.*—Meadows and thickets; not common.

3. **G. macrophyl'lum,** Willd. Bristly-hairy, stout. Root-leaves interruptedly pinnate, *with a very large round-heart-shaped terminal leaflet.* Stem-leaves with 2-4 minute lateral leaflets, the terminal 3-cleft, with *wedge-form rounded lobes. Petals yellow,* longer than the calyx. *Receptacle nearly naked.*—Atl. Prov. chiefly.

4. **G. strictum,** Ait. (YELLOW A.) Stem 2-3 feet high, rather hairy. Root-leaves interruptedly pinnate; stem-leaves 3-5-foliolate, leaflets obovate or ovate. *Petals yellow,* longer than the calyx. Receptacle of the fruit downy. Achenes tipped with the hooked style.—Dry thickets.

5. **G. riva'le,** L. (WATER or PURPLE AVENS.) *Petals purplish-yellow; calyx brown-purple.* Flowers nodding, but the fruiting heads upright. *The upper joint of the style feathery, persistent.* Stem simple, 2 feet high. Root-leaves lyrate; stem-leaves few, 3-foliolate, lobed.—Bogs and wet places.

6. **G. triflo'rum,** Pursh. Stem about a foot high, soft-hairy. Flowers 3 or more, on long peduncles, purple. Styles not jointed, feathery, *at least 2 inches long in the fruit.* —Dry hills and thickets. Not common.

8. WALDSTEI'NIA, Willd. BARREN STRAWBERRY.

W. fragarioi'des, Tratt. A low plant, 4-6 inches high. Leaflets 3, broadly wedge-form, crenately toothed. Scapes several-flowered. Petals yellow, longer than the calyx.— Dry woods and hill-sides.

9. POTENTILLA, L. CINQUE-FOIL. FIVE-FINGER.

* Styles thickened and glandular towards the base. Inflorescence cymose.

    ← *Style arising from near the base of the carpel.*

1. **P. argu'ta,** Pursh. Stem stout, 1–2 feet high, brownish-hairy. Leaves pinnate, of 7–11 oval serrate leaflets, downy underneath. Flowers in dense cymose clusters. Petals yellowish or cream-coloured, deciduous. Stamens about 30. Plant clammy above.—Dry thickets.

    ← ← *Style arising from the top of the carpel.*

2. **P. Norve'gica,** L. (NORWAY CINQUE-FOIL.) Stem *erect, hairy,* branching above. *Leaves palmate, of 3 leaflets;* leaflets obovate-oblong, coarsely serrate. Flowers in cymose clusters. *Calyx large.* Petals pale yellow, small, *not longer than the sepals.* Stamens about 15.—Fields and low grounds.

3. **P. riva'lis,** Nutt., var. **millegra'na,** Watson. *More slender and branching* than the last, softly villous. Leaves all of 3 serrate leaflets. *Cymes loose. Calyx small.* Petals yellow, minute. Stamens 10–20, rarely 5. — N.W.

4. **P. paradox'a,** Nutt., (*P. supi'na,* L.) a plant of spreading or decumbent habit, with *pinnate leaves of 5–11 leaflets, loose leafy cymes, small petals,* as long as the sepals, and achenes with an appendage at the base, occurs along the western shore of Lake Ontario.

5. **P. Pennsylvan'ica,** L. Perennial; more or less white-woolly. Leaflets 5–9, *white-woolly beneath,* less so above, cut-pinnatifid, the segments linear. Stamens about 25.— Chiefly eastward and N.W.

Var. **strigo'sa,** Pursh, is silky-woolly throughout, 6–12 inches high. Leaflets deeply pinnatifid, the margins of the narrow lobes revolute. Cyme short and close.—N. W.

6. **P. recta,** L. Perennial, tall, sparingly villous, glandular-puberulent. Leaves *palmate,* of 5–7 leaflets, the latter cut-pinnatifid. Flowers yellow, large, in a broad cyme.— Not common.

** *Styles filiform, not glandular at the base. Inflorescence cymose.*

+ *Style arising from the top of the carpel.*

7. **P. Hippia'na,** Lehm. *Densely white-tomentose throughout.* Stem slender. Leaves pinnate, of 5-11 leaflets, diminishing regularly down the petiole. Leaflets sharply toothed at least at the apex. Stamens 20.—N.W.

Var. **pulcher'rima,** Watson, has the upper surfaces of the crowded leaflets *green and pubescent.*—N.W.

8. **P. effu'sa,** Dougl. *Tomentose throughout, with scattered villous hairs.* Stems diffusely branched above, 4-12 inches high. Leaflets 5-11, interruptedly pinnate. Carpels 10.—N.W.

9. **P. Platten'sis,** Nutt. Stems decumbent. Pubescence appressed silky-villous throughout. Leaflets 7-13, crowded, deeply cut-pinnatifid into linear segments. Flowers few. Carpels many.—N.W.

10. **P. gra'cilis,** Dougl. Stems 2-3 feet high. *Leaves palmate,* of about 7 leaflets, the latter serrate, tomentose beneath, green above. Carpels many.—N.W.

Var. **flabellifor'mis,** Torr. and Gray, has very deeply pinnatifid leaflets.—N.W.

Var. **rig'ida,** Watson, is villous, but not tomentose, and usually tall and stout.—N.W.

11. **P. pilo'sa,** Willd., is distinguished from the preceding species by its tall rigid stems and light yellow flowers.—Niagara Falls, not common.

12. **P. argen'tea,** L. (SILVERY C.) Stem ascending, branched at the summit, *white-woolly.* Leaves palmate, of 5 leaflets, the latter deeply serrate towards the apex, *with revolute margins, and woolly beneath.* Petals yellow, longer than the sepals.—Dry fields and roadsides.

13. **P. humifu'sa,** Nutt. Stems decumbent, 2-4 inches long, slender. *Densely white-tomentose.* Leaflets 5, serrate at the rounded or truncate apex with 3 or 5 teeth.—N.W.

\+ \+ *Style arising from the side of the carpel.*

14. **P. frutico'sa,** L. (SHRUBBY C.) Stem erect, *shrubby*, 1–3 feet high, much branched. Leaves pinnate, of 5–7 leaflets, closely crowded, *entire*, silky, especially beneath. Flowers numerous, large, yellow, terminating the branches. —Bogs.

15. **P. tridenta'ta,** Ait., (THREE-TOOTHED C.) is common eastward towards the sea-coast. Stem 4–6 inches high. Leaves rigid, palmate, of 3 wedge-shaped leaflets, *3-toothed at the apex. Petals white.*

16. **P. palustris,** Scop. (MARSH FIVE-FINGER.) Stem ascending. Leaves pinnate, of 5–7 lanceolate, crowded, deeply serrate leaflets, whitish beneath. *Calyx an inch broad, dark purple inside. Petals purple.*—Bogs.

*** *Styles filiform, lateral. Peduncles axillary, solitary, 1-flowered.*

17. **P. Anseri'na,** L. (SILVER-WEED.) A low plant, *creeping with slender runners.* Leaves all radical, interruptedly pinnate; leaflets 9–19, serrate, green above, *silvery-silky beneath. Flowers solitary, on long scape-like peduncles,* bright yellow.—River and lake margins.

18. **P. Canaden'sis,** L. (CANADA C.) Stem prostrate or ascending, silky-hairy. *Leaves palmate, of 5 leaflets,* the latter serrate towards the apex. Flowers solitary. Petals yellow, *larger than the sepals.*—Dry soil.

10. CHAMÆ'RHODOS, Bunge.

C. erecta, Bunge. Stem slender, 2–12 inches high, branching above. Radical leaves many-cleft, forming a rosette; cauline ones 3–5-cleft.—N.W. prairies.

11. FRAGA'RIA, Tourn. STRAWBERRY.

1. **F. Virginia'na,** Ehrhart. *Achenes deeply imbedded in pits* on the surface of the fleshy receptacle; calyx erect after flowering. Leaflets firm.

2. **F. ves'ca,** L. *Achenes not sunk in pits,* but merely on the surface of the receptacle; calyx spreading. Leaflets thin.

12. DALIBAR'DA, L. DALIBARDA.

D. repens, L. (*Rubus Dalibarda*, L., in Macoun's Catalogue.) Stems tufted, downy. Whole plant with something of the aspect of a violet.—Low woods.

13. RUBUS, Tourn. BRAMBLE.

1. **R. odora'tus**, L. (PURPLE FLOWERING - RASPBERRY.) Shrubby 3-5 feet high. Branches, peduncles, and calyx *clammy with glandular hairs. Flowers large and handsome, rose-purple*. Leaves large, broadly ovate, 3-5-lobed, the lobes acute, minutely toothed. *Fruit flat.*

2. **R. Nutka'nus**, Mocino, is distinguished from No. 1 by its *oval white petals*, and its 5-lobed coarsely toothed leaves. —N.W.

3. **R. Chamæmo'rus**, L. (CLOUD-BERRY. BAKED-APPLE BERRY.) A low herb with *diœcious* flowers. *Stem simple*, without prickles, 2-3-leaved, bearing one large white flower. Leaves simple, kidney-form, 5-lobed, serrate.—Sphagnous swamps, chiefly eastward.

4. **R. triflo'rus**, Richardson. (DWARF RASPBERRY.) Stems ascending or trailing, a foot high, not prickly. Leaflets 3-5, nearly smooth, rhombic-ovate, acute at both ends, doubly serrate. Peduncle usually 3-flowered. Petals white; sepals reflexed. Fruit red.—Cedar-swamps.

5. **R. strigo'sus**, Michx. (WILD RED RASPBERRY.) *Stems upright, beset with stiff straight bristles.* Leaflets 3-5, oblong-ovate, pointed, cut-serrate, whitish beneath. *Fruit light red.*—Hillsides and thickets.

6. **R. occidenta'lis**, L. (BLACK RASPBERRY.) *Stem glaucous, recurved, armed with hooked prickles.* Leaflets 3, ovate, pointed, coarsely serrate, white-downy beneath. Fruit purplish-black.—Borders of fields, especially where the ground has been burned over.

7. **R. villo'sus**, Ait. (HIGH BLACKBERRY.) Stem shrubby, furrowed, erect or reclining, armed with hooked prickles. Leaflets 3-5, unequally serrate, the terminal one conspicu-

ously stalked. Lower surface of the leaflets *hairy and glandular*. Flowers racemed, numerous, large and white. *Fruit oblong, black.* Var. **frondosus** is smoother and less glandular. Var. **humifusus** is trailing and smaller, and the flowers are less numerous.—Borders of thickets.

8. **R. Canaden'sis**, L. (Low Blackberry. Dewberry.) Stem shrubby, *extensively trailing, slightly prickly*. Leaflets chiefly 3, oval or ovate-lanceolate, *nearly smooth*, sharply serrate. Flowers in racemes.—Thickets and rocky hills.

9. **R. his'pidus**, L., (Running Swamp-Blackberry) occurs occasionally in low meadows. Stem prostrate, with small reflexed prickles, sending up at intervals the short flowering shoots. Leaflets mostly 3, smooth and shining. Fruit of few grains, red or purple.

14. ROSA, Tourn. Rose.

* *Styles cohering in a protruding column, as long as the stamens.*

1. **R. setig'era**, Michx. *Stem climbing.* Prickles nearly straight. Leaflets 3-5, ovate. Petals deep rose-coloured, changing to white.—Borders of thickets and along fences; south-western Ontario.

** *Styles separate, included within the calyx-tube; sepals spreading after flowering, and deciduous.*

2. **R. Caroli'na**, L. (Swamp Rose.) Stem 1-7 feet high, erect, armed with *stout straight or somewhat curved prickles*. Leaflets 5-9, *very finely serrate*, mostly narrowly oblong. Stipules long and very narrow. *Flowers in corymbs, numerous. Calyx and globular calyx-tube beset with glandular bristles.*—Wet places, Ontario.

3. **R. lu'cida**, Ehrhart. Stem 1-6 feet high, armed with stout more or less hooked prickles. Leaflets 5-9, *coarsely serrate, smooth and often shining above*, rather thick. *Peduncles 1-3-flowered.*—Dry soil, or borders of swamps, chiefly eastward.

4. **R. hu'milis**, Marsh. Low, more slender and less leafy than the last, with *straight slender spines. Stipules narrow.* Leaflets thin and pale. *Outer sepals always more or less lobed.*—Mostly in sandy soil.

5. **R. nit'ida,** Willd. Low. Stem and branches usually thickly covered with *prickles interspersed with straight slender spines*. Stipules mostly dilated. Leaflets bright green and shining, mostly narrowly oblong. Flowers generally solitary. *Sepals entire.*—Margins of swamps, Atl. Prov.

6. **R. rubigino'sa,** L. (SWEET - BRIER.) Stem tall. Prickles numerous, the larger hooked, the smaller awl-shaped. Leaflets 5-7, doubly serrate, *glandular beneath, aromatic*. *Flowers mostly solitary*. Fruit *pear-shaped* or obovate.—Roadsides and fields.

*** *Styles separate; sepals erect and connivent after flowering, persistent.*

← *Fruit globose.*

7. **R. blanda,** Ait. (EARLY WILD ROSE.) Stem 1-3 feet high. *Prickles* (if any) *few and scattered, straight*. Leaflets 5-7, mostly oblong-lanceolate, cuneate at the base, not resinous, simply serrate. Sepals hispid, not lobed. *Peduncles 1-3-flowered.*—Rocks and rocky shores, mostly eastward.

8. **R. Say'i,** Schwein. Stems 1-2 feet high, *very prickly*. Leaflets 3-7, broadly elliptical to oblong-lanceolate, *resinous, the teeth serrulate*. Flowers large, mostly solitary. Outer sepals usually lobed, not hispid.—Our most northern rose.

9. **R. Arkansa'na,** Porter. Stems low, *very prickly*. Stipules narrow. Leaflets 7-11, broadly elliptical to oblong-lanceolate, wedge-shaped at the base, simply serrate, not resinous. Outer sepals lobed, rarely hispid.—N.W. prairies.

← *Fruit oblong-ovate to oblong.*

10. **R. Engelman'ni,** Watson. Stems 3-4 feet high, often very prickly. Leaflets 5-7, the teeth serrulate. Flowers solitary. Sepals not lobed. Fruit ½-1 inch long.—Shores of Lake Superior and westward.

15. CRATÆ'GUS, L. HAWTHORN.

1. **C. coccin'ea,** L. (SCARLET - FRUITED THORN.) A low tree with reddish branches, and stout chestnut-brown

spines. Peduncles and calyx glandular. Leaves rather *thin*, roundish-ovate, *on slender petioles;* acutely glandular-toothed, sometimes cut-lobed. Fruit bright red, globose or obovate, half an inch broad.—Thickets, common.

Var. **macracantha,** Dudley, (*C. tomentosa,* L. in part) has longer spines and thicker leaves, wedge-shaped at the base, on stout petioles and often deeply cut. The cymes also are broader and the flowers and fruit rather large.—Thickets.

Var. **mollis,** Torr. and Gray, (*C. tomentosa,* L., var. *mollis,* Gray) has densely pubescent shoots and large slender-petioled leaves, usually with acute narrow lobes. Fruit bright scarlet with a slight bloom, an inch broad.—Queenston and westward along Lake Erie.

2. **C. puncta'ta,** Jacq. Not glandular. Branches horizontal. Leaves rather small, wedge-obovate, tapering and entire below, unequally toothed above, villous-pubescent when young, not shining. Fruit globose, about an inch broad.

3. **C. Crus'galli,** L. (Cockspur Thorn.) A shrub or low tree, *glabrous*. Leaves thick, *shining above, wedge-obovate,* finely serrate. *Petioles very short.* Fruit globular, dull red, $\frac{1}{3}$ of an inch broad. Thorns very long.—Thickets, south-western Ontario.

4. **C. Douglasii,** Lindl., has few short and stout spines, ovate thin doubly serrate leaves, and small black-purple fruit.—N.W. prairies.

16. PYRUS, L. Pear. Apple.

1. **P. corona'ria,** L. (American Crab-Apple.) A small tree, with ovate serrate *simple leaves,* tomentose beneath. Flowers in umbel-like cymes. Styles woolly and cohering at the base. *Fruit a greenish apple.*—Toronto and westward.

2. **P. arbutifo'lia,** L. (Choke-berry.) A shrub, with oblong or oblanceolate finely serrate *simple leaves,* tomentose

beneath. Flowers in compound cymes. *Fruit berry-like*, nearly globular, *dark red or purple.*—Swamps.

Var. **melanocar'pa,** Hook., is nearly smooth throughout, and has large *black* fruit.—Swamps.

3. **P. America'na,** DC. (AMERICAN MOUNTAIN-ASH.) A small tree with *odd-pinnate leaves of 13-15 leaflets*, the latter *lanceolate, taper-pointed, sharply serrate*, bright green. *Fruit scarlet, berry-like*, not larger than peas. Flowers in flat cymes.—Swamps and cool woods, northward.

4. **P. sambucifo'lia,** Cham. and Schlecht., differs from the last in having oblong, oval or lance-ovate, obtuse leaflets, smaller cymes, and larger flowers and berries.—N.W.

17. **AMELAN'CHIER,** Medic. JUNE-BERRY. SASKATOON-BERRY.

1. **A. Canadensis,** Torr. and Gray. (SHADBUSH. SERVICE-BERRY.) A tree 10-30 feet high, with a purplish, berry-like edible fruit. Leaves ovate to ovate-oblong, slightly cordate, pointed, very sharply serrate, nearly or soon glabrous. Bracts and stipules silky-ciliate. Flowers large, in drooping nearly glabrous racemes. Petals oblong. Pedicels of the berries long.—Open woodlands.

Var. **rotundifolia,** Torr. and Gray, has broader leaves.

Var. **oblongifolia,** Torr. and Gray, is a small tree (6-10 feet), with the young leaves and racemes densely white-tomentose. Leaves broader and generally rounded at the base. Flowers in shorter and denser racemes, and pedicels of the fruit shorter. Petals oblong-spathulate. — Low grounds.

2. **A. oligocarp'a,** Rœm. A shrub 2-4 feet high. Leaves thin, oblong, *acute at both ends.* Petals oblong-obovate. Fruit dark purple with a dense bloom.—Cold swamps.

3. **A. alnifolia,** Nutt. A shrub 3-8 feet high. Leaves mostly broadly elliptical and obtuse, coarsely toothed towards the apex. Raceme short and rather dense. Petals wedge-shaped-oblong.—N.W. prairies.

ORDER XXXIII. SAXIFRAGA'CEÆ. (SAXIFRAGE F.)

Herbs or shrubs, distinguished from Rosaceæ chiefly in having opposite as well as alternate leaves, and usually no stipules; stamens only as many or twice as many as the (usually 5) petals; and the carpels fewer than the petals (mostly 2), and usually more or less united with each other. Stamens and petals generally inserted on the calyx.

Synopsis of the Genera.

1. **Ri'bes.** Shrubs, sometimes prickly, with alternate and palmately-veined and lobed leaves, which are plaited in the bud. Calyx 5-lobed, the tube adherent to the ovary (superior). Petals 5, small, inserted on the calyx. Stamens 5. Styles 2. *Fruit a many-seeded berry.*

2. **Parnas'sia.** Smooth herbs, with entire and chiefly radical leaves, and solitary flowers terminating the long scapes. Petals 5, large, *veiny, each with a cluster of sterile filaments at the base.* Proper stamens 5. Stigmas 4. Pod 4-valved. Calyx free from the ovary.

3. **Saxifraga.** Herbs with *clustered root-leaves.* Flowers in close cymes. Calyx-lobes hardly adherent to the ovary. Petals 5. Stamens 10. *Fruit a pair of follicles*, slightly united at the base.

4. **Mitel'la.** Low and slender herbs, with round-heart-shaped radical leaves, *those on the scape (if any) opposite.* Flowers in terminal racemes. Calyx 5-lobed, adherent to the base of the ovary. Petals 5, slender, *pinnatifid.* Stamens 10, *short.* Styles 2. Pod 2-beaked, but 1-celled.

5. **Tiarel'la.** Slender herbs, with radical heart-shaped leaves, and *leafless* scapes, bearing a simple raceme of flowers. Calyx bell-shaped, 5-parted. Petals 5, *entire.* Stamens 10, *long and slender.* Pod 2-valved, *the valves unequal.*

6. **Heu'chera.** Perennial herbs with round-heart-shaped radical leaves; those on the stem (if any) *alternate.* Greenish or purple flowers, clustered in a long narrow panicle. Petals 5, small, spathulate, entire. *Stamens 5.* Styles 2. Pod 2-beaked.

7. **Chrysosple'nium.** Small and smooth herbs, with mostly opposite roundish leaves. Calyx-tube adherent to the ovary. *Petals none.* Stamens twice as many as the calyx-lobes (8-10), inserted on a conspicuous disk. Pod 2-lobed.

1. RI'BES, L. CURRANT. GOOSEBERRY.

1. **R. Cynos'bati,** L. (WILD GOOSEBERRY.) Stem with small thorns at the bases of the leaves, the latter downy. on

slender petioles, roundish heart-shaped, 3-5-lobed. *Peduncles slender*, 2-3-flowered. *Berry covered with long prickles.*—Open woods and clearings.

2. **R. oxyacanthoi'des**, L. (*R. hirtellum*, Michx.) (SMALL WILD GOOSEBERRY.) Stems with very short thorns or none. *Peduncles very short*, 1-2-flowered. Stamens shorter than the broadly oblong calyx-lobes. *Berry small, smooth.*—Low grounds.

3. **R. rotundifo'lium**, Michx. Like the last, but *the stamens are longer than the narrowly oblong-spathulate calyx-lobes.*—Chiefly eastward.

4. **R. lacus'tre**, Poir. (SWAMP GOOSEBERRY.) Shrubby. Young stems prickly, and thorny at the bases of the leaves. Leaves cordate, *deeply 3-5-lobed, the lobes deeply cut*. *Racemes 4-9-flowered, slender, nodding*. Fruit bristly.—Swamps and wet woods.

5. **R. flor'idum**, L. (WILD BLACK CURRANT.) *Stems and fruit without prickles or thorns. Leaves resinous-dotted*, sharply 3-5-lobed, doubly serrate. Racemes many-flowered, drooping. Calyx bell-shaped. *Fruit black*, smooth.—Woods.

6. **R. rubrum**, L. (WILD RED CURRANT.) A low shrub with straggling stems. Leaves obtusely 3-5-lobed. *Racemes from lateral buds separate from the leaf-buds*, drooping. Calyx flat. *Fruit red*, smooth.—Bogs and wet woods.

7. **R. prostra'tum**, L'Her. (FETID CURRANT.) Stems reclined. Leaves deeply cordate, 5-7-lobed, smooth, the lobes ovate, acute, doubly serrate. Racemes erect, slender, the flowers greenish. Fruit pale red, glandular-bristly.—Cold damp woods and rocks.

8. **R. Hudsonia'num**, Richards. Like the last, but with white flowers crowded in the erect raceme, and darker and *smooth* fruit.—N. W. Also in central Ontario.

2. PARNAS'SIA, Tourn. GRASS OF PARNASSUS.

1. **P. Carolinia'na**, Michx. Petals sessile, very veiny *Sterile filaments 3 in each set.* Leaves ovate or rounded

usually only one low down on the stalk. Flower an inch across, the petals much longer than the calyx.—Beaver meadows and wet banks.

2. **P. palustris,** L. *Sterile filaments 9–15 in each set.* Scapes 3–10 inches high. Flower nearly an inch across, the petals not much longer than the sepals. Leaves cordate.— Sandy banks.

3. **P. parviflo'ra,** DC. *Petals but little longer than the sepals.* Sterile filaments *about 7 in each set.* Leaves ovate or oblong.—Atl. Prov. and northward.

3. SAXIF'RAGA, L. SAXIFRAGE.

1. **S. Virginien'sis,** Michx. (EARLY SAXIFRAGE.) Stem 4–9 inches high. Scape clammy. Leaves obovate, crenately toothed. Petals white, oblong, twice as long as the sepals. —Rocks and hillsides.

2. **S. tricuspida'ta,** Retz. Stems tufted, 4–8 inches high. Leaves alternate, oblong or spathulate, *with 3 rigid sharp teeth at the end.* Petals yellow.—Chiefly N. and N.W.

3. **S. Aizo'on,** Jacq. Scape 5–10 inches high. Leaves *thick, spathulate, with white finely-toothed margins.* Petals cream-colour, obovate, often spotted at the base. — Moist rocks, Atl. sea-coast and northward.

4. MITEL'LA, Tourn. MITRE-WORT. BISHOP'S-CAP.

1. **M. diphyl'la,** L. (TWO-LEAVED MITRE-WORT.) Stem hairy. Leaves cordate, 3–5-lobed, *those on the scape 2, opposite,* nearly sessile. Flowers white, oblong.- Rich woods.

2. **M. nuda,** L. (NAKED-STALKED M.) Stem small and delicate. Leaves kidney-shaped, *doubly crenate. Scape leafless,* few-flowered. *Flowers greenish.*—Deep woods, on moss-covered logs, etc.

5. TIAREL'LA, L. FALSE MITRE-WORT.

T. cordifo'lia, L. Scapes leafless, 5–12 inches high. Leaves heart-shaped, sharply toothed, sparsely hairy above, downy beneath. Petals white, oblong.—Rich woods.

6. HEU'CHERA, L. ALUM-ROOT.

1. **H. America'na,** L. (COMMON ALUM-ROOT.) Stems 2-3 feet high, glandular and short-hairy. Flowers small, in a loose panicle. Petals not longer than the calyx-lobes. Stamens and style exserted.—S. W. Ontario.

2. **H. his'pida,** Pursh, has larger flowers in a very narrow panicle, and taller stems, with long spreading hairs. Stamens short, but soon exserted.—N. W. prairies.

7. CHRYSOSPLE'NIUM, Tourn. GOLDEN SAXIFRAGE.

C. America'num, Schwein. A low and delicate smooth herb, with spreading and forking stems. Flowers greenish-yellow, inconspicuous, nearly sessile in the forks.—Shady wet places.

ORDER XXXIV: **CRASSULA'CEÆ.** (ORPINE FAMILY.)

Succulent herbs (except in one genus), chiefly differing from Saxifragaceæ in having *symmetrical flowers*, the sepals, petals and carpels being the same in number, and the stamens either as many or twice as many.

Synopsis of the Genera.

1. **Pen'thorum.** Not succulent. The carpels united, forming a 5-celled pod.
2. **Sedum.** Succulent. Carpels distinct.

1. PEN'THORUM, Gronov. DITCH STONE-CROP.

P. sedoi'des, Gronov. Not succulent. Sepals 5. Petals 5, if any; sometimes wanting. Stamens 10. *Pod 5-angled, 5-horned, and 5-celled.* Leaves scattered, lanceolate, acute at both ends. A homely weed, with greenish-yellow flowers in a loose cyme.—Wet places. (Parts of the flowers occasionally in sixes or sevens.)

2. SEDUM, Tourn. STONE-CROP. ORPINE.

1. **S. acre,** L. (MOSSY STONE-CROP.) Leaves very thick and succulent, crowded, very small. Petals yellow. A spreading moss-like plant, which has escaped from cultivation in many places.—Roadsides.

2. **S. Tele'phium,** L. (LIVE-FOR-EVER.) Stems tall and stout. Leaves oval, toothed. Flowers in compound cymes, *petals purple.* *Sepals, petals, and carpels 5 each.* Stamer 10. (Int. from Eu.)

3. **S. Rhodi'ola,** DC. (ROSE-ROOT.) Stems 5-10 inches high. Flowers diœcious, greenish-yellow or purplish. *Stamens mostly 8,* other parts in *fours.*—Rocky shores, Atl. Prov.

ORDER XXXV. **HAMAMELA'CEÆ.** (WITCH-HAZEL F.)

Tall shrubs, with alternate simple leaves, and deciduous stipules. Flowers in clusters or heads, often monœcious. Calyx 4-parted, adherent to the base of the ovary, the latter of 2 united carpels. Fruit a 2-beaked, 2-celled, woody pod, opening at the top. Petals 4, strap-shaped, inserted on the calyx. Stamens 8, 4 of them anther-bearing, the remainder reduced to scales. The only genus with us is

HAMAME'LIS, L. WITCH-HAZEL.

H. Virginia'na, L. Leaves obovate or oval, crenate or wavy-toothed, pubescent. Flowers yellow, appearing late in the autumn.—Damp woods, chiefly west of Toronto.

ORDER XXXVI. **HALORA'GEÆ.** (WATER-MILFOIL F.)

Aquatic or marsh plants, with small inconspicuous flowers, sessile in the axils of the leaves or bracts. Calyx-tube adherent to the ovary (but calyx and corolla wanting in Callitriche), the latter 2-4-celled (in Hippuris of a single carpel). Limb of the calyx minute or none. Petals small or none. Stamens 1-8. Fruit indehiscent, a single seed in each cell.

Synopsis of the Genera.

1. **Myriophyl'lum.** Flowers monœcious or polygamous, with the parts in fours. Stamens 4 or 8. Immersed leaves pinnately dissected into capillary divisions.
2. **Hippu'ris.** Flowers perfect. Stamen, style, and carpel *only one.* Leaves *entire*, linear, acute; in whorls of 8 or 10.

3. **Proserpina'ca.** Flowers perfect, the parts in threes. Petals none. Stems creeping at the base. Leaves alternate, the immersed ones pinnately dissected.

4. **Callit'riche.** Flowers monœcious. Calyx and corolla wanting. Leaves entire, opposite. Staminate flower of a single stamen; pistillate flower a single 4-celled ovary. Fruit compressed, 4-lobed. 4-celled, breaking up into 4 one-seeded pieces.

1. MYRIOPHYL'LUM, Vaill. WATER-MILFOIL.

1. **M. spica'tum,** L. *Stamens 8.* Bracts ovate, entire, *shorter than the flowers.* Leaves in whorls of 3 or 4. Flowers greenish, in terminal spikes. Stems very long.—Deep water.

2. **M. verticilla'tum,** L. *Stamens 8.* Leaves finely dissected and whorled as in No. 1. Bracts pectinate-pinnatifid, much longer than the flowers, and the spikes therefore leafy. Stem 2-4 feet long.—Stagnant water.

3. **M. heterophyl'lum,** Michx. *Stamens 4.* Lower leaves dissected, in whorls of 4 or 5. Bracts ovate or lanceolate, finely serrate, crowded, the lower ones pinnatifid. Stem stout.—Stagnant or slow water.

4. **M. tenel'lum,** Bigel. Flowering stems nearly leafless. Bracts small, entire. Flowers alternate, monœcious. Stamens 4.—Borders of ponds.

2. HIPPU'RIS, L. MARE'S TAIL.

H. vulga'ris, L. A perennial aquatic, with jointed erect stem.—Muddy margins of ponds and streams.

3. PROSERPINA'CA, L. MERMAID-WEED.

P. palustris, L. (MERMAID-WEED.) Low herb. Stem creeping at base. Leaves alternate, lanceolate, sharply serrate. Petals none. Stamens 3. Fruit nut-like, 3-seeded. —Wet swamps.

4. CALLIT'RICHE, L. WATER-STARWORT.

1. **C. ver'na,** L. Amphibious. Floating leaves obovate, tufted; submersed leaves linear. Flowers monœcious, axillary, usually between a pair of bracts. Sterile flower *a single stamen;* fertile flower *a single pistil* with a 4-celled ovary. *Leaves beset with stellate scales.*

2. **C. autumna'lis.** Growing under water. May be distinguished from **C. verna** by its leaves being *retuse and all linear from a broader base*, and its flowers *without bracts*.

Order XXXVII. **ONAGRA'CEÆ.** (Evening-Primrose F.)

Herbs with perfect and symmetrical flowers, the parts of the latter in twos or fours. Calyx-tube adherent to the ovary, and usually prolonged above it. Petals and stamens inserted on the calyx. Style 1. Stigmas 2 or 4 or capitate. (See Part I., sections 44–47, for description of a typical plant.)

Synopsis of the Genera.

1. **Circæ'a.** *Petals 2, obcordate. Stamens 2.* Stigma capitate. Fruit *bur-like*, 1-2-seeded, beset with hooked bristles. Delicate low plants with opposite leaves and very small white flowers in racemes.
2. **Epilo'bium.** *Petals 4. Stamens 8.* Calyx-tube hardly prolonged beyond the ovary. Fruit a linear pod, many-seeded, *the seeds provided with tufts of downy hairs.*
3. **Œnothe'ra.** *Petals 4. Stamens 8.* Stigma 4-lobed or discoid. *Flowers yellow* (white in one species). Calyx-tube much prolonged. Pods cylindrical or club-shaped. *Seeds without tufts.*
4. **Gau'ra.** *Petals 4*, clawed and unequal. *Stamens 8.* Stigma 4-lobed, with a cup-like border. Calyx-tube much prolonged. Fruit hard and nut-like, ribbed, *indehiscent or nearly so*, few-seeded. Flowers rose-coloured or white, turning scarlet.
5. **Ludwig'ia.** *Petals 4, or none. Stamens 4.* Calyx-tube not prolonged. Stigma capitate.

1. CIRCÆ'A, Tourn. Enchanter's Nightshade.

1. **C. Lutetia'na,** L. Stem 1-2 feet high. Leaves opposite, ovate, slightly toothed. *No bracts under the pedicels.* Fruit roundish, *bristly-hairy, 2-celled.*—Rich woods.

2. **C. alpi'na,** L. Stem *low* and delicate (3-8 inches). Leaves cordate, coarsely toothed. *Minute bracts under the pedicels.* Fruit *club-shaped, soft-hairy, 1-celled.*—Deep low woods.

2. EPILO'BIUM, L. Willow-herb.

1. **E. angustifo'lium,** L. (*E. spica'tum,* Lam.) (Great Willow-herb.) Stem 3-6 feet high, simple. Leaves lanceolate, scattered. Flowers purple, very showy, *in a terminal*

raceme or spike. Stamens and style deflexed. Stigma of 4 long lobes.—Newly-cleared land.

Var. **canescens,** Wood, has *white* flowers and silvery pods.

2. **E. hirsu'tum,** L., is found about Niagara Falls. The stem is tall, erect, and densely soft-hairy, with opposite lance-oblong leaves. Flowers smaller than in No. 1, rose-purple, forming a leafy short raceme. Stamens and style erect.

3. **E. panicula'tum,** Nutt. Glabrous, or pubescent above. Stem varying from 1 to 10 feet high, branching above. Leaves narrowly linear, mostly alternate and fascicled. Flowers few, small, terminating the spreading slender and almost leafless branches. Stamens and style erect. Stigma club-shaped.—Colpoy's Bay, Lake Huron; and N.W.

4. **E. linea're,** Muhl. (*E. palustre,* L., var. *lineare,* Gray.) Stem 1-2 feet high, erect, slender, branching above, *hoary-pubescent.* Leaves linear-lanceolate, nearly entire. Flowers *small, corymbed* at the ends of the branches, purplish or white. Petals erect. Stigma club-shaped.—Bogs.

5. **E. strictum,** Muhl., (*E. molle,* Torr.) is occasionally met with. It differs from No. 4 chiefly in having the leaves crowded, broader, and their points more obtuse. The petals are rose-coloured.—Bogs.

6. **E. palustre,** L. Stem low, slender, and simple (about a foot high), finely pubescent. Leaves erect or ascending, sessile, linear to linear-lanceolate, obtuse, with revolute margins. Seeds roughened with points.—Atl. seacoast and N. W.

7. **E. colora'tum,** Muhl. Stem 1-2 feet high, nearly smooth, but with 2-4 hairy lines decurrent from the leaves, the latter *lanceolate* or *ovate-lanceolate, serrate,* with conspicuous petioles. Flowers small, more or less nodding, corymbed. Petals pale, deeply notched. Coma of the seed brownish.—Not so common as the next.

8. **E. adenocau'lon,** Haussk. More glandular-pubescent than the last. Leaves abruptly contracted into short

petioles. Flowers erect. Coma of the seed much lighter in colour.—Common in wet places.

3. ŒNOTHE'RA, L. Evening Primrose.
* Stigma-lobes linear.

1. **Œ. bien'nis,** L. (Common Evening Primrose.) *Stem 2-4 feet high*, hairy. Leaves ovate-lanceolate. Flowers yellow, odorous, in a leafy spike, opening in the evening or in cloudy weather. Pods oblong, narrowing towards the top.—Waste places.

Var. **grandiflo'ra,** Lindl., has petals as long as the calyx-tube.

Var. **murica'ta,** Lindl. has rough-bristly stem and pods, and petals rather longer than the stamens.

2. **Œ. albicau'lis,** Nutt. Flowers white, changing to rose-colour, nodding in the bud. Stem white, and commonly glabrous.—N. W.

3. **Œ. pu'mila,** L. (Small E.) Stem low, *5-12 inches high*, smooth or nearly so. Leaves lanceolate or oblanceolate. Pods nearly sessile, club-shaped, *4-angled*. Flowers pale yellow, opening in sunshine.—River and lake margins.

4. **Œ. chrysan'tha,** Michx. Distinguished from the preceding by the orange-yellow flowers, and *pedicelled pods*, the latter scarcely wing-angled.—Drier ground than the preceding.

** Stigma discoid.

5. **Œ. serrula'ta,** Nutt., var. **Douglasii,** Torr. and Gray. A low and slender plant, with linear to lanceolate leaves. *Calyx-tube broadly funnel-form.* Petals obovate.—N.W.

4. GAURA, L.

G. coccin'ea, Nutt. Hoary and very leafy, 6-12 inches high. Flowers small, in simple spikes.—N. W.

5. LUDWIG'IA, L. False Loosestrife.

1. **L. palustris,** Ell. (Water Purslane.) Stems creeping in the mud of ditches or river margins, smooth. *Leaves opposite*, tapering into a slender petiole. Flowers sessile, solitary, usually without petals. Pod 4-sided.

2. **L. alternifo'lia,** L. (SEED-BOX.) Stem branching, about 3 feet high, nearly smooth. Flowers *yellow, conspicuous,* peduncled in the upper axils. *Leaves alternate,* pointed at both ends. Capsules wing-angled.—S. W. Ontario.

3. **L. polycarp'a,** Short and Peter. Stem 1-3 feet high, erect and branching, but producing runners at the base. Flowers small, sessile in the axils, with greenish petals or none. Leaves acute at both ends, those of the runners oblong-spathulate. Bractlets at the base of the capsules linear-awl-shaped.—S. W. Ontario, not common.

ORDER XXXVIII. **MELASTOMA'CEÆ.** (MELASTOMA F.)

Low herbs with opposite 3-5-ribbed leaves. Calyx-tube adherent to the ovary, the limb 4-cleft. Petals 4, showy, convolute in the bud. Stamens 8, with 1-celled anthers opening by a pore at the apex; these and the petals inserted on the calyx. Style and stigma 1. Pod 4-celled, many-seeded; seeds coiled. The only representative with us is

RHEXIA, L. DEER-GRASS. MEADOW-BEAUTY.

R. Virgin'ica, L. Stem square, wing-angled. Leaves oval-lanceolate. Petals purple.—Shores of the Muskoka Lakes.

ORDER XXXIX. **LYTHRA'CEÆ.** (LOOSESTRIFE F.)

Herbs, or slightly woody plants, with opposite or whorled entire leaves, without stipules. Calyx enclosing, *but free from,* the ovary. Petals and stamens inserted on the calyx, Flowers axillary or whorled. Style 1. Stigma capitate.

Synopsis of the Genera.

1. **Lyth'rum.** Petals mostly 6. Stamens mostly 6 or 12. Flowers purple, solitary in the axils, or forming an interrupted spike. Calyx-teeth with projections in the sinuses. Pod oblong, 2-celled.
2. **Nesæ'a.** Petals 5 (rarely 4). Stamens twice as many, in two sets. Pod 3-5-celled.

1. LYTH'RUM, L. LOOSESTRIFE.

1. **L. ala'tum,** Pursh. Tall and wand-like. Flowers solitary in the upper axils. Calyx-teeth often shorter than

the projections between them. Petals deep-purple.—S. W. Ontario.

2. **L. Salica'ria,** L. (SPIKED LOOSESTRIFE.) More or less downy and tall. Flowers purple, crowded, and forming an interrupted spike. Stamens twice as many as the petals, in two sets.—Atl. Prov. Also at Ottawa and opposite Montreal.

2. NESÆ'A, Commerson, Juss. SWAMP LOOSESTRIFE.

N. verticilla'ta, H. B. K. (*Dec'odon verticillatus,* Ell.) Stems curving, 2-6 feet long, 4-6-sided. Leaves lanceolate, mostly whorled. Flowers purple, in the axils of the upper leaves. Calyx bell-shaped, with 5-7 erect teeth, with supplementary projections between them. Stamens 10, exserted, 5 longer than the rest.—Swamps.

ORDER XL. CUCURBITA'CEÆ. (GOURD FAMILY.)

Herbs, climbing by tendrils. Flowers monœcious. Calyx-tube adherent to the 1-3-celled ovary. Corolla commonly more or less gamopetalous. Stamens usually 3, united by their tortuous anthers, and often also by the filaments. Leaves alternate, palmately lobed or veined.

Synopsis of the Genera.

1. **Sl'cyos.** Flowers greenish-white, small; the staminate corymbed, the pistillate clustered in a head on a long peduncle. *Corolla 5-cleft,* with a spreading border. Style slender; stigmas 3. Ovary 1-celled. Fruit *dry and indehiscent,* prickly, bur-like in appearance.
2. **Echinocys'tis.** Flowers whitish, small; the staminate in long compound racemes, the pistillate in small clusters from the same axils. *Corolla 6-parted.* Stigma broad, almost sessile. Ovary 2-celled, 4-seeded. Fruit fleshy, becoming dry, clothed with weak prickles.

1. SI'CYOS, L. STAR CUCUMBER.

S. angula'tus, L. A clammy-hairy weed in damp yards. Leaves roundish heart-shaped, 5-angled or lobed.

2. ECHINOCYS'TIS, Torr. and Gray. WILD BALSAM APPLE.

E. loba'ta, Torr. and Gray. Climbing high about dwellings. Leaves deeply and sharply 5-lobed. The oval fruit 2 inches long.

Order XLI. CACTA'CEÆ. (Cactus Family.)

Very fleshy and commonly leafless plants; the stems globose or columnar and angled, or of flattened joints, mostly prickly. Flowers solitary, sessile. The numerous sepals and petals adherent to the 1-celled ovary. Style 1, stigmas many. Fruit a 1-celled berry with many seeds on the walls.

Synopsis of the Genera.

1. **Mamilla'ria.** Stems globose or oval, covered with spine-bearing tubercles, the flowers between these.
2. **Opun'tia.** Stems of flattened joints, bearing very small awl-shaped leaves with clusters of bristles in their axils.

1. MAMILLA'RIA, Haw.

M. vivip'ara, Haw. Stems 1-5 inches high, the tubercles bearing bundles of 5-8 reddish-brown spines, surrounded by numerous grayish ones. Flowers purple.—N. W. plains.

2. OPUN'TIA, Tourn. PRICKLY PEAR.

O. Rafines'quii, Engel. Stem prostrate, deep green. Bristles reddish-brown. Spines few. Flowers yellow, sometimes with reddish centre.—Point Pelee.

Order XLII. FICOI'DEÆ. (Ice-Plant Family.)

A miscellaneous group, embracing plants formerly included in Caryophyllaceæ and Portulacaceæ; differing, however, from true representatives of these in having *partitions in the ovary*. Petals wanting in our genus.

MOLLU'GO, L. CARPET-WEED.

M. verticilla'ta, L. A prostrate much-branched herb, growing in patches. Leaves spathulate, apparently verticillate. Flowers on long axillary pedicels, clustered into a sort of umbel. Sepals 5, white inside. Petals none. Stamens mostly 3. Styles 3. Pod 3-celled, 3-valved, loculicidal, the partitions breaking away from the many-seeded axis.—Mostly in south-western Ontario.

ORDER XLIII. **UMBELLIF'ERÆ.** (PARSLEY FAMILY.)

Herbs with small *flowers mostly in compound umbels.* Calyx-tube grown fast to the surface of the ovary; calyx-teeth minute or none. The 5 petals and 5 stamens inserted on a disk which crowns the ovary. Styles 2. Fruit dry, 2-seeded. Stems hollow. Leaves usually much cut. (See Part I., Chapter VII., for description of a typical flower.)

Synopsis of the Genera.

§ 1. *Seeds flat (not hollow) on the inner face.*

1. **Hydrocot'yle.** *Umbels simple, or one springing from the summit of another,* axillary. Flowers white. Stem slender and creeping. Leaves round-kidney-shaped.

2. **Sanic'ula.** *Umbels irregular* (or compound), the *greenish flowers capitate* in the umbellets. Leaves palmately lobed or parted. Fruit globular, covered with hooked prickles.

(*In the Genera which follow, the umbels are regularly compound.*)

3. **Dau'cus.** Stem bristly. Leaves twice or thrice-pinnate, or pinnatifid. Bracts of the involucre pinnatifid, very long. Fruit ribbed, the *ribs bristly.*

4. **Heracle'um.** *Stem 3-4 feet high, woolly* and grooved. Leaves 1-2-ternately compound. *Flowers white;* the outer corollas larger than the others. *Fruit wing-margined at the junction of the carpels,* very flat. Carpels 5-ribbed on the back.

5. **Pastina'ca.** *Stem smooth,* grooved. Leaves pinnate. *Flowers yellow, all alike.* Fruit as in No. 4.

6. **Peuced'anum.** Low acaulescent herbs, with spindle-shaped roots and twice-pinnate leaves. No involucre; involucels of scarious-margined (often purplish) lanceolate bractlets. Flowers white or yellow. Fruit round, winged on the sides.—N. W. only.

7. **Archem'ora.** Stem smooth. Leaves pinnate, of 3-9 rather narrow leaflets. *Flowers white.* Fruit broadly winged, flat, 5-ribbed on the back.

8. **Archangel'ica.** Stem smooth, stout, purple. Leaves 2-3-ternately compound. *Flowers greenish-white.* Fruit smooth, flattish on the back, *double-wing-margined,* each carpel with 3 ribs on the back.

9. **Coniosell'num.** Stem smooth. Leaves 2-3-pinnately compound, the petioles inflated. *Flowers white.* Fruit *doubly wing-margined,* and with *3 narrow wings on the back of each carpel.*

10. **Thaspium.** Stem smooth. Leaves 1-3-ternately divided. *Flowers yellow.* Fruit not flattened, *10-winged or ribbed.*

11. **Æthu'sa.** Leaves twice or thrice ternately compound, the divisions finely dissected. No involucre: involucels of long narrow bractlets. Flowers white. Fruit ovate-globose, the carpels with 5 thick sharp ribs.

12. **Ligus'ticum.** Stem smooth, from large aromatic roots. Leaves twice ternate, coarsely toothed. Flowers white. Fruit with prominent acute ribs, having broad spaces between them.

13. **Zizia.** Stem slender, smooth and glaucous. Leaves 2-3-ternately compound. *Flowers yellow.* Rays of the umbel long and slender. *Fruit contracted at the junction of the carpels; the carpels narrowly 5-ribbed.*

14. **Cicu'ta.** Stem streaked with purple, stout. Leaves *thrice compound.* *Flowers white.* Fruit a little contracted at the sides, *the carpels strongly 5-ribbed.*

15. **Carum.** Stem erect, slender, smooth. Leaves pinnately decompound, the divisions filiform. *Roots tuberous.* Flowers white. Ribs of the ovate or oblong fruit inconspicuous.

16. **Sium.** Stem grooved. Leaves simply *pinnate.* *Flowers white.* Fruit as in No. 10.

17. **Cryptotæ'nia.** Stem smooth. *Leaves 3-foliolate. The umbels with very unequal rays.* Flowers white. Fruit nearly as in Nos. 10 and 11.

§ 2. *Inner face of each seed hollowed lengthwise.*

18. **Bupleu'rum.** Stem smooth. *Leaves perfoliate,* ovate, entire. No involucre: involucels of 5 very conspicuous ovate mucronate bractlets. Flowers *yellow.*

19. **Muse'nium.** Stem short, branching from the base. Flowers *yellow.* Leaves twice-pinnatifid. Roots spindle-shaped. No involucre; involucels 1-sided, of a few narrow bractlets. Ribs of the ovate fruit slightly prominent.—N. W. only.

20. **Osmorrhi'za.** Leaves large, 2-3-ternately compound. Flowers white. Fruit linear-oblong, angled, tapering downwards into a stalk-like base. Ribs of the carpels bristly upwards.

21. **Co'nium.** Leaves large, decompound. Flowers white. Fruit ovate, flattened at the sides, 5-ribbed, *the ribs wavy.*

§ 3. *Inner face of each seed curved inwards at top and bottom.*

22. **Erige'nia.** Stem low and smooth. Leaves 2-3-ternately divided. *Fruit twin.* Carpels nearly kidney-form. Umbels 3-rayed, small. Flowers white.

UMBELLIFERÆ.

1. HYDROCOTYLE, Tourn. WATER PENNYWORT.

H. America'na, L. Stem spreading and creeping, very slender. Leaves kidney-shaped, crenate, slightly lobed. Umbels 3–5-flowered, inconspicuous, in the axils of the leaves.—Shady wet places.

2. SANIC'ULA. Tourn. SANICLE. BLACK SNAKEROOT.

S. Marilan'dica, L. *Leaves 5–7-parted.* Staminate flowers very numerous, and *on slender pedicels.* Styles long, recurved.—Rich woods.

Var. Canaden'sis, L. *Leaves 3–5-parted. A few staminate flowers* among the perfect ones, and on *very short pedicels.* Styles shorter than the prickles of the fruit.—Low rich woods, not so common.

3. DAU'CUS, Tourn. CARROT.

D. Caro'ta, L. (COMMON CARROT.) Found wild occasionally in old fields. In fruit the umbel becomes hollow like a bird's nest.

4. HERACLE'UM L. COW-PARSNIP.

H. lana'tum, Michx. Umbels large and flat. Petioles of the leaves spreading and sheathing. Leaves very large; leaflets broadly heart-shaped, deeply lobed. Low wet meadows.

5. PASTINA'CA Tourn. PARSNIP.

P. sati'va, L. (COMMON PARSNIP.) Found wild in old fields and along roadsides. Leaflets shining above.

6. PEUCED'ANUM, L.

1. P. nudicau'le. Nutt. Pubescent. Peduncles 3–8 inches high. Flowers white, in early spring.—N.W.

2. P. villo'sum, Nutt., has *yellow* flowers.—N.W.

7. ARCHEM'ORA, D.C. COWBANE.

A. rig'ida, DC. Calyx 5-toothed. Involucre almost none; involucels of many small bractlets.—Sandy swamps, southwestern Ontario.

8. ARCHANGEL'ICA, Hoffm. ARCHANGELICA.

1. **A. atropurpu'rea,** Hoffm. (*Angelica atropurpurea,* L.) (GREAT ANGELICA.) Stem very tall (4-6 feet) and stout, dark purple. Whole plant strong-scented. Petioles much inflated at the base.—Marshes and low river-banks.

2. **A. Gmel'ini,** DC. Stem slightly downy at the top. Involucels about as long as the umbellets. Plant but little aromatic.—Rocky coasts; Atl. Prov.

9. CONIOSELI'NUM, Fischer. HEMLOCK-PARSLEY.

C. Canadense, Torr. and Gr. Stem 2-4 feet high. Petioles much inflated. Leaflets of the involucels awl-shaped.—Swamps.

10. THASPIUM, Nutt. MEADOW-PARSNIP.

1. **T. au'reum,** Nutt. Stem 1-2 feet high, angular-furrowed. Leaflets oblong-lanceolate, sharply serrate. Flowers *deep yellow*. *Fruit with 10 winged ridges*, or in var. **apterum** *with 10 ribs*.—Dry or rich woods.

Var. **trifoliatum,** Coult. and Rose, is a far western form. Leaflets *crenate*.

2. **T. barbino'de,** Nutt. Loosely branched, pubescent on the joints. Leaflets *ovate to lanceolate*, coarsely cut-serrate. Flowers *light yellow*.—S. W. Ontario.

11. ÆTHU'SA, L. FOOL'S PARSLEY.

Æ. Cyna'pium, L. A poisonous annual, with an unpleasant odour, found occasionally in cultivated grounds.

12. LIGUS'TICUM, L. LOVAGE.

L. Scot'icum, L. (SCOTCH LOVAGE.) Stem 1-2 feet high, simple. Leaflets ovate. Fruit narrowly oblong.—Salt marshes, Atl. sea-coast.

13. ZIZIA, DC. ZIZIA.

Z. integer'rima, DC. (*Pimpinella integerrima,* Benth. and Hook., in Macoun's Catalogue.) Stem slender, 1-2 feet high, from a stout root stock. Involucels none. Plant strong-scented.—Rocky hill-sides.

14. CICU'TA, L. WATER-HEMLOCK.

1. C. macula'ta, L. (SPOTTED COWBANE. BEAVER POISON.) Stem 3-6 feet high, purplish, smooth. Leaflets *ovate-lanceolate, coarsely serrate,* pointed.—Swamps and low grounds.

2. C. bulbif'era, L., is easily distinguished from No. 1 by bearing *clusters of bulblets* in the axils of the upper leaves. The leaflets, also, are *linear.*—Swamps and low grounds.

15. CARUM, L. CARAWAY.

C. Car'ui, L. (CARAWAY.) Escaped from cultivation in many places.

16. SIUM, L. WATER-PARSNIP.

S. linea're, Michx. (*S. cicutæfolium*, Gmelin, in Macoun's Catalogue.) Stem 2-3 feet high, smooth, furrowed. Leaflets varying from linear to oblong, sharply pointed and serrate. Fruit oblong or ovate, with prominent ribs.—Borders of marshes usually in the water.

17. CRYPTOTÆ'NIA, DC. HONEWORT.

C. Canadensis, DC. Stem 1-2 feet high, slender. Leaflets large, ovate, doubly serrate. No involucre.—Rich woods and thickets.

18. BUPLEU'RUM, L. THOROUGH-WAX.

B. rotundifo'lium, L. Ballast-heaps.—Atl. Prov.

19. MUSE'NIUM, Nutt.

M. divarica'tum, Nutt. Decumbent. Leaves glabrous and shining, the divisions confluent with the winged rhachis. Rather ill-smelling herbs.—N.W.

20. OSMORRHI'ZA, Raf. SWEET CICELY.

1. O. longis'tylis, DC. (SMOOTHER SWEET CICELY.) Stem reddish, nearly smooth. Leaflets sparingly pubescent, short-pointed. *Styles slender, nearly as long as the ovary,* recurved.—Rich woods.

2. O. brevis'tylis, DC. (HAIRY SWEET CICELY.) Whole plant hairy. Leaflets taper-pointed. *Styles very short, conical.*—Rich woods.

21. CO'NIUM, L. POISON HEMLOCK.

C. macula'tum, L. Stem smooth, spotted. Leaflets lanceolate, pinnatifid, pale green, with an offensive odour when bruised. Involucels one-sided. Inner face of the seed marked with a deep groove.—Waste places.

22. ERIGE'NIA, Nutt. HARBINGER-OF-SPRING.

E. bulbo'sa, Nutt. Stem 4-6 inches high, from a tuber deep in the ground, producing 2 leaves, the lower radical. Leaflets much incised. Flowers few.—Alluvial soil.

ORDER XLIV. ARALIA'CEÆ. (GINSENG FAMILY.)

Herbs (with us) differing from the last Order chiefly in having, as a rule, *more than 2 styles*, and the *fruit a drupe*. The umbels, also, are either single, or corymbed, or panicled. Flowers often polygamous. The only Canadian genus is

ARA'LIA, Tourn. GINSENG. WILD SARSAPARILLA.

* Umbels corymbed or panicled. Petals, stamens, and styles each 5. Fruit black or dark-purple.

1. A. racemo'sa, L. (SPIKENARD.) *Umbels in a large compound panicle.* Stem 2-3 feet high, widely branching. Leaves very large and decompound; leaflets ovate-cordate, doubly serrate. Roots aromatic.—Rich woods.

2. A. his'pida, Michx. (BRISTLY SARSAPARILLA. WILD ELDER.) Stem 1-2 feet high, *bristly*, leafy, somewhat shrubby at the base. *Umbels 2-7, corymbed.* Leaves twice-pinnate. Leaflets sharply serrate. Fruit black.—Rocky or sandy woods.

3. A. nudicau'lis, L. (WILD SARSAPARILLA.) True stem very short, sending up a naked scape bearing 3 or 4 long-peduncled umbels at the summit, and one long-petioled leaf, ternately divided, and with 5 leaflets on each division. Root horizontal, aromatic.—Rich woods.

* * *Umbel single, on a long peduncle. Styles 2 or 3.*

4. **A. quinquefo'lia**, Decaisne. (GINSENG.) Leaves in a whorl of 3 at the summit of the stem, the latter a foot high. *Leaflets mostly 5, long-stalked.*—Rich woods.

5. **A. trifo'lia**, Decaisne. Stem 4-6 inches high. Leaves in a whorl of 3 at the summit, but *the leaflets usually only 3, and sessile.*—Rich woods.

ORDER XLV. **CORNA'CEÆ.** (DOGWOOD FAMILY.)

Shrubs or trees (rarely herbs) with simple leaves. Calyx-tube adherent to the 1-2-celled ovary, the limb of the calyx inconspicuous. Petals and stamens all epigynous. Style 1 ; stigma flat or capitate. Fruit a 1-2-seeded drupe.

Synopsis of the Genera.

1. **Cornus.** Flowers perfect (the parts in fours), either forming cymes, or in close heads surrounded by a showy involucre resembling a corolla. Leaves mostly opposite.
2. **Nyssa.** Flowers diœciously polygamous, the parts in fives. Leaves alternate, mostly crowded at the ends of the branchlets.

1. CORNUS, Tourn. CORNEL. DOGWOOD.

**Flowers in a close head, surrounded by a showy involucre of 4 white bracts. Fruit red.*

1. **C. Canadensis**, L. (BUNCH-BERRY.) Stem simple, 5 or 6 inches high. Upper leaves crowded and apparently whorled, ovate, the lower scale-like. Leaves of the involucre ovate. Rich woods.

2. **C. flor'ida**, L. (FLOWERING DOGWOOD.) A small tree, with opposite ovate pointed leaves. Leaves of the involucre notched at the apex, very showy.—Rocky woods. South-west Ontario.

* * *Flowers (white) in flat cymes. No involucre. Fruit blue or white.*

3. **C. circina'ta**, L'Her. (ROUND-LEAVED DOGWOOD.) A shrub 4-6 feet high, with *greenish warty dotted branches.* Leaves opposite, *broadly oval*, white-woolly beneath. Fruit light blue.—Rich woods.

4. **C. seric'ea,** L. (SILKY CORNEL.) A large shrub with *purplish branches*. Leaves opposite, narrowly ovate or oblong, silky beneath. Branchlets often rusty. Fruit light blue. Distinguished from No. 3 by the colour of the branches and the much smaller leaves.—Low wet grounds.

5. **C. stolonif'era,** Michx. (RED-OSIER DOGWOOD.) A shrub forming clumps by the production of suckers or stolons, 3-6 feet high. *Branches bright red-purple, smooth.* Leaves opposite, ovate, roughish, whitish beneath. Fruit white or whitish.—Low wet grounds.

6. **C. panicula'ta,** L'Her. (PANICLED CORNEL.) A shrub 4-8 feet high, with erect, gray, and smooth branches. Flowers white, very numerous. Leaves opposite, ovate-lanceolate, taper-pointed. Cymes convex. Fruit white.—Thickets and river-banks.

7. **C. asperifo'lia,** Michx., (ROUGH-LEAVED DOGWOOD) is reported by Macoun as common on Point Pelee. Branches brownish, the branchlets rough-pubescent. Leaves opposite, rather small, oblong or ovate; rough above, downy beneath. Fruit bluish or white.

8. **C. alternifo'lia,** L. (ALTERNATE-LEAVED CORNEL.) A large shrub or small tree, with *alternate greenish branches streaked with white*. Leaves mostly alternate, oval, acute at each end, crowded at the ends of the branches. Flowers yellowish, in loose cymes. Fruit deep blue, on reddish stalks.—Thickets.

2. NYSSA, L. PEPPERIDGE. SOUR-GUM TREE.

N. sylvat'ica, Marsh. A middle-sized tree with horizontal branches, reported from Niagara and south-western Ontario. Sterile flowers in dense axillary clusters, with small calyx, and small fleshy petals or none. Stamens mostly 10, but sometimes fewer. Fertile flowers in clusters of 3-8, at the summit of a slender axillary peduncle, larger than the staminate ones. Style revolute. Ovary 1-celled. Drupe ovoid, bluish-black, about half an inch long.

II. GAMOPET'ALOUS DIVISION.

Embracing plants with both calyx and corolla, the latter with the petals united (in however slight a degree.)

ORDER XLVI. **CAPRIFOLIA'CEÆ.** (HONEYSUCKLE F.)

Shrubs, rarely herbs, with the calyx-tube adherent to the ovary, the corolla borne on the ovary, and the stamens on the tube of the corolla. Leaves opposite and without stipules, but some species of **Vibur'num** have appendages resembling stipules. Fruit a berry, drupe, or pod.

Synopsis of the Genera.

* *Corolla tubular, sometimes 2-lipped. Style slender.*

1. **Linnæ'a.** A trailing or creeping herb, with evergreen oval crenate leaves and slender scape-like peduncles which fork at the top into 2 pedicels, each of which bears a nodding narrowly bell-shaped purplish flower. Stamens 4, 2 shorter than the others.

2. **Symphoricar'pus.** Upright branching shrubs, with oval entire short-petioled leaves. Flowers in interrupted spikes at the ends of the branches, rose-coloured. Corolla bell-shaped, 4-5-lobed, with as many stamens. Berries large and white, 4-celled, but only 2-seeded.

3. **Lonice'ra.** Upright or twining shrubs, with entire leaves. Corolla funnel-form, more or less irregular, *often with a projection on one side at the base.* Berry several-seeded.

4. **Diervil'la.** Low upright shrubs with ovate pointed serrate leaves. Calyx-tube tapering towards the top, the teeth slender. Flowers light yellow, peduncles mostly 4-flowered. Corolla funnel-form, nearly regular. Pod slender-pointed.

5. **Trios'teum.** Coarse herbs. Lobes of the calyx leaf-like. Flowers brownish-purple, sessile in the axils of the leaves. Corolla bulging at the base. Fruit a 3-seeded orange-coloured drupe.

** *Corolla rotate or urn-shaped, regular, 5-lobed. Flowers white, in broad cymes.*

6. **Sambu'cus.** Upright shrubs with pinnate leaves, the leaflets serrate. Stigmas 3. Fruit purple or red, a juicy berry-like drupe, with 3 seed-like stones.

7. **Vibur'num.** Upright shrubs with simple leaves, and white flowers in compound cymes. Fruit a 1-seeded drupe.

1. LINNÆ'A, Gronov. TWIN-FLOWER.

L. borea'lis, Gronov.—Cool mossy woods and swamps.

2. SYMPHORICAR'PUS, Dill. SNOWBERRY.

1. **S. racemo'sus,** Michx. (SNOWBERRY.) Corolla bearded inside. Flowers in a rather loose spike. Var. **pauciflo'rus,** Robbins, is low, diffusely branched, and spreading, with two or three flowers only, in the axils of the uppermost leaves.—Dry rocky hill-sides.

2. **S. occidenta'lis,** Hook. (WOLFBERRY.) Flowers in denser spikes than the last, and with larger and more funnel-form corolla and longer stamens. *Corolla much bearded within.*—N.W.

3. LONICE'RA, L. HONEYSUCKLE. WOODBINE.

1. **L. parviflo'ra,** Lam. (*L. glauca,* Hill, in Macoun's Catalogue.) (SMALL HONEYSUCKLE.) *Twining shrub,* 2–4 feet high, with smooth leaves which are glaucous beneath, the upper ones connate-perfoliate ; corolla yellowish-purple.—Rocky banks.

2. **L. hirsu'ta,** Eaton. (HAIRY HONEYSUCKLE.) Stem *twining high.* Leaves *not glaucous, very large, downy-hairy,* the upper ones connate-perfoliate. Flowers in close whorls ; corolla greenish-yellow, clammy-pubescent.—Deep thickets.

3. **L. Sullivan'tii,** Gray. *Twining.* At length *glaucous-whitened.* Leaves oval and ovate-oblong, mostly connate on the flowering stems. Corolla pale yellow.—N.W.

4. **L. cilia'ta,** Muhl. (FLY-HONEYSUCKLE.) A branching, straggling shrub, with *thin oblong-ovate ciliate leaves.* Peduncles axillary, filiform, shorter than the leaves, *each 2-flowered at the top.* Corolla greenish-yellow, *almost spurred at the base.* The two berries *separate.*—Damp woods.

5. **L. cæru'lea,** L. (MOUNTAIN F.) Smaller (1–2 feet high), and with upright branches. Leaves oval. *Ovaries united into one berry.*—Eastward and northward.

6. **L. oblongifo'lia,** Muhl. (SWAMP FLY-HONEYSUCKLE). A shrub with upright branches, and *oblong leaves. Peduncles long and slender, 2-flowered.* Corolla deeply 2-lipped. Berries united at the base.—Swamps and low grounds.

7. **L. involucra'ta,** Banks, is at once recognized by the involucre of four leaf-like bracts under the two flowers. Corolla yellowish, viscid-pubescent. Berries dark-purple.— Woods and banks of streams, Atl. Prov. and N.W.

4. DIERVIL'LA, Tourn. BUSH-HONEYSUCKLE.

D. trif'ida, Mœnch.—Rocky woods and clearings.

5. TRIOS'TEUM, L. FEVER-WORT.

T. perfolia'tum, L. A coarse herb, 2-4 feet high, soft-hairy. Leaves oval, narrowed at the base. Fruit orange-coloured.—Old clearings and thickets.

6. SAMBU'CUS, Tourn. ELDER.

1. **S. Canadensis,** L. (COMMON ELDER.) Shrub 5-10 feet high, in clumps. Leaflets 5-11, oblong. Cymes flat. Fruit black-purple.—Open grounds, and along streams.

2. **S. racemo'sa,** L. (*S. pubens,* Michx.) (RED-BERRIED ELDER) may be distinguished from No. 1 by its warty bark, brown pith, 5-7 leaflets, convex or pyramidal cymes, and red berries.—Rocky woods.

7. VIBUR'NUM, L. ARROW-WOOD. LAURESTINUS.

1. **V. Lenta'go,** L. (SWEET VIBURNUM. SHEEP-BERRY.) A small tree, with *ovate finely-serrate pointed leaves,* with long and margined petioles. *Cyme sessile.* Fruit black.— Along streams.

2. **V. cassinoides,** L. (WITHE-ROD.) A smooth shrub with somewhat scurfy shoots and tall straight stems. Leaves thickish, entire or wavy-toothed, dotted beneath. *Cymes with short peduncles,* about 5-rayed. Fruit black.—Cold swamps.

3. **V. pubes'cens,** Pursh. (DOWNY ARROW-WOOD.) A straggling shrub, not more than 4 feet high. with small ovate

coarsely serrate leaves, *the lower surface soft-downy*. Cymes small. Fruit oblong, dark-purple.—Rocky places.

4. **V. acerifo'lium,** L. (Maple-leaved A. Dockmackie.) A shrub 3–6 feet high, with greenish bark. Leaves 3-lobed, 3-ribbed, soft-downy beneath. Stipular appendages bristle-shaped. Cymes small, on long peduncles. Fruit red, becoming black.—Thickets and river-banks.

5. **V. Op'ulus,** L. (Cranberry-tree.) An upright shrub, 5–10 feet high, with strongly 3-lobed leaves, broader than long, the lobes spreading and pointed. Cymes peduncled. Marginal flowers of the cyme very large and neutral. Stipular appendages conspicuous. Fruit red, pleasantly acid.—Low grounds.

6. **V. pauciflo'rum,** Pylaie. A low shrub. Leaves 5-ribbed at the base, serrate, with 3 short lobes at the summit. Cyme few-flowered. Stamens shorter than the corolla. Fruit red, sour, with a very flat stone.—Cold woods, Atl. Prov. chiefly.

7. **V. lantanoi'des,** Michx. (Hobble-bush.) A straggling shrub with reclining branches. Leaves large, round-ovate, heart-shaped at the base, serrate, many-veined, the veins underneath and the stalks and branchlets very rusty-scurfy. Stipular appendages conspicuous. Cymes sessile, very broad and flat, with very conspicuous neutral flowers on the margin. —Moist woods.

Order XLVII. **RUBIA'CEÆ.** (Madder Family.)

Herbs or shrubs, chiefly distinguished from the preceding Order by the presence of stipules between the opposite entire leaves, or by the leaves being in whorls without stipules. Calyx superior. Stamens alternate with the (mostly 4) lobes of the corolla, and inserted on its tube. Ovary 2–4-celled.

RUBIACEÆ. 99

Synopsis of the Genera.

1. **Ga'lium.** *Leaves in whorls. Slender weak herbs with square stems.* Calyx-teeth inconspicuous. Corolla 4-parted, wheel-shaped. Styles 2. Fruit twin, separating into two 1-seeded carpels.

2. **Sherard'ia.** *Leaves in whorls.* Stems square, slender, procumbent. Corolla funnel-form, 4-5-lobed. Calyx-lobes lanceolate. Flowers blue or pinkish, with a gamophyllous involucre.

3. **Cephalan'thus.** *Leaves opposite. Shrubs with the flowers in a globular peduncled head.* Lobes of calyx and corolla each 4. Style very slender, much protruded. Stigma capitate.

4. **Mitchel'la.** *Leaves opposite. Shining trailing evergreen herbs,* with flowers in pairs, *the ovaries united.* Lobes of calyx and corolla each 4, the corolla bearded inside. Style 1. Stigmas 4. Fruit a red 2-eyed berry.

4. **Housto'nia.** *Leaves opposite. Low and slender erect herbs,* with the flowers in small terminal clusters. Lobes of calyx and corolla each 4. Style 1. Stigmas 2.

1. GA'LIUM, L. BEDSTRAW. CLEAVERS.

1. **G. Apari'ne**, L. (CLEAVERS. GOOSE-GRASS.) *Leaves about 8 in a whorl*, lanceolate, rough-margined. Peduncles 1-2-flowered, *axillary. Fruit covered with hooked prickles.* Low grounds.

2. **G. triflo'rum**, Michx. (SWEET-SCENTED BEDSTRAW.) *Leaves chiefly 6 in a whorl*, elliptical-lanceolate, bristle-pointed. Peduncles 3-flowered, *terminating the branches. Fruit covered with hooked prickles.*—Woods.

3. **G. pilo'sum**, Ait. Leaves in whorls of 4, *hairy, oval.* Peduncles twice- or thrice-forked.—Southwestern Ontario.

4. **G. lanceola'tum**, Torr. (WILD LIQUORICE.) *Leaves all in whorls of 4 each*, lanceolate, *tapering at the apex*, more or less 3-nerved. Peduncles mostly once-forked. Flowers few or several, remote. Fruit covered with hooked prickles.

5. **G. circæ'zans**, Michx., is similar to No. 4, but the leaves are *obtuse* instead of tapering.—Woods.

6. **G. Kamtschat'icum,** Steller, has leaves *orbicular to oblong-ovate*, and corolla *yellowish-white* and *glabrous.*—Mountain woods, Atl. Prov.

7. **G. asprel'lum,** Michx. (ROUGH BEDSTRAW.) Leaves in whorls of 6, or 4 or 5 on the branchlets, elliptical-lanceolate, very rough on the edges and midrib. Stem weak, 3-5 feet high, leaning upon and clinging to bushes by its rough edges. *Flowers numerous in panicled clusters.* Fruit not rough. Thickets.

8. **G. trif'idum,** L. (SMALL BEDSTRAW.) Leaves in whorls of 4-6. Stem 6-18 inches high, roughened on the edges, as are the leaves usually. *Flowers few, not panicled. Parts of the flowers generally in threes.* Fruit smooth. Var. **latifolium,** Torr., is easily known by its broad leaves and widely branching stems.—Low grounds and swamps.

Var. **tincto'rium,** Torr. and Gray. Stem taller and stouter, with nearly smooth angles. Peduncles 3-7-flowered. Corolla-lobes and stamens 4.

9. **G. borea'le,** L. (NORTHERN BEDSTRAW.) *Leaves in whorls of 4,* linear-lanceolate, 3-nerved. Flowers very numerous, crowded in a narrow and compact terminal panicle. Stem erect and rigid, 1-3 feet high.— Rocky thickets and river-banks.

2. SHERARD'IA, Dill.

S. arven'sis, L. Sparingly naturalized.

3. CEPHALAN'THUS, L. BUTTON BUSH.

C. occidenta'lis, L. A smooth shrub growing in swamps, with ovate petioled pointed leaves, which are opposite or in whorls of 3. Easily recognized by the globular head of white flowers.

4. MITCHEL'LA, L. PARTRIDGE BERRY.

M. repens, L.—Common in dry woods. Leaves round-ovate, shining, sometimes with whitish lines.

5. HOUSTO'NIA, L. HOUSTONIA.

1. **H. purpu'rea,** L. Stems tufted, 3-6 inches high. Leaves varying from roundish-ovate to lanceolate, 3-5-ribbed, sessile.—Woodlands.

Var. longifo'lia, Gray, has thinner oblong-lanceolate to linear leaves, and is lower.—Chiefly N.W.

2. **H. cæru'lea,** L. (BLUETS. INNOCENCE.) A slender herb with erect stems. A single flower on each slender peduncle. Leaves oblong-spathulate. Corolla light blue to nearly white, with a yellowish eye and a long tube.—Moist grassy places, Atl. Prov.

ORDER XLVIII. VALERIANA'CEÆ. (VALERIAN F.)

Herbs with opposite exstipulate leaves, and small cymose flowers. Calyx-tube adherent to the ovary, the latter 3-celled, but *only one of these fertile.* Stamens 1-3, *fewer than the lobes of the corolla.* Style slender. Stigmas 1-3. The only common genus is

VALERIA'NA. Tourn. VALERIAN.

1. **V. sylvat'ica,** Banks. Not uncommon in cedar-swamps. *Root fibrous. Calyx-limb consisting of several bristles rolled inwards in the flower, but expanding in fruit.* Corolla gibbous at the base. Stamens 3. Root-leaves ovate or oblong, entire; stem-leaves pinnate, leaflets 5-11. Stem erect, striate, 1-2 feet high.

2. **V. ed'ulis,** Nutt. *Root spindle-shaped, large.* Flowers in a long and narrow interrupted-panicle, nearly diœcious. Stem-leaves deeply pinnatifid.—Low grounds, western Ontario.

ORDER XLIX. DIPSA'CEÆ. (TEASEL FAMILY.)

Herbs with the flowers in heads, surrounded by a many-leaved involucre, as in the next Family, *but the stamens are distinct.* Leaves opposite. Represented in Canada by the genus

DIP'SACUS, Tourn. TEASEL.

D. sylves'tris, Mill. (WILD TEASEL.) A stout, coarse, prickly plant, not unlike a thistle in appearance. Flowers in oblong very dense heads, bluish. Corolla 4-cleft. Stamens 4, on the corolla. Bracts among the flowers terminating in a long awn. Leaves generally connate.—Roadsides and ditches. Very common in the Niagara district, but found also elsewhere.

ORDER L. **COMPOS'ITÆ.** (COMPOSITE FAMILY.)

Flowers in a dense head on a common receptacle, and surrounded by an involucre. Calyx-tube adherent to the ovary, its limb either obsolete or forming a pappus of few or many bristles or chaffy scales. Corolla either tubular or with one side much prolonged (strap-shaped or ligulate). Stamens usually 5, on the tube of the corolla, their anthers united (syngenesious). Style 2-cleft. (See Part I., sections 60-62, for examination of a typical flower.)

The heads of flowers present some variety of structure. *All* the flowers of a head may be tubular ; or only the central ones or *disk-flowers*, as they are then called, may be tubular, whilst those around the margin, then known as *ray-flowers*, are ligulate or strap-shaped. Or again, *all* the flowers may be strap-shaped. It is not unusual also to find a mixture of perfect and imperfect flowers in the same head.

The bracts which are often found growing on the common receptacle among the florets are known as the *chaff*. When these bracts are entirely absent the receptacle is said to be *naked*. The leaves of the involucre are called its *scales*.

Artificial Synopsis of the Genera.

SUBORDER I. **TUBULIFLO'RÆ.**

Heads either altogether without strap-shaped corollas, or the latter, if present, forming only the outer circle (the *ray*). Ray-flowers, when present, *always without stamens*, and often without a pistil also.

COMPOSITÆ. 103

A. Ray-flowers entirely absent.

Scales of the involucre in many rows, *bristly-pointed, or fringed.*

+ *Florets all perfect.*

1. **Cni'cus.** *Leaves and scales of the involucre prickly.* Pappus of long plumose bristles. Receptacle with long soft bristles among the florets. Flowers reddish-purple.

2. **Car'duus.** Resembling Cnicus, but the bristles of the pappus are *not plumose.* Atl. Prov.

3. **Onopor'don.** *Leaves and scales of the involucre prickly.* Heads much as in Cnicus, but the *receptacle naked,* and deeply honeycombed. Pappus of long bristles, *not* plumose. Stem winged by the decurrent bases of the leaves. Flowers purple.

4. **Arc'tium.** *Leaves not prickly, but the scales of the globular involucre tipped with hooked bristles.* Pappus of many short rough bristles. Receptacle bristly. Flowers purple.

   Verno'nia. One species has scales of the involucre somewhat bristly. See No. 15.

++ *Marginal florets often sterile, and much larger than the others, forming a kind of false ray.*

5. **Centaure'a.** Leaves not prickly. *Scales of the involucre fringed.* Pappus double and bristly, or very short or none. Receptacle bristly.

+++ *Sterile and fertile florets in separate heads,* i.e., *monœcious. Fruit a completely closed involucre (usually bristly) containing only one or two florets, these heads sessile in the axils of the bracts or upper leaves. Sterile heads with more numerous florets in flattish involucres, and forming racemes or spikes. Pappus none.*

6. **Xan'thium.** Fertile florets only 2 together in burs with hooked prickles, clustered in the axils. Sterile heads in short spikes above them, the scales of their involucres in one row only, *but not united together.*

7. **Ambro'sia.** Fertile florets single, in a closed involucre armed with a few spines at the top. Sterile heads in racemes or spikes above, the scales of their involucres in a single row and *united into a cup.*

8. **Franse'ria.** Inflorescence much like that of Ambrosia, but the fruiting involucre is armed with a number of *long flat and thin spines.*—N.W.

* * Scales of the involucre without bristles of any kind.

+ *Marginal florets without stamens.*

++ *Pappus none or minute. Receptacle naked. Very strong scented herbs.*

9. **Tanace'tum.** Flowers yellow, in numerous corymbed heads. Scales of the involucre dry, imbricated. Pappus 5-lobed. Leaves dissected.

10. **Artemis'ia.** Flowers yellowish or dull purplish, in numerous small heads which are panicled or racemed. Scales of the involucre with dry and scarious margins, imbricated. Achenes with narrow top.

++++ *Pappus of all the florets bristly. Receptacle naked.*

11. **Erechti'tes.** Flowers whitish. *Scales of the involucre in a single row, linear, with a few bractlets at the base.* Corolla of the marginal florets very slender. Pappus copious, of fine soft white hairs. Heads corymbed. Erect and coarse herbs.

12. **Gnapha'lium.** Flowers whitish or yellowish. Scales of the involucre yellowish-white, in many rows, dry and scarious, woolly at the base. Outer corollas slender. Pappus a single row of rough bristles. Flocculent-woolly herbs.

13. **Antenna'ria.** Very much like Gnaphalium in appearance, being white-woolly, *but the heads are usually diœcious*, and the bristles of the pappus thicker in the sterile florets.

+ + *All the florets in the head perfect.*

13. **Antenna'ria,** with diœcious heads, may be looked for here. See previous paragraph.

Bidens. One or two species have no rays. See No. 41.

Sene'cio. One species is without rays. See No. 19.

14. **Lia'tris.** Flowers handsome, rose-purple. Receptacle naked. Pappus of long and slender bristles, plumose or rough. Achenes slender, 10-ribbed. Lobes of the corolla slender. Stem wand-like, leafy, from a corm or tuber. *Leaves narrow or grass-like.*

15. **Verno'nia.** Flowers purple. Scales of the involucre (with us) *with slender bristly tips.* Receptacle naked. Pappus double, the outer minute, the inner capillary. Branches of the style *long and slender, minutely bristly.*

16. **Eupato'rium.** Flowers white or purple. Receptacle naked. Pappus of slender hair-like bristles, smooth or nearly so. Achenes 5-angled. Heads in corymbs. Leaves whorled, or connate, or opposite.

+ + + *Marginal florets pistillate and fertile, those in the centre of the head staminate and sterile. Receptacle chaffy. Pappus wanting.*

17. **Caca'lia.** Flowers white or whitish. *Heads 5-flowered*, rather large, in flat corymbs. *Involucre 5-leaved.* Receptacle naked, bearing a scale-like pointed appendage in the centre. Corolla deeply 5-cleft. Pappus of numerous capillary bristles.

18. **Iva.** Marginal pistillate florets 1-5, with very small tubular corolla or none. Staminate florets with funnel-form 5-toothed corolla. Anthers nearly separate. Achenes very short. Coarse plants with small greenish-white nodding heads. N.W.

B. **Rays or strap-shaped corollas round the margin of the head.**

* *Pappus of hair-like bristles. Receptacle naked.*

19. **Sene'cio.** Rays yellow, or in one species *none*. Scales of the involucre in a single row, or with a few bractlets at the base. Pappus very fine and soft. Heads corymbose. Leaves alternate.

20. **In'ula.** Rays yellow, numerous, very narrow, in a single row. Outer scales of the involucre leaf-like. *Anthers with two tails at the base.* Stout plants, with large alternate leaves which are woolly beneath.

21. **Chrysop'sis.** Rays yellow, many. Disk yellow. *Pappus double*, the outer of very small chaffy bristles, the inner capillary. Low hairy herbs.—N.W. only.

22. **Aplopap'pus.** Rays yellow, many. Disk yellow. *Pappus simple, of unequal bristles.*—N.W. only.

23. **Solida'go.** Rays yellow, few, as are also the disk-florets. Involucre oblong, scales of unequal lengths, appressed. Achenes many-ribbed. Heads small in compound racemes, or corymbs. Stems usually wand-like. Leaves alternate.

24. **Ar'nica.** Rays yellow. Disk yellow. Scales of the bell-shaped involucre in two rows, lanceolate, erect. Receptacle flat, minutely fringed. Achenes slender. Leaves opposite. Heads showy, with long peduncles. Stem simple.

25. **Tussila'go.** Rays yellowish, in several rows, very narrow. Involucre nearly simple. Pappus copious, soft. A low perennial, with creeping rootstocks, sending up a scaly scape in early spring bearing one head, and producing later heart-shaped angled or toothed leaves, woolly when young.

26. **Petasi'tes.** Rays whitish or purplish. Heads in a corymb, fragrant. Scales of the involucre in a single row. Heads somewhat diœcious, the staminate with one row of pistillate ray-flowers, the pistillate with ray-flowers in many rows. *Woolly herbs, with large leaves, all radical*, and sheathing scaly bracts on the scape.

27. **Aster.** Rays white, purple, or blue, *never yellow*, but the *disk* generally yellow. Pappus of numerous fine roughish bristles, (in one species *double*). Achenes flattish. Heads corymbed or racemose. Flowering in late summer.

28. **Erig'eron.** Rays and disk as in Aster, *but the rays very narrow, and usually in more than one row.* Scales of the involucre in one or two rows, nearly of equal length. Pappus of long bristles *with shorter ones intermixed.* Heads corymbed or solitary. Leaves generally sessile.

***Pappus not of hair-like bristles, but either altogether wanting or consisting of a few chaffy scales or teeth, or only a minute crown.*

+ *Receptacle naked.*

29. **Grinde'lia.** Rays yellow, many. Disk yellow. Heads large. Pappus of 2 or 3 rigid awns, falling off early. Scales of the involucre with spreading tips. Coarse herbs with sessile serrate leaves.—N.W. chiefly.

30. **Gutierre'zia.** Rays yellow, few. Disk-florets 3 or 4. Heads numerous, small, crowded. Pappus of a few short chaffy scales. A low partly shrubby plant, with narrowly linear entire alternate leaves. N.W. only.

31. **Hele'nium.** Rays yellow, wedge-shaped, 3-5-cleft at the summit. Scales of the involucre reflexed, awl-shaped. Pappus of 5-8 chaffy scales, 1-nerved, the nerve usually extending to a point. Leaves alternate, decurrent on the angled stem. Heads corymbed, showy.

32. **Chrysan'themum.** Rays white; disk yellow. Disk-corollas with a flattened tube. *Pappus none.* Heads single, or corymbed. Involucre broad and flat, the scales with scarious margins. Receptacle *flattish.*

33. **Bellis.** Rays white or pink, numerous. Scales of the involucre in about 2 rows. Receptacle conical, naked. Low herbs with solitary heads on scapes. Escaped from cultivation.

34. **Matrica'ria.** Rays white, disk yellow. Receptacle *conical.* Achenes 3-5-ribbed. Pappus a short crown or border.

+ + *Receptacle chaffy.*

35. **An'themis.** Rays white. Disk yellow. *Ray-florets pistillate or neutral.* Pappus none or a minute crown. Receptacle conical, more or less chaffy. Herbs with strong odour, and pinnately dissected leaves. Heads terminal.

36. **Rudbeck'ia.** Rays yellow, usually long; *disk dark-purple, or in one species greenish-yellow.* Scales of the involucre leaf-like. Receptacle conical. Pappus none, or only a minute crown. Ray-florets neutral.

37. **Helian'thus.** Rays yellow, neutral. Receptacle flattish or convex. Chaff persistent, and *embracing the 4-sided achenes.* Pappus *deciduous*, of 2 thin scales. Stout coarse herbs.

38. **Actinom'eris.** Rays 2-8, irregular, yellow, neutral. Receptacle small, convex. Scales of the involucre few, soon deflexed. Achenes flat, obovate, *spreading when ripe.* Pappus of 2 or 3 smooth awns, *persistent.* Tall branching perennials, with mostly decurrent leaves.

39. **Coreop'sis.** Rays mostly 8, yellow (sometimes brown-tinted), neutral. Receptacle flat. Involucre double, each of about 8 scales, the outer leaf-like and spreading, the inner rather membranaceous and appressed. Achenes flat, *compressed parallel with the scales of the involucre*, 2-toothed or sometimes naked. Heads corymbed.

40. **Gaillard'ia.** Rays yellow (sometimes purplish), numerous 3-cleft. Receptacle convex, the chaff bristly or awl-shaped. Achenes top-shaped, 5-ribbed. Pappus of several long thin scales, awn-tipped. Heads showy, fragrant, terminal, on scape-like stems.—N.W.

41. **Bidens.** Rays yellow, few; but 2 species are without rays. Scales of the involucre in 2 rows, the outer large and leaf-like. Ray-florets neutral. *Achenes crowned with 2 or more stiff awns which are barbed backward.*

42. **Heliop'sis.** Rays yellow, 10 or more, pistillate. Scales of the involucre in 2 or 3 rows, the outer leaf-like. Receptacle conical; chaff linear. Achenes smooth, 4-angled. Pappus none.

43. **Achille'a.** Rays white (occasionally pinkish), few. Receptacle flattish. Pappus none. Achenes margined. Heads small, in flat corymbs. Leaves very finely dissected.

44. **Polym'nia.** Rays whitish-yellow, wedge-form, shorter than the involucre, few in number. Scales of the involucre in 2 rows, the outer leaf-like, the inner small, and partly clasping the achenes. Pappus none. Coarse clammy herbs with an unpleasant odour.

45. **Sil'phium.** Easily known by its stout square stem, and the upper connate leaves forming a sort of cup. Flowers yellow. Achenes broad and flat.

Suborder II. LIGULIFLO'RÆ.

Corolla strap-shaped in all the florets of the head. All the florets perfect. Herbs with milky juice, and alternate leaves.

46. **Krig'ia.** Flowers yellow. Pappus double, the outer short, of many minute chaffy scales, the inner of many long capillary bristles. Low perennials branching from the base, with single small heads on scapes.

47. **Lamp'sana.** Flowers yellow, 8-12 in a head. Scales of the involucre 8, in a single row. *Pappus none.* Stem slender. Heads small, in loose panicles.

48. **Ciclio'rium.** Flowers bright blue, showy. Scales of the involucre in 2 rows, the outer of 5 short scales, the inner of 8-10 scales. *Pappus chaffy.* Heads sessile, 2 or 3 together.

49. **Leon'todon.** Flowers yellow. Involucre with bractlets at the base. *Pappus of plumose bristles*, these broader at the base. Heads borne on branching scapes. Leaves radical.

50. **Hiera'cium.** Flowers yellow. Scales of the involucre more or less imbricated. *Pappus a single row of tawny hair-like rough bristles.* Heads corymbose.

51. **Crepis.** Like Hieracium, but the pappus is of copious white and soft capillary bristles.

52. **Prenan'thes.** Stems leafy. Flowers yellowish or greenish-white often tinged with purple; heads nodding. Involucre of 5-14 scales in a single row, with a few bractlets below. *Pappus copious, of brownish or yellowish rough bristles.* Achenes short, blunt.

53. **Lygodes'mia.** Stems nearly leafless. Flowers white or cream-colour or pinkish. Involucre with bractlets below. Pappus white. Heads erect. Achenes long, tapering.—N.W. only.

54. **Trox'imon.** Flowers yellow. Head large, solitary, borne on a scape. Leaves radical, elongated, linear, tufted. Achenes 10-ribbed. Involucre loosely imbricated in 2 or 3 rows.—N.W. only.

55. **Tarax'acum.** Flowers yellow, *on slender naked hollow scapes.* Achenes prolonged into a slender thread-like beak. Leaves all radical. (See Part I., Chapter viii.)

56. **Lactu'ca.** Flowers pale yellow or purplish. Florets few (about 20) in the head. Scales of the involucre in 2 or more rows of

unequal length. *Achenes with long thread-form beaks*, and a pappus of very soft white bristles. Heads numerous, panicled. Tall smooth herbs with runcinate leaves.

57. **Mulge′dium.** Flowers chiefly blue. Structure of the heads and general aspect of the plant as in Lactuca, *but the beak of the achenes short and thick, and the pappus tawny.* Heads in a dense panicle.

58. **Son′chus.** Flowers pale yellow. Heads many-flowered, enlarging at the base. Achenes without beaks. Pappus very soft and white. Tall glaucous herbs with *spiny-toothed leaves.*

59. **Tragopo′gon.** Flowers yellow or purple. Heads large. Involucre of about 12 lanceolate rather fleshy scales in one row, somewhat united at the base. Achenes with long tapering beaks. Pappus of plumose bristles, 5 of these longer and naked at the summit. Leaves entire, straight-veined, clasping.

1. CNI′CUS, Tourn. COMMON THISTLE.

1. **C. lanceola′tus,** Hoff. (*Cirsium lanceolatum,* Scop.) (COMMON THISTLE.) *All the scales of the involucre prickly-pointed.* Leaves decurrent, pinnatifid, the lobes prickly-pointed, rough above, woolly with webby hairs beneath.—Fields and roadsides everywhere.

2. **C. undula′tus,** Gray. (*Cirsium undulatum,* Spreng.) *White-woolly* throughout. Low and stout, leafy. Scales tipped with short prickles. Leaves *undulate-pinnatifid.* —N.W. prairies.

4. **C. pu′milus,** Torr. Differs from the last in being pubescent but *not woolly.* The leaves, also, are not pinnatifid. Stem 1-flowered.—Lake Huron.

5. **C. altiss′imus,** Willd., var. **dis′color,** Gray. (*Cirsium discolor,* Spreng.) The *inner* scales of the involucre not prickly. Stem grooved. Leaves prickly, green above, *white-woolly beneath.* Flowers pale purple. Whole plant with a whitish aspect.—Dry thickets.

3. **C. mu′ticus,** Pursh. (*Cirsium muticum,* Michx.) (SWAMP THISTLE.) *Scales of the webby involucre hardly prickly,*

and not spreading. *Stem very tall*, *and smoothish, and sparingly leafy. Heads single or few.*—Swamps and low woods.

4. C. arvensis, Hoff. (*Cirsium arvense*, Scop.) (CANADA THISTLE.) Scales of the involucre with reflexed points. Leaves prickly, smooth both sides, or slightly woolly beneath. Roots extensively creeping. Heads small and numerous.—Fields and roadsides.

2. CAR'DUUS, Tourn. PLUMELESS THISTLE.

1. C. nu'tans, L. (MUSK THISTLE.) Leaves decurrent, sinuate, spiny. Heads solitary, drooping. Flowers purple. —Atl. Prov.

3. ONOPOR'DON, Vaill. SCOTCH THISTLE.

O. Acan'thium, L. A coarse branching herb, 2–4 feet high, with woolly stem and leaves. Bristles of the pappus united at the base into a ring.—Roadsides and old fields; not common.

4. ARCTIUM, L. BURDOCK.

A. Lappa, L. (*Loppa officinalis*, All., var. *major*, Gray.) A coarse plant with very large cordate petioled leaves, and numerous small globular heads of purple flowers. The involucre forms a bur which clings to one's clothing, or to the hair of animals.—Near dwellings, mostly in manured soil. Varies somewhat as to size and pubescence of the heads.

5. CENTAURE'A, L. STAR-THISTLE.

1. C. Cy'anus, L. (BLUE-BOTTLE.) An old garden plant, found occasionally along roadsides. False rays very large. Scales of the involucre fringed. Leaves linear, entire or nearly so. Stem erect. Heads single at the ends of the branches, the flowers bluish. Pappus very short.

2. C. ni'gra, L. (KNAPWEED.) *No false rays.* Scales of the globular involucre *black-fringed.* Leaves lanceolate, mostly entire, rough. Flowers purple. Pappus short or none.—Waste places, chiefly eastward.

3. **C. benedic'ta,** L. (*Cnicus benedictus*, L.) *Flowers yellow.* Pappus double, of 10 long outer bristles and 10 short inner ones. A low branching annual, with clasping cut leaves, and large sessile leafy-bracted heads.—Atl. Prov.

6. XANTHIUM, Tourn. CLOTBUR.

1. **X. Canadense,** Mill., var. **echina'tum,** Gray. (COMMON COCKLEBUR.) Stem rough, not prickly or spiny. Leaves broadly triangular, and somewhat heart-shaped, long-petioled. Fruit a hard 2-celled bur, nearly an inch long, clothed with stiff hooked prickles, the two beaks of the fruit long and usually incurved.—Low river-banks and waste places.

2. **X. spino'sum,** L. (SPINY CLOTBUR.) Stem armed with conspicuous straw-coloured triple slender spines, at the bases of the lanceolate short-petioled leaves, the latter white-woolly beneath.—Town of Dundas, Ontario; the seeds having been brought in wool from South America.

7. AMBRO'SIA, Tourn. RAGWEED.

1. **A. artemisiæfo'lia,** L. (HOG-WEED.) Stem erect, 1–3 feet high, branching, hairy. *Leaves twice-pinnatifid,* the lobes linear, paler beneath.—Waste places everywhere, but not so common northward.

2. **A. trif'ida,** L., (GREAT RAGWEED) is found in low grounds in the south-west of Ontario; also at Montreal and Ottawa. Stem stouter than No. 1, 2–4 feet high. *Leaves opposite, deeply 3-lobed,* the lobes oval-lanceolate and serrate.

8. FRANSE'RIA, Cav.

F. **Hookeria'na,** Nutt. Low and diffuse, hairy. Leaves bipinnatifid, at least the lower ones.—N.W.

9. TANACE'TUM, L. TANSY.

1. **T. vulga're,** L. (COMMON TANSY.) A very strong-scented herb, 2–4 feet high, smooth. Leaves twice-pinnate, the lobes serrate, as are also the wings of the petiole. Heads densely corymbed. Var. **crispum,** DC., is easily distinguished by its crisper and more incised leaves.—Old gardens and roadsides near dwellings.

2. **T. Huronen'se**, Nutt. Hairy when young. Heads usually few, and much larger than in T. vulgare. *Pistillate flowers flattened*, instead of tereté as in T. vulgare.—River-margins, Atl. Prov.

10. ARTEMIS'IA, L. WORMWOOD.

1. **A. Canadensis**, Michx. Stem smooth or sometimes hoary with silky down, erect, usually brownish. Lower leaves twice-pinnatifid, the lobes linear.—Shores of the Great Lakes.

2. **A. cauda'ta**, Michx., has the lobes of the dissected leaves *filiform*, and the heads *small*, in a wand-like long panicle. Disk-flowers perfect *but sterile*; marginal florets fertile.—Atl. Prov. and N.W.

3. **A. dracunculoi'des**, Pursh. Glabrous or slightly hoary. Stems 2-4 feet high, branching. Leaves mostly *entire*, linear, sometimes 3-cleft. Heads very numerous, in a compound panicle.—N.W. plains.

4. **A. glau'ca**, Pall. A smaller plant than the last, of stricter aspect. Leaves linear to oblong-lanceolate.—N.W. plains.

5. **A. vulga'ris**, L. (COMMON MUGWORT.) Stem tall, and branching above. *Leaves green and smooth above*, white-woolly beneath, pinnatifid, the lobes linear-lanceolate. Heads small, erect, in panicles. Flowers purplish.—Old fields near dwellings.

6. **A. Ludovicia'na**, Nutt. *White-woolly* throughout. Leaves lanceolate, the upper entire, the lower lobed or toothed, the upper surface sometimes green and smoothish. Heads small, mostly sessile in narrow panicles.—N.W.

Var. **gnaphalo'des**, Torr. and Gr., has the leaves serrate at the tips.—N.W.

7. **A. bien'nis**, Willd. *Glabrous;* stem *strict*, 1-3 feet high. Leaves once- or twice-pinnately parted, the lobes linear, acute. Heads in very short axillary spikes or clusters, crowded in a narrow leafy panicle. Florets all fertile.—N.W., and spreading through the Eastern Provinces.

8. **A. Absin'thium,** L. (COMMON WORMWOOD.) Somewhat shrubby. Whole plant silky hoary. Stem angular, branched, the branches with drooping extremities. Leaves 2-3-pinnately divided, the lobes lanceolate. Heads nodding. —Escaped from gardens in some places.

9. **A. frig'ida,** Willd. (PASTURE SAGE-BRUSH.) A *low* plant, growing in tufts, white-silky. Leaves dissected into narrowly linear divisions. Heads globose, in racemes.— N.W.

10. **A. ca'na,** Pursh. (SAGE-BRUSH.) Slightly shrubby, 1-2 feet high, much branched, silvery-hoary. Leaves narrow, tapering to both ends, small, mostly entire. Heads clustered in a leafy narrow panicle, few-flowered.—N. W. plains.

11. ERECHTI'TES, Raf. FIREWEED.

E. **hieracifo'lia,** Raf. Stem tall, grooved. Leaves sessile, lanceolate, cut-toothed, upper ones clasping.— Common in places recently over-run by fire.

12. GNAPHA'LIUM, L. CUDWEED.

1. **G. decur'rens,** Ives. (EVERLASTING.) Stem erect, 2 feet high, *clammy-pubescent*, white-woolly on the branches. Heads corymbed. Leaves linear-lanceolate, partly clasping, *decurrent.*—Fields and hillsides.

2. **G. polyceph'alum,** Michx. (COMMON EVERLASTING.) Stem erect, 1-2 feet high, white-woolly. Heads corymbed. Leaves lanceolate, tapering at the base, *not decurrent.*— Old pastures and woods.

3. **G. uligino'sum,** L. (LOW CUDWEED.) Stem spreading, 3-6 inches high, white-woolly. Leaves linear. Heads small *in crowded terminal clusters* subtended by leaves.— Low grounds.

4. **G. sylvat'icum,** L.—Erect, usually 9-12 inches high. *Leaves linear.* Heads axillary, nearly sessile, forming an erect leafy spike. Scales obtuse *with a brown bar across each near the top.*—Atl. Prov.

13. ANTENNA'RIA, Gærtn. EVERLASTING.

1. **A. margarita'cea,** R. Brown. (*Anaphalis margaritacea*, Benth. and Hook., in Macoun's Catalogue.) (PEARLY EVERLASTING.) Stem in clusters, downy. Leaves linear-lanceolate, taper-pointed, sessile. Scales of the involucre pearly-white. Heads in corymbs.—Along fences and in open woods.

2. **A. plantaginifo'lia,** Hook. (PLANTAIN - LEAVED E.) *Stem scape-like, 4–6 inches high.* Radical leaves spathulate or obovate; stem-leaves few, linear. Heads small, in a crowded corymb. Involucre white or purplish.—Old pastures and woods.

14. LIA'TRIS, Schreb. BLAZING-STAR.

1. **L. cylindra'cea,** Michx. Stem wand-like, 6–18 inches high. Leaves linear, rigid, generally 1-nerved. Heads few, cylindrical. Scales with abruptly mucronate tips. *Pappus plumose.*—Sandy fields and thickets.

2. **L. spica'ta,** Willd. Stem stout and rigid, 2–5 feet high, very leafy. Leaves linear, erect, the lowest 3–5-nerved. Heads crowded in a long spike. Scales obtuse. Pappus not obviously plumose.—Low grounds, south-western Ontario.

3. **L. scario'sa,** Willd. *Heads many-flowered,* large, few or many. *Pappus not obviously plumose.* Stem 2–5 feet high, stout, pubescent or hoary. Leaves lanceolate. Scales of the involucre obovate or spathulate, often with coloured tips or edges.—S. W. Ontario, and rather common N. W.

4. **L. puncta'ta,** Hook. *Heads 4-6-flowered,* usually many in a dense spike. *Pappus plumose.* Stem 10–30 inches high, stout. Scales of the involucre acuminate. Leaves narrowly linear.—N.W.

15. VERNO'NIA, Schreb. IRON-WEED.

V. altiss'ima, Nutt. Tall. Leaves lanceolate to lance-oblong. Heads in an open cyme, *the involucre purplish.*—Fields and roadsides, S. W. Ontario.

16. EUPATO'RIUM, Tourn. THOROUGHWORT.

1. **E. purpu'reum,** L. (JOE-PYE WEED. TRUMPET-WEED.) Stem tall and simple. *Leaves petioled, 3-6 in a whorl.* Flowers purplish or flesh-coloured. Heads in dense corymbs. —Low grounds.

2. **E. perfoli'atum,** L. (BONESET.) Stem short, hairy. *Leaves rugose, connate-perfoliate, tapering.* Flowers whitish. Corymbs very large.—Low grounds.

3. **E. ageratoi'des,** L. (WHITE SNAKE-ROOT.) Stem very smooth, commonly branching, 2-3 feet high. *Leaves opposite, petioled, broadly ovate, pointed, coarsely serrate.* Flowers white, in corymbs.—Low rich woods.

17. CACA'LIA, L. INDIAN PLANTAIN.

C. tubero'sa, Nutt. Stem angled and grooved, tall, from a thick root. *Leaves 5-7-nerved*, the lower lance-ovate or oval, tapering into long petioles.—Western Ontario.

18. IVA, L. MARSH ELDER.

1. **I. axilla'ris,** Pursh. Stem 1-2 feet high, the nodding heads mostly solitary in the axils of the upper leaves. Leaves small, obovate to oblong-linear, entire, sessile.— N. W.

2. **I. xanthiifo'lia,** Nutt. Stem 3-5 feet high, the narrow spike-like clusters of heads forming a naked panicle. Leaves large, broadly ovate, long-petioled, serrate.—N. W.

19. SENE'CIO, L. GROUNDSEL.

1. **S. vulga'ris,** L. (COMMON GROUNDSEL.) *Ray-florets wanting.* Stem low, branching. Leaves pinnatifid and toothed, clasping. Flowers yellow, terminal.—Cultivated and waste grounds.

2. **S. visco'sus,** L. *Ray-florets very minute.* Stem *viscid-pubescent* and strong-scented. Leaves twice-pinnatifid.— Atl. Prov.

3. **S. au'reus,** L. (GOLDEN RAGWORT. SQUAW-WEED.) *Rays 8-12.* Stem smooth, or woolly when young, 1-2 feet

high. Root-leaves simple, *rounded, usually cordate, crenately-toothed, long-petioled.* Stem leaves sessile, lanceolate, deeply pinnatifid. Heads in a corymb nearly like an umbel. —Swamps, often in gardens.

Var. **obova'tus,** Torr. and Gr., has root-leaves round-obovate with a cuneate or truncate base.

Var. **Balsam'itæ,** Torr. and Gr., has root-leaves oblong, spathulate, or lanceolate, serrate.

Var. **lanceola'tus,** Oakes, has thin lance-oblong root-leaves on long petioles.—Atl. Prov.

4. **S. integer'rimus,** Nutt. Rays conspicuous. Stem woolly-pubescent when young, *soon smoothish and green.* Leaves oblong-lanceolate or oblong, the upper bract-like. Heads in a naked corymb, the scales green-tipped.—N.W.

5. **S. canus,** Hook. Low, *persistently woolly.* Leaves small, spathulate to oblong. Rays conspicuous.—N.W.

6. **S. Jacobæa,** L. (COMMON RAGWORT.) Flowers golden yellow, the heads forming a spreading corymb. Stem erect, 2-3 feet high, branching, glabrous or somewhat cottony. Leaves numerous, lyrate, bipinnatifid, the lower with broad segments, the upper with linear divisions, all glabrous.— A troublesome weed in the Atl. Prov. Rare westward.

20. IN'ULA, L. ELECAMPANE.

I. **Hele'nium,** L. (COMMON ELECAMPANE.) Stem stout, 2-5 feet high. Root-leaves very large, ovate, petioled. Stem-leaves clasping. Rays numerous, narrow.—Roadsides.

21. CHRYSOP'SIS, Nutt. GOLDEN ASTER.

C. **villo'sa,** Nutt. Hirsute and villous-pubescent. Leaves narrowly oblong, hoary, bristly-ciliate towards the base. Stem branching, the branches terminating in the single heads.—Dry plains, N.W.

22. APLOPAPPUS, Cass.

1. **A. spinulo'sus,** DC. Low, perennial, *branches* very minutely hoary-pubescent. Leaves narrow, pinnately or tripinnately lobed, the lobes and teeth bristly, as are also the scales of the involucre.—Dry plains, N. W.

2. **A. lanceola'tus,** Torr. and Gr. Stem *simple*, with a tuft of coriaceous radical leaves, and bearing 3 to 15 heads. Rays 20 to 50. Scales of the involucre in 3 or 4 unequal series, lanceolate, with greenish tips and whitish base.—Marshy plains, N.W.

3. **A. acau'lis,** Gray, var. **glabra'tus,** Eaton. Stems *depressed-tufted*, from a woody rootstock, glabrous or nearly so. Leaves rigid, persistent, crowded below, a few only on the scape-like flowering stems, which occasionally branch above and bear 2 or 3 heads.—Dry gravelly ridges, N.W.

23. SOLIDA'GO, L. GOLDEN-ROD.

* *Heads clustered in the axils of the feather-veined leaves.*

1. **S. squarro'sa,** Muhl. Stem stout, 2–5 feet high, simple, hairy above. *Scales of the involucre with reflexed herbaceous tips.* Leaves large, oblong, serrate, veiny; the lower tapering into a long winged petiole, the upper sessile and entire. Heads in racemose clusters, *the whole forming a dense, leafy, interrupted, compound spike.*—Rocky woods.

2. **S. bi'color,** L. Stem hoary-pubescent, usually simple. Leaves oval-lanceolate, acute at both ends; the lower oval and tapering into a petiole, serrate. Heads in short racemes in the upper axils, the whole forming an interrupted spike or compound raceme. *Ray-florets whitish.* The variety **con'color** has *yellow* rays.—Dry banks and thickets.

3. **S. latifo'lia,** L. Stem smooth, angled, zigzag, 1–3 feet high. Leaves broadly ovate or oval, strongly and sharply serrate, *pointed at both ends.* Heads in very short axillary clusters. Rays 3 or 4.—Cool woods.

4. **S. cæ'sia,** L., var. **axilla'ris,** Gray. Stem smooth, terete, glaucous, slender, usually branching above. Leaves smooth, *lanceolate*, pointed, serrate, sessile. Heads in very short clusters in the axils of the leaves.—Rich woods and hillsides.

** *Racemes terminal, erect, loosely thyrsoid, not one-sided. Leaves feather-veined,*

5. **S. hu'milis,** Pursh. (*S. virgaurea,* L., var. **humilis,** Gray.) Stem low, 6–12 inches high, usually smooth; the

heads, peduncles, etc., mostly glutinous. Leaves lanceolate or oblanceolate, serrate or entire, the radical ones petiolate, obtuse, and serrate at the apex.—Rocky banks, not common.

6. **S. macrophyl'la,** Pursh. (*S. thyrsoidea*, E. Meyer.) Stems stout, simple, pubescent near the summit. Leaves thin, ovate, with *sharp projecting teeth*, the lower ones *abruptly contracted into long margined petioles. Heads large*, in an oblong raceme, loose and thin, long-pointed. Rays 8–10, *long*.—Wooded hillsides, chiefly Atl. Prov.

*** *Heads in a compound corymb terminating the simple stem, not at all racemose.*

7. **S. Ohioen'sis,** Riddell. *Very smooth* throughout. Stem slender, reddish, leafy. Radical leaves very long (often a foot), slightly serrate towards the apex, tapering into long margined petioles; stem-leaves oblong-lanceolate, entire, sessile.—Wet grassy shores of Red Bay, Lake Huron.

8. **S. rig'ida,** L. *Rough and rather hoary.* Stem stout, 2–5 feet high, very leafy. Leaves *oval or oblong*.—Dry soil, western Ontario, and N. W.

**** *Heads in one-sided racemes, spreading or recurved. Leaves not 3-ribbed, but sometimes obscurely 3-nerved.*

9. **S. semper'virens,** L. Stem stout. Leaves *long*, lanceolate, *thickish*, smooth, entire, obscurely 3-nerved. Racemes short, in a terminal panicle. Heads *large, showy*. —Salt marshes and sea-shores.

10. **S. puber'ula,** Nutt. *Stem and panicle minutely hoary.* Stem-leaves lanceolate, acute, tapering to the base, somewhat smooth, the lowermost spathulate, sparingly toothed. Heads not large, crowded in compact short racemes, which form a long, dense, terminal panicle. Rays 10–14.—Barren soil, Atl. Prov.

11. **S. uligino'sa,** Nutt. (*S. stricta*, Ait.) Smooth. Stem simple, strict. Leaves lanceolate, pointed, smooth, the lower tapering into winged petioles, finely but sparingly serrate, or entire. Racemes crowded and appressed in a close wand-like panicle. Heads middle-sized. Rays 5–6.

small.—Peat-bogs and wet places, Atl. Prov., westward and northward.

12. **S. specio'sa,** Nutt., somewhat resembles the last, but the leaves are oval or ovate, and the stem is taller and stouter (3-6 feet). The rays, also, are larger.—Copses, Atl. Prov.

13. **S. jun'cea,** Ait. (*S. arguta*, Torr. and Gray.) Whole plant smooth, 1-4 feet high, rigid, branching above. Lower leaves oval or elliptical-lanceolate, serrate with spreading teeth, pointed, tapering into winged and ciliate petioles; upper ones lanceolate. Racemes very dense, naked, at length elongated and recurved.—Woods and banks.

14. **S. argu'ta,** Ait. (*S. Muhlenbergii*, Torr. and Gray.) Stem smooth, angled or furrowed. Leaves large and thin, ovate; the upper elliptical-lanceolate. Racemes much shorter and looser than in No. 13, and the rays much larger.—Moist woods and thickets.

15. **S. rugo'sa,** Mill. (*S. altissima*, Torr. and Gray.) Stem rough-hairy, *less than a foot high.* Leaves ovate-lanceolate or oblong, coarsely serrate, veiny, *often rugose.* Racemes panicled, spreading.—Borders of fields and copses.

16. **S. neglecta,** Torr. and Gr. Stem smooth, 2-3 feet high, stout. Leaves thickish, smooth both sides, the upper oblong-lanceolate, nearly entire, the lower ovate-lanceolate or oblong, sharply serrate, tapering into a petiole. Heads rather large. Racemes short and dense, at first erect and scarcely one-sided, at length spreading.—Swamps.

* * * * * *Racemes one-sided and recurved, and the leaves plainly 3-ribbed.*

17. **S. nemora'lis,** Ait. Stem minutely and closely hoary-pubescent, simple or corymbed. Leaves more or less hoary, obscurely serrate or entire; the lower oblanceolate, somewhat crenate, and tapering into a petiole. Racemes numerous, dense, at length recurved, forming a large panicle.—Dry fields.

18. **S. Missouriensis,** Nutt. Stem *smooth.* Leaves linear-lanceolate, tapering to both ends, with rough margins. Racemes densely crowded.—Dry prairies, N.W.

Var. **monta'na,** Gray, is only 6-15 inches high, with a small and compact panicle, not more than 2 or 3 inches long. —N.W.

19. **S. Canadensis,** L. Stem *rough-hairy*, tall and stout. Leaves lanceolate, serrate, pubescent beneath, rough above. Panicle exceedingly large.—Very common along fences and in moist thickets.

Var. **scabra,** Torr. and Gray, has the leaves very rough above and whitish-woolly beneath.—N. W.

20. **S. sero'tina,** Ait. (*S. gigantea,* Torr. and Gray.) Stem *smooth*, stout. Leaves lanceolate, taper-pointed, sharply serrate, except at the base, smooth both sides, *rough-ciliate*. Rays 7-14, rather long. Panicle large, *pubescent*.— Open thickets and meadows.

Var. **gigante'a,** Gray, is very tall and the leaves more or less pubescent beneath.—Thickets and low grounds.

* * * * * * *Inflorescence a flat-topped corymb.*

21. **S. lanceola'ta,** L. Stem pubescent above, much branched. Leaves linear-lanceolate, the nerves (3-5) and margins rough-pubescent. Heads in dense corymbed clusters, giving a decidedly characteristic aspect to this species. —Low lake- and river-margins.

24. AR'NICA, L.

1. **A. Chamisso'nis,** Less. Soft-hairy. Stem leafy to the top, bearing 1-5 heads. Leaves thin and veiny, toothed; the upper ovate-lanceolate, sessile; the lower narrow, tapering to a margined petiole.—Atl. Prov.

2. **A. folio'sa,** Nutt., has lanceolate leaves, the upper partly clasping and the lower with tapering bases *connate*. Stem strict.—N. W.

3. **A. alpi'na,** Olin. Less leafy, low, yellowish-pubescent or villous. Stem simple, *bearing a single head* as a rule. Leaves thickish, narrowly oblong to lanceolate, *the cauline only 1 or 2 pairs*.—N. W.

25. TUSSILA'GO, Tourn. COLTSFOOT.

T. Far'fara, L. Wet places, chiefly eastward, but also in Ontario.

26. PETASI'TES, Tourn. SWEET COLTSFOOT.

1. **P. palma'ta**, Gray. Leaves rounded, somewhat kidney-shaped, palmately 5-7-lobed, the lobes toothed and cut.—Cedar-swamps and bogs; wet banks of streams.

2. **P. sagitta'ta**, Gray. Leaves deltoid-oblong to reniform-hastate, repand-dentate.—Swamps, N.W.

27. ASTER, L. STARWORT. ASTER.

* *Leaves, at least the lower ones, heart-shaped and petioled.*

1. **A. corymbo'sus**, Ait. Rays 6-9, white or nearly so. Heads in corymbs. Stems slender, 1-2 feet high, zigzag. Leaves thin, smoothish, sharp-pointed, coarsely serrate, all the lower ones on slender naked petioles.—Woodlands.

2. **A. macrophyl'lus**, L. Rays white or bluish. Stem stout, 2-3 feet high. Leaves thickish, rough, finely serrate, the lower long-petioled. Heads in closer corymbs than in No. 1.—Woodlands.

3. **A. azu'reus**, Lindl. Rays 10-20, bright blue. Heads racemed or panicled. Stem roughish, erect, racemose-compound above. Leaves entire or nearly so, rough; the lower ovate-lanceolate, on long petioles; the upper lanceolate or linear, sessile. The latest flowering of our Asters.—Dry soil.

4. **A. undula'tus**, L. Rays bright blue. Heads racemed or panicled. Stem hoary with close pubescence, spreading. Leaves with somewhat wavy margins, entire or nearly so, ovate or ovate-lanceolate, roughish above, downy beneath; the lowest cordate, on margined petioles; the upper with *winged short petioles* clasping at the base, or sessile.—Dry woods.

5. **A. cordifo'lius**, L. Rays pale blue or nearly white. Heads *small*, profuse, panicled. Stem much branched. Leaves thin, *sharply serrate*, the lower on slender ciliate

petioles. Scales of the inversely conical involucre all appressed, and tipped with very short green points.—Woods and along fences.

6. **A. sagittifo'lius,** Willd. Rays pale blue or purple. Heads *small*, in dense compound racemes or panicles. Stem smooth or nearly so, erect, with *ascending* branches. Leaves conspicuously serrate, ovate-lanceolate, pointed, pubescent, the lowest on long margined petioles, the upper lanceolate or linear, pointed at both ends. Scales of the oblong involucre linear, tapering into awl-shaped, slender and loose tips.—Thickets and along fences.

7. **A. Lindleya'nus,** Torr. and Gr. Rays pale violet. Stem rather stout, smooth or sparsely pubescent. Leaves conspicuously serrate. Root-leaves and lowest stem-leaves ovate, more or less cordate, with margined petioles; uppermost *sessile, and pointed at both ends. Heads rather small, in a loose thyrse or panicle.* Scales linear, green-tipped.—Open barren grounds.

** *Upper leaves all sessile or clasping by a heart-shaped base; lower ones not heart-shaped.*

8. **A. lævis,** L. *Rays large*, sky-blue. Very smooth throughout. Heads in a close panicle. Leaves lanceolate or ovate-lanceolate, chiefly entire, rough on the margins, *the upper ones clasping by an auricled base.*—Dry woods.

9. **A. Novæ-An'gliæ,** L. Rays many, narrow, violet-purple, showy; *heads large. Involucre of many slender equal scales, apparently in a single row, clammy.* Stem stout, 3-8 feet high, hairy, corymbed above. Leaves very numerous, lanceolate, entire, clasping by an auricled base, pubescent. —River-banks and borders of woods.

10. **A. puni'ceus,** L. *Rays long,* lilac-blue. *Scales of the involucre narrowly linear, loose,* in about two rows. Stem 3-6 feet high, stout, rough-hairy, *usually purple below.* Leaves oblong-lanceolate, clasping by an auricled base, sparingly serrate in the middle, rough above, smooth beneath, pointed.—Swamps; usually clustered.

11. A. tardiflo'rus, L. Rays pale violet. Stem glabrous or nearly so, 1-2 feet high. Leaves lanceolate or oblong-lanceolate, acuminate, with tapering and somewhat auricled base. Heads few, as a rule. Involucre loose, the outer scales leaf-like.—Atl. Prov.

* * * *None of the leaves heart-shaped; those of the stem sessile, tapering at the base (except in No. 11).*

12. A. rad'ula, Ait. Rays light violet. Radical leaves all tapering into margined petioles. Stem-leaves sessile oblong-lanceolate, pointed, sharply serrate in the middle, very rough both sides, rugose. Scales of the bell-shaped involucre oblong, appressed, with slightly spreading herbaceous tips.—Low grounds, Atl. Prov.

13. A. ericoi'des, Ait. Rays white or nearly so. Stem smooth or nearly so, 1-3 feet high, the branchlets or peduncles racemose on the upper side of the spreading branches. Lowest leaves oblong-spathulate, the others linear-lanceolate or linear-awl-shaped.—Barren soil, and margins of lakes and rivers. Var. **villosus**, Torr. and Gray, has hairy stems and leaves.

14. A. multiflo'rus, Ait. Rays white, 10-20. Stem pale or hoary with minute pubescence, 1 foot high, bushy. *Leaves crowded, linear*, with rough margins; the upper partly clasping. Heads small, crowded on the racemose branches. Scales of the involucre with spreading green tips.—Dry soil.

15. A. Tradęscan'ti, L. (*A. tenuifolius.*) Rays white or purplish. Scales of the involucre narrowly linear, in 3 or 4 rows. Heads small, very numerous, in 1-sided *close* racemes on the branches. Stem 2-4 feet high, much branched, *smooth*. Leaves linear-lanceolate, *with long slender point*, the larger ones with a *few remote teeth* in the middle.—Low grounds.

16. A. diffu'sus, Ait. (*A. miser*, L.) Rays pale blue or whitish. Involucre nearly as in No. 15. Stem *more or less* pubescent, much branched. Heads small, in *loose* 1-sided racemes on the spreading branches. Leaves lanceolate, acute at each end, *sharply serrate in the middle.*—Low grounds.

17. **A. panicula'tus,** Lam. (*A. simplex,* Willd.) Rays pale blue or whitish. Scales of the involucre linear-awl-shaped. Stem stout, smooth or nearly so, with numerous leafy branches. Heads medium-sized, scattered, loosely paniculate. Leaves smooth, oblong to lanceolate, tapering at both ends, the lower serrate.—Moist and shady banks.

18. **A. jun'ceus,** Ait. Rays light purple. Stem slender, 1-3 feet high, simple, with few small heads, or loosely branching. Leaves narrow, entire, or the lower sparingly denticulate. Scales of the involucre small, narrow, in 2 or 3 rows.—Bogs and wet places.

19. **A. vimin'eus,** Lam. Rays white or nearly so. Stem 2-5 feet high, smooth, bushy. Leaves linear or narrowly lanceolate, the larger ones sparingly serrate in the middle with fine teeth. Heads very numerous, in 1-sided racemes on short branchlets. Scales of the involucre narrowly linear, in 3 or 4 rows.—Moist banks.

20. **A. nemora'lis,** Ait. Rays lilac-purple, elongated. Stem slender and leafy, the upper branches terminating in 1-flowered nearly naked peduncles. Leaves small, rigid, narrowly lanceolate, nearly entire, *with revolute margins.*—Swamps, Atl. Prov. and Muskoka.

21. **A. ptarmicoi'des,** Torr. and Gr. Rays *pure white.* Stems clustered, generally a foot high, each bearing a flat corymb of small heads. Leaves linear-lanceolate, acute, rigid, entire, mostly 1-nerved, with rough margins.—Dry or gravelly hills. Our earliest Aster.

22. **A. acumina'tus,** Michx. Rays white or faintly purple. Stem about a foot high, somewhat hairy, zigzag, panicled-corymbose at the top. Leaves large, thin, oblong-lanceolate, *pointed,* coarsely toothed towards the apex, entire at the base.—Cool sandy woods; mostly eastward.

23. **A. umbella'tus,** Mill. (*Diplopappus umbellatus,* Torr. and Gr.) *Pappus double,* the inner of long capillary bristles,

the outer of short and rigid bristles. *Rays white.* Stem smooth, leafy to the top, tall, simple. Leaves lanceolate, long-pointed. Heads small, very numerous, in compound flat corymbs.—Moist thickets.

28. ERIG'ERON, L. FLEABANE.

1. **E. Canaden'sis,** L. (HORSE-WEED. BUTTER-WEED.) Rays white, *but very inconspicuous,* shorter than their tubes. Heads very small, numerous, in panicled racemes. Stem 1-5 feet high, erect and wand-like, bristly-hairy. Leaves linear, mostly entire.—Common in burnt woods and new clearings.

2. **E. acris,** L. Rays purplish or bluish, about the same length as the copious simple pappus. Heads several or many, small, at length corymbose, hirsute. Stem 10-20 inches high, pubescent or smoothish. Leaves mostly lanceolate, entire. *A set of pistillate flowers within the circle of ray-flowers.*—Atl. Prov. and N. W.

3. **E. bellidifo'lius,** Muhl. (ROBIN'S PLANTAIN.) Rays *bluish-purple*, numerous. Heads medium-sized, few, on slender corymbose peduncles. Stem hairy, *producing offsets from the base.* Radical leaves spathulate or obovate, toothed above the middle; stem-leaves oblong, *few*, sessile or partly clasping, entire.—Thickets.

4. **E. hyssopifo'lius,** Michx. Rays rose-purple or whitish. Stem slightly pubescent, slender, 6-12 inches high, *from slender rootstocks.* Leaves very many, short, linear. Heads small, *terminating the slender naked branches.* Pappus simple.—Atl. sea-coast and northward.

5. **E. cæspito'sus,** Nutt. Stem *dwarf, tufted*, from a stout rootstock, more or less *hoary-pubescent.* Rays white, 40-50, narrow.—N. W.

6. **E. Philadel'phicus,** L. (COMMON FLEABANE). Rays *rose-purple, very numerous and narrow.* Heads small, few, in corymbs. Stem hairy, with *numerous* stem-leaves. Radical leaves spathulate and toothed; the upper ones clasping by a heart-shaped base, entire.—Moist grounds.

7. **E. strigo'sus,** Muhl. (DAISY FLEABANE.) Rays white, conspicuous, numerous. *Pappus plainly double.* Stem and leaves roughish with minute appressed hairs, or nearly smooth. Lower leaves spathulate and slender-petioled, *entire or nearly so,* the upper lanceolate, scattered.—Dry fields and meadows.

8. **E. glabell'us,** Nutt. Rays purple, very many, much longer than the *hoary-hispid* involucre. Stem 6-15 inches high, stout, smooth below, bearing 1-7 large heads on the naked summit. *Pappus double.* Leaves *smooth* but ciliate, the upper oblong-lanceolate and pointed, sessile or clasping; the lower petiolate, spathulate.—N.W.

Var. **as'perus,** has very rough leaves and stem.

9. **E. an'nuus,** Pers. (LARGER DAISY FLEABANE.) Rays white, tinged with purple. *Pappus double.* Stem rough with spreading hairs. Leaves *coarsely toothed;* the lower ovate, tapering into a margined petiole; the upper ovate-lanceolate. Heads corymbed.—Fields and meadows.

29. GRINDE'LIA, Willd.

G. squarro'sa, Dunal. Leaves spathulate to linear-oblong. Heads large, terminating the leafy branches.—Dry prairies, N.W.; also at Ottawa.

30. GUTIERRE'ZIA, Lag.

G. Eutha'miæ, Torr. and Gr. Not more than 8 inches high.—Dry plains, N.W.

31. HELE'NIUM, L. SNEEZE-WEED.

H. autumna'le, L. (SNEEZE-WEED.) Stem nearly smooth. Leaves lanceolate, toothed. Disk globular.—Low river- and lake-margins.

32. CHRYSAN'THEMUM, Tourn. OX-EYE DAISY.

1. **C. Leucan'themum,** L. (*Leucanthemum vulgare,* Lam.) (OX-EYE DAISY. WHITE-WEED.) Stem erect, naked above, bearing a single large head. Leaves pinnatifid or cut-toothed, the lowest spathulate, the others partly clasping.—Pastures and old fields.

2. **C. Parthe'nium,** Pers. (FEVERFEW.) Stem branching, leafy. Leaves twice-pinnately divided, the divisions *ovate, cut.* Heads corymbed.—Escaped from gardens.

3. **C. Balsam'itæ,** L. (COSTMARY.) A garden-escape, smooth, with pleasant odour. Leaves pale, oblong, somewhat toothed. Heads small, pale yellow.

33. BELLIS, Tourn. DAISY.

B. perennis, the true Daisy, a native of the Old World, is a low stemless herb. It is an uncommon garden escape. The heads are many-flowered with numerous pistillate rays. The scales of the involucre equal, in about 2 rows, herbaceous. Receptacle conical. Pappus wanting.

34. MATRICA'RIA, Tourn. WILD CHAMOMILE.

M. inodo'ra, L. Leaves twice-pinnately divided into very narrow lobes. Heads large, naked-peduncled, the rays many and long.—Chiefly Atl. Prov.

35. AN'THEMIS, L. CHAMOMILE.

1. **A. Cot'ula,** DC. (*Maruta Cotula.*) (MAY-WEED.) Stem branching. Leaves *thrice-pinnate*, finely dissected. Odour disagreeable. Rays soon reflexed.—Roadsides everywhere.

2. **A. arven'sis,** L. (CORN CHAMOMILE.) Resembling the last, but the leaves are not so finely dissected, and the odour not so unpleasant.—Atl. Prov., rare.

36. RUDBECK'IA, L. CONE-FLOWER.

1. **R. lacinia'ta,** L. Rays linear, 1–2 inches long, drooping. *Disk greenish-yellow.* Stem tall, *smooth*, branching. Lowest leaves pinnate, of 5–7 lobed leaflets; upper ones 3–5-parted, or the uppermost undivided and generally ovate. Heads terminal, long-peduncled.—Swamps.

2. **R. hir'ta,** L. Rays bright yellow. *Disk purplish-brown.* Stem very *rough-hairy*, naked above, bearing single large heads. Leaves 3-ribbed, the lowest spathulate, narrowed into a petiole, the upper ones sessile.—Meadows.

3. **R. columna'ris,** Pursh. (*Lepachys columnaris,* Torr.

and Gr.) Rays yellow or purplish ; disk grayish. Receptacle oblong. Chaff thickened and bearded at the tip. Pappus none or of 2 teeth. Stem branching from the base, 1-2 feet high. Leaves pinnately divided, the divisions 5-9, narrow. Heads single, on the naked branches.—N.W., and at Ottawa.

37. HELIAN'THUS, L. Sun-flower.

Annuals. Leaves alternate. Receptacle flat. Disk brownish.

1. **H. an'nuus**, L., (Common Sunflower.) Tall, rough. Leaves 3-ribbed, ovate, serrate. Scales of the involucre long-pointed, ciliate.—Escaped from cultivation.

2. **H. petiola'ris**, Nutt. More slender, 1-3 feet high. Leaves narrow, mostly entire. Scales seldom ciliate.—N.W.

**Perennials. Receptacle convex. Lower leaves usually opposite.*

— Disk dark.

3. **H. rig'idus**, Desf. Stem tall and stout, rough. Leaves very thick and rigid, rough both sides, oblong-lanceolate, pointed at both ends, the lowest oval, 8-nerved. Rays 20-25. Pappus of 2 large, and often several small scales.—N. W.

— — Disk yellow.

4. **H. Nuttall'ii**, Torr. and Gr. Stem slender, smooth, simple. Leaves lanceolate or linear. Pappus-scales long and narrow.—N.W.

5. **H. strumo'sus**, L. Stem 3-6 feet high, smooth below. Leaves broadly lanceolate, rough above and *whitish beneath*, pointed, serrate with small appressed teeth, short-petioled. Rays about 10.—Moist copses and low grounds.

6. **H. divarica'tus**, L. Stem 1-4 feet high, *smooth*, simple or forking above. Leaves all opposite, widely spreading, *sessile, rounded or truncate at the base*, ovate-lanceolate, *3-nerved*, long-pointed, serrate, *rough on both sides*. Heads few, on short peduncles. Rays about 12.—Open thickets and dry plains.

7. **H. decapet'alus**, L. Stem 3-6 feet high, branching, smooth below, rough above. Leaves *thin, green on both sides, ovate, coarsely serrate*, pointed, abruptly contracted

into short margined petioles. Rays usually 10.—Thickets and river-banks.

8. **H. gigante'us,** L. Stem tall, *hairy or rough*, branching above. Leaves lanceolate, pointed, serrate, very rough above, hairy below, narrowed and ciliate at the base. Heads somewhat corymbed, not large. Disk yellow; rays pale yellow, 15-20. — Low grounds, western and southwestern Ontario.

9. **H. tubero'sus,** L., (JERUSALEM ARTICHOKE) has escaped from cultivation in some places. It is at once recognized by its tubers.

38. ACTINOM'ERIS, Nutt.

A. squarro'sa, Nutt. Stem hairy, tall, commonly winged above. Leaves oblong to ovate-lanceolate, pointed at both ends.—S. W. Ontario.

39. COREOP'SIS, L. TICKSEED.

1. **C. tincto'ria,** Nutt. Stem smooth, 2-3 feet high. Leaves once- or twice-pinnately divided, the lobes narrow. Pappus none. Achenes *oblong*. Rays yellow, *brown-tinted*. —N. W.

2. **C. trichosperm'a,** Michx., var. **tenuilo'ba,** Gray. (TICKSEED SUNFLOWER.) Stem smooth, branching. Rays *golden-yellow*. Leaves short-petioled, pinnately divided, the segments serrate. Achenes *narrowly wedge-oblong*, 2-toothed. —Swamps, S. W. Ontario.

3. **C. verticilla'ta,** L. Glabrous. *Leaves sessile, divided into 3 sessile leaflets* (appearing whorled), these pinnately dissected into very narrow divisions.—Damp places, S. W. Ontario.

4. **C. trip'teris,** L. Smooth. Stem tall, corymbed above. Leaves petioled, pinnately divided, the divisions lanceolate, acute, entire. Disk turning brownish.—Damp places, S. W. Ontario.

40. GAILLARD'IA, Foug.

G. arista'ta, Pursh. Rough-hairy, about 2 feet high. Leaves lanceolate to oblanceolate, entire to coarsely pinnatifid.—Dry soil, N. W.

41. BIDENS, L. BUR-MARIGOLD.

1. **B. frondo'sa,** L. (COMMON BEGGAR-TICKS.) *Rays none. Achenes flat*, wedge-obovate, *ciliate on the margins with bristles, pointing upwards*, 2-awned. Stem tall, branched. Leaves thin, long-petioled, pinnately 3-5-divided, the leaflets ovate-lanceolate, pointed, serrate.

2. **B. conna'ta,** Muhl. (SWAMP BEGGAR-TICKS.) *Rays none.* Achenes flat, narrowly wedge-shaped, 2-4-awned, *ciliate with minute bristles, pointing downwards.* Stem 1-2 feet high, smooth. Leaves lanceolate, pointed, serrate, tapering and connate at the base, the lowest often 3-parted and decurrent on the petiole.—In shallow water and low grounds.

3. **B. cer'nua,** L. (SMALLER BUR-MARIGOLD.) *Rays short, pale yellow.* Achenes flat, wedge-obovate, *4-awned, ciliate with bristles pointing downwards.* Stem nearly smooth, 5-10 inches high. Leaves all simple, lanceolate, unequally serrate, hardly connate. Heads nodding.—Wet places.

4. **B. chrysanthemoï'des,** Michx. (LARGE BUR-MARIGOLD.) *Rays an inch long, showy, golden yellow.* Achenes wedge-shaped, 2-4-awned, *bristly downwards.* Stem smooth, 6-30 inches high, erect or ascending. Leaves lanceolate, tapering at both ends, connate, regularly serrate.—Swamps and ditches.

5. **B. Beck'ii,** Torr. (WATER MARIGOLD.) *Aquatic.* Stems long and slender. Immersed leaves dissected into fine hair-like divisions; those out of water lanceolate, slightly connate, toothed. Rays showy, golden yellow, larger than the involucre. *Achenes linear, bearing 4-6 very long awns barbed towards the apex.*—Ponds and slow streams.

42. HELIOP'SIS, Pers. OX-EYE.

1. **H. lævis,** Pers. Stem smooth, slender, branching. Leaves ovate-lanceolate, acute, sharply serrate, on slender petioles. Heads showy; peduncles elongated.—Dry open thickets; London and westward.

2. **H. scabra,** Dunal. *Roughish,* especially the leaves. Pappus of 2 or 3 teeth, or a mere chaff-like border.—Niagara Falls and N.W.

43. ACHILLE'A, L. YARROW.

1. **A. millefo'lium,** L. (MILFOIL.) Stems simple. Leaves dissected into fine divisions. Corymb flat-topped. Rays only 4 or 5, short.—Fields and along fences; very common.

2. **A. Ptar'mica,** L. (SNEEZE-WORT.) Leaves simple, lance-linear, serrate. Corymb loose. Rays 8-12, much longer than the involucre.—Atl. Prov.

44. POLYM'NIA, L. LEAF-CUP.

P. Canadensis, L. A coarse clammy-hairy herb. Lower leaves opposite, petioled, pinnatifid; the upper alternate, angled or lobed. Heads small; rays pale yellow.—Shaded ravines; south-westward.

45. SIL'PHIUM, L. ROSIN-PLANT.

1. **S. perfolia'tum,** L., (CUP-PLANT) is found in south-western Ontario. Stem stout, *square*, 4-8 feet high. Leaves ovate, coarsely toothed, the upper ones united by their bases.

2. **S. terebinthina'ceum,** L. (PRAIRIE DOCK.) Stem tall, *round*, naked above, smooth. Radical leaves sometimes 2 feet long, rough-hairy, coarsely serrate, on slender petioles. Heads small, loosely panicled.—Open woods and grassy banks, south-western Ontario.

46. KRIG'IA, Schreber. DWARF DANDELION.

K. amplexicau'lis, Nutt. (*Cynthia Virginica,* Don.) Roots fibrous. Stem-leaves 1-2, oblong or lanceolate-spathulate, clasping, mostly entire, the radical ones on short winged petioles. Peduncles 2-5.—South-western Ontario.

47. LAMP'SANA, Tourn. NIPPLE-WORT.

L. commu'nis, L. Very slender and branching. Leaves angled or toothed. Heads small, loosely panicled.—Borders of springs; common at Queenston Heights.

48. CICHO'RIUM, Tourn. SUCCORY. CICHORY.

C. In'tybus, L. Stem-leaves oblong or lanceolate, partly clasping; radical ones runcinate.—Roadsides and waste places.

49. LEON'TODON, L. FALL DANDELION.

L. autumna'le, L. (FALL DANDELION.) Leaves lanceolate, laciniate-toothed or pinnatifid. *Scape branched.*—Roadsides and waste places; not common westward.

50. HIERA'CIUM, Tourn. HAWKWEED.

1. **H. Canadensis,** Michx. (CANADA HAWKWEED.) *Heads large.* Stem simple, leafy, corymbed, 1-3 feet high. *Peduncles downy.* Leaves ovate-oblong, with a few coarse teeth, somewhat hairy, sessile, or the uppermost slightly clasping. Achenes tapering towards the base.—Dry banks and plains.

2. **H. scabrum,** Michx. (ROUGH H.) *Heads small.* Stem stout, 1-3 feet high, *rough-hairy*, corymbose. *Peduncles or involucre densely clothed with dark bristles.* Achenes not tapering.—Sandy woods and thickets.

3. **H. Grono'vii,** L. (HAIRY H.) Heads small. Stem wand-like, leafy and very hairy below, *naked above*, forming a long and narrow panicle. *Achenes with a very taper summit.*—Dry soil, western Ontario.

4. **H. veno'sum,** L., (RATTLESNAKE-WEED) with a smooth naked scape (or bearing one leaf), and a loose corymb of very slender peduncles, is found in the Niagara region and south-westward.

5. **H. panicula'tum,** L. Stem slender, *leafy*, diffusely branched, hairy only below. Heads *very small*, in a loose panicle, on slender diverging pedicels, 10-20-flowered. Achenes short, not tapering above. Leaves lanceolate, acute at both ends.—Open woods, S.W. Ontario.

6. **H. auranti'acum,** L., a low hirsute species, with clustered heads of deep-orange flowers on a simple peduncle, leafy at the base, is reported from the neighbourhood of London, Ont.

51. CREPIS, L.

C. runcina'ta, Torr. and Gr. Stem 1-2 feet high, glabrous. Radical leaves obovate-oblong to oblong-lanceolate, with short lobes or teeth. Cauline leaves absent or very small. Involucre pubescent, sometimes it and the upper part of the scape glandular.—N.W.

52. PRENAN'THES, Vaill. RATTLESNAKE-ROOT.

1. **P. alba,** L. (*Nabalus albus,* Hook.) (WHITE LETTUCE.) Heads 8-12-flowered. *Pappus deep cinnamon-coloured.* Stem 2-4 feet high, smooth and glaucous, corymbose-paniculate. Leaves triangular-halberd-shaped, or 3-5-lobed, the uppermost oblong and undivided.—Rich woods.

2. **P. altiss'ima,** L. (*Nabalus altissimus.*) (TALL WHITE LETTUCE.) Heads 5-6-flowered. *Pappus pale straw-coloured.* Stem taller but more slender than in No. 1, *with a long, leafy panicle at the summit.*—Rich woods.

3. **P. racemo'sa,** Michx. (*Nabalus racemosus,* Hook.) Heads about 12-flowered. Involucre and peduncles *hairy.* Stem wand-like, smooth. Leaves oval or oblong-lanceolate, slightly toothed. Heads crowded in a long and narrow interruptedly spiked panicle. Pappus straw-colour; flowers flesh-colour.—Shore of Lake Huron and south-westward.

4. **P. Mainen'sis,** Gray. Leaves as in the last, but the radical ones ovate and more abruptly narrowed to the short petiole. Heads 8-12-flowered, persistently drooping on slender pedicels.—Atl. Prov.

53. LYGODES'MIA. Don.

L. jun'cea, Don. Much branched from the base, the branches closely erect and rigid, rush-like, terminating in erect heads of pinkish flowers. Leaves small, the lower lance-linear, the upper scale-like.—N.W. plains.

54. TROX'IMON, Nutt.

1. **T. cuspida'tum,** Pursh. Scape a foot high. Leaves lanceolate, tapering to a sharp point, entire, woolly on the margins. Achenes beakless.—N.W. prairies.

2. **T. glau'cum,** Nutt. Scape 1-2 feet high. Leaves varying from entire to dentate or laciniate. Achenes long-beaked.—N.W. prairies.

55. TARAX'ACUM, Haller. DANDELION.

T. officina'le, Weber. (*T. dens-leonis,* Desf.) (COMMON DANDELION.) Outer involucre reflexed. Leaves runcinate. —Fields everywhere.

56. LACTU'CA, Tourn. LETTUCE.

1. **L. Canaden'sis,** L. (WILD LETTUCE.) Heads numerous, in a long and narrow naked panicle. Stem stout, smooth, hollow, 4-9 feet high. Leaves mostly runcinate, partly clasping, pale beneath; the upper entire. Achenes longer than their beaks.—Borders of fields and thickets.

2. **L. scari'ola,** L. (PRICKLY LETTUCE.) Stem below sparingly bristly. Leaves vertical, spinulose-denticulate, oblong or lanceolate, prickly on the midrib below. Heads small, 6-12 flowered.—Waste places.

3. **L. integrifo'lia,** L. Stem 3-6 feet high; leaves all undivided, entire or slightly toothed. Flowers pale yellow, cream-colour, or purple.—Dry soil.

4. **L. hirsu'ta,** Muhl. Leaves runcinate, the midrib beneath often sparingly bristly-hairy. Flowers yellowish-purple, rarely white.—Dry soil.

57. MULGE'DIUM, Cass. FALSE OR BLUE LETTUCE.

M. leucophæ'um, DC. (*Lactuca leucophœa.* Gray, in Macoun's Catalogue.) Stem tall and very leafy. Heads in a dense compound panicle.—Borders of damp woods, and along fences.

58. SON'CHUS, L. SOW-THISTLE.

1. **S. olera'ceus,** L. (COMMON SOW-THISTLE.) Stem-leaves runcinate, slightly toothed with soft spiny teeth, clasping; *the auricles acute.*—Manured soil about dwellings.

2. **S. asper,** Vill. (SPINY-LEAVED S.) Leaves *hardly lobed,* fringed with soft spines, clasping; *the auricles rounded.* Achenes *margined.*—Same localities as No. 1.

3. **S. arven'sis**, L., (Field S.) with bright yellow flowers and bristly involucres and peduncles, is found eastward.

59. TRAGOPO'GON, L. Goat's Beard. Salsify.

1. **T. praten'sis**, L. (Yellow Goat's Beard.) Spreading westward along the railway lines. Flowers *yellow*. Peduncle little thickened below the head.

2. **T. porrifo'lius**, L. (Salsify.) Stem 2-3 feet high. Peduncle thickened below the head. Flowers *purple*.

Order LI. LOBELIA'CEÆ. (Lobelia Family.)

Herbs with milky acrid juice, alternate leaves, and loosely racemed flowers. Corolla irregular, 5-lobed, *the tube split down one side*. Stamens 5, syngenesious, and commonly also monadelphous, *free from the corolla*. Calyx tube adherent to the many-seeded ovary. Style 1. The only genus is

LOBE'LIA, L. Lobelia.

1. **L. cardina'lis**, L. (Cardinal Flower.) Corolla *large, deep red*. Stem simple, 2-3 feet high, smooth. Leaves oblong-lanceolate, slightly toothed. Bracts of the flowers leaf-like. —Low grounds.

2. **L. syphilit'ica**, L. (Great Lobelia.) Corolla *rather large, light blue*. Stem hairy, simple, 1-2 feet high. Leaves thin, acute at both ends, serrate. Calyx-lobes half as long as the corolla, the tube hemispherical. Flowers in a dense spike or raceme.—Low grounds.

3. **L. infla'ta**, L. (Indian Tobacco.) Flowers *small, ⅛ of an inch long, pale blue*. Stem leafy, *branching*, 8-18 inches high, pubescent. Leaves ovate or oblong, toothed. Pods *inflated*. Racemes leafy.—Dry fields.

4. **L. spica'ta**, Lam. Flowers small, ⅛ *of an inch long, pale blue*. Stem slender, erect, *simple*, 1-3 feet high, minutely pubescent below. Leaves barely toothed, the lower spathulate or obovate, the upper reduced to linear bracts.— *Racemes long and naked*.—Sandy soil.

5. **L. Kal'mii,** L. Flowers small, ⅛ of an inch long, *light blue. Stem low*, 4–18 inches high, *very slender*. Pedicels filiform, as long as the flowers, with 2 minute bractlets above the middle. Leaves mostly linear, the radical ones spathulate and the upper ones reduced to bristly bracts.— Wet rocks and banks, chiefly northward.

6. **L. Dortman'na,** L., (WATER LOBELIA) with small leaves, all tufted at the root, and a scape 5 or 6 inches long with a few small light-blue pedicelled flowers at the summit, occurs in the shallow borders of ponds in Muskoka.

ORDER LII. **CAMPANULA'CEÆ.** (CAMPANULA F.)

Herbs with milky juice, *differing from the preceding Order chiefly in having a regular 5-lobed corolla (bell-shaped or wheel-shaped), separate stamens (5), and 2 or more (with us, 3) stigmas.*

Synopsis of the Genera.

1. **Campan'ula.** Calyx 5-cleft. Corolla nearly wheel-shaped, 5-lobed. Pod short.
2. **Specula'ria.** Calyx 5-cleft. Corolla nearly wheel-shaped, 5-lobed. Pod prismatic or oblong.

1. CAMPAN'ULA, Tourn. BELL-FLOWER.

1. **C. rotundifo'lia,** L. (HAREBELL.) Flowers blue, loosely panicled, on long slender peduncles, nodding. Stem slender, branching, several-flowered. Root-leaves round-heart-shaped; stem-leaves linear. Calyx-lobes awl-shaped. —Shaded banks.

2. **C. aparinoi'des,** Pursh. (MARSH BELL-FLOWER.) Flowers *white* or nearly so, about ⅛ of an inch long. *Stem very slender and weak*, few-flowered, *angled, roughened backwards*. Leaves linear-lanceolate. Calyx-lobes triangular. —Wet places in high grass. This plant has the habit of a Galium.

3. **C. America'na,** L. (TALL BELL-FLOWER.) Flowers light blue, about an inch across, *crowded in a leafy spike*. Corolla deeply 5-lobed. *Style long and curved*. Stem 3–6

feet high, simple. Leaves ovate or ovate-lanceolate, taper-pointed, serrate.—Moist rich soil.

4. **C. rapunculoi'des**, L. Flowers nodding, single in the axils of bracts, forming a raceme. Stem-leaves pointed, lanceolate, serrate; the lower cordate, long-petioled.—Atl. Prov.; int. from Europe.

5. **C. Scheuchz'eri**, Vill. Stem *low*, from a filiform rootstock, bearing usually *a single erect flower*. Leaves 1-2 inches long, lanceolate to linear-lanceolate, all more or less dentate. Calyx-lobes much longer than the tube, and exceeding the tube of the. *shallow reddish-purple corolla*.—N.W.

2. **SPECULA'RIA**, Heister. VENUS'S LOOKING-GLASS.

S. perfolia'ta, A. DC. Flowers purplish-blue, only the latter or upper ones expanding. Stem hairy, 3-20 inches high. Leaves roundish or ovate, clasping. Flowers solitary or 2 or 3 together in the axils.—Sterile open ground, chiefly south-westward.

ORDER LIII. **ERICA'CEÆ.** (HEATH FAMILY.)

Chiefly shrubs, *distinguished by the anthers opening, as a rule, by a pore at the top of each cell.* Stamens (as in the two preceding Orders) free from the corolla, as many or twice as many as its lobes. Leaves simple and usually alternate. Corolla in some cases polypetalous.

Synopsis of the Genera.

SUBORDER I. **VACCINIEÆ.** (WHORTLEBERRY FAMILY.)

Calyx-tube adherent to the ovary. Fruit a berry crowned with the calyx-teeth.

1. **Gaylussa'cia.** Stamens 10, the anthers opening by a pore at the apex. Corolla tubular, ovoid, the border 5-cleft. *Berry 10-celled, 10-seeded.* Flowers white with a red tinge. *Leaves covered with resinous dots.* Branching shrubs.
2. **Vaccin'ium.** Stamens 8 or 10, the anthers prolonged upwards into tubes with a pore at each apex. Corolla deeply 4-parted and revolute, or cylindrical with the limb 5-toothed. Berry 4-celled, or more or less completely 10-celled. Flowers white or reddish, solitary or in short racemes. Shrubs.

3. **Chiog'enes.** Stamens 8, *each anther 2-pointed at the apex.* Corolla bell-shaped, *deeply 4-cleft,* Limb of the calyx 4-parted. Flowers very small, nodding from the axils, with 2 bractlets under the calyx. *Berry white,* 4-celled. A trailing slender evergreen.

SUBORDER II. **ERICINEÆ.** (HEATH FAMILY PROPER.)

Calyx free from the ovary. Shrubs or small trees. Corolla gamopetalous, except in No. 11.

4. **Arctostaph'ylos.** Corolla urn-shaped, the limb 5-toothed, revolute. Stamens 10, the anthers each with 2 reflexed awns on the back. Fruit a berry-like drupe, 5-10-seeded. A trailing thick-leaved evergreen, with nearly white flowers.

5. **Epigæ'a.** Corolla salver-shaped, hairy inside, rose-coloured. Stamens 10; filaments slender; anthers awnless, *opening lengthwise.* Calyx of 5 pointed and scale-like nearly distinct sepals. A trailing evergreen, bristly with rus'y hairs.

6. **Gaulthe'ria.** Corolla ovoid or slightly urn-shaped, 5-toothed, nearly white. Stamens 10, the anthers 2-awned. Calyx 3-cleft, closing the pod and *becoming fleshy and berry-like in fruit.* Stems low and slender, leafy at the summit.

7. **Cassan'dra.** Corolla cylindrical, 5-toothed. Stamens 10, the anther-cells tapering into beaks with a pore at the apex, awnless. Calyx of 5 overlapping sepals, and 2 similar bractlets. Pod with a double pericarp, the outer of 5 valves, the inner cartilaginous and of 10 valves. A low shrub, with rather scurfy leaves, and white flowers.

8. **Androm'eda.** Corolla globular-urn-shaped, 5-toothed. Calyx of 5 nearly distinct valvate sepals, without bractlets. Stamens 10; the filaments bearded; the anther-cells each with a slender awn. A low shrub, with white flowers in a terminal umbel.

9. **Callu'na.** Corolla bell-shaped, 4-parted, persistent, becoming scarious. A low evergreen shrub, with numerous minute opposite leaves. Flowers rose-coloured or white, in mostly 1-sided racemes.

10. **Kal'mia.** Corolla broadly bell-shaped, *with 10 pouches receiving as many anthers.* Shrubs with showy rose-purple flowers.

11. **Le'dum.** Calyx 5-toothed, very small. Corolla of 5 obovate and spreading *distinct petals.* Stamens 5-10. Leaves evergreen, *with revolute margins, covered beneath with rusty wool.*

12. **Rhododen'dron.** Corolla irregular (in our species), nearly an inch long, 2-lipped, the upper lip 3-lobed, the lower of 2 oblong-linear curved nearly or quite distinct petals. Stamens 10, as long as the rose-coloured corolla. A shrub with alternate oblong somewhat pubescent leaves.—Atl. Prov.

ERICACEÆ.

SUBORDER III. **PYROLEÆ.** (PYROLA FAMILY.)

Calyx free from the ovary. Corolla polypetalous. More or less herbaceous evergreens.

13. **Py'rola.** Calyx 2-parted. Petals 5, concave. Stamens 10. Stigma 5-lobed. Leaves evergreen, *clustered at the base of an upright scaly-bracted scape which bears a simple raceme of nodding flowers.*

14. **Mone'ses.** Petals 5, orbicular, spreading. Stamens 10. Stigma large, peltate, with 5 narrow radiating lobes. Plant having the aspect of a Pyrola, *but the scape bearing a single terminal flower.*

15. **Chimaph'ila.** Petals 5, concave, orbicular, spreading. Stamens 10. Stigma broad and round, the border 5-crenate. Low plants with running underground shoots, and thick, shining, sharply serrate, somewhat whorled leaves. Flowers corymbed or umbelled on a terminal peduncle.

SUBORDER IV. **MONOTROPEÆ.** (INDIAN-PIPE FAMILY.)

16. **Monot'ropa.** A smooth *perfectly white* plant, parasitic on roots. bearing scales instead of leaves, and a single flower at the summit of the stem.

17. **Pteros'pora.** A purplish-brown clammy-pubescent plant, *parasitic* on the roots of pines. Stem simple. Flowers numerous, nodding, white, forming a raceme.

18. **Hypop'itys.** A tawny or reddish *parasitic* plant, with several flowers in a scaly raceme, the terminal one generally with 5 petals and 10 stamens, and the others with 4 petals and 8 stamens.

1. GAYLUSSA'CIA, H.B.K. HUCKLEBERRY.

1. **G. resino'sa,** Torr and Gr. (BLACK HUCKLEBERRY.) Fruit *black, without a bloom.* Racemes short, 1-sided, in clusters. Leaves oval or oblong. Branching shrub, 1-3 feet high.—Low grounds.

2. **G. dumo'sa,** Torr. and Gr. (DWARF HUCKLEBERRY.) Fruit black, insipid. Racemes long, with leaf-like, persistent bracts. Leaves obovate, oblong, mucronate.—Sandy low ground, Atl. Prov.

2. VACCIN'IUM, L. CRANBERRY. BLUEBERRY.

1. **V. Oxycoc'cus,** L. (*Oxycoccus vulgaris,* Pursh, in Macoun's Catalogue.) (SMALL CRANBERRY.) A creeping or

trailing very slender shrubby plant, with ovate acute evergreen leaves only ¼ of an inch long, the margins revolute. Corolla rose-coloured, 4-parted, the lobes reflexed. Anthers 8. *Stem 4–9 inches long. Berry only about ¼ of an inch across*, often speckled with white.—Bogs.

2. **V. macrocar'pon,** Ait. (*Oxycoccus macrocarpus*, Pursh, in Macoun's Catalogue.) (LARGE or AMERICAN CRANBERRY.) Different from No. 1 in having prolonged stems (1–3 feet long) and the flowering branches lateral. The leaves also are nearly twice as large, and the *berry is fully ½ an inch broad.*—Bogs.

3. **V. Vitis-Idæa,** L. A low plant with erect branches from tufted creeping stems. Leaves evergreen, *obovate*, with revolute margins, shining above, dotted with blackish bristly points beneath. Corolla *bell-shaped*, 4-lobed. Anthers 8–10. Flowers in a short bracted raceme.—Northward and eastward.

4. **V. Pennsylvan'icum,** Lam. (DWARF BLUEBERRY.) Stem 6–15 inches high, the branches green, angled and warty. Corolla cylindrical, bell-shaped, 5-toothed. Anthers 10. Flowers in short racemes. Leaves lanceolate or oblong, serrulate with bristly-pointed teeth, smooth and shining on both sides. Berry blue or black, with a bloom —Dry plains and woods.

5. **V. Canaden'se,** Kalm. (CANADIAN BLUEBERRY.) Stem 1–2 feet high. Leaves oblong-lanceolate or elliptical, *entire, downy both sides*, as are also the branchlets.—A very common Canadian species.

6. **V. vacil'lans,** Solander. (Low BLUEBERRY.) Stem 1–2 feet high, *glabrous*, with yellowish-green branchlets. Leaves obovate or oval, *very pale or dull* and *glaucous*, at least beneath. Corolla between bell-shaped and cylindraceous, the mouth somewhat contracted.—Dry soil; western Ontario mostly.

7. **V. corymbo'sum,** L., (SWAMP BLUEBERRY) is a tall shrub (8–10 feet) growing in swamps and low grounds, with

leaves varying from ovate to elliptical-lanceolate, and flowers and berries very much the same as those in No. 4, but the berries ripen later.

Var. **amœ'num,** Gray, has the leaves bristly-ciliate, green both sides, shining above.—S.W. Ontario.

Var. **pal'lidum,** Gray, has the leaves mostly glabrous, pale or whitish, glaucous at least underneath, and serrulate with bristly teeth.—Atl. Prov.; also Niagara River.

Var. **atrococ'cum,** Gray, has the leaves entire, downy, or woolly underneath, as well as the branchlets.—Chiefly eastward.

8. **V. stamin'eum,** L. (DEER-BERRY. SQUAW HUCKLEBERRY.) Stem diffusely branching, 2-3 feet high. Leaves ovate or oval, pale, whitish beneath. Corolla open-bell-shaped, 5-lobed, greenish-white or purplish. Anthers 2-awned on the back, much exserted. Flowers slender-pedicelled. Berries greenish or yellowish, large. — Niagara River, above Queenston.

3. CHIOG'ENES, Salisb. CREEPING SNOWBERRY.

C. hispid'ula, Torr. and Gr. Leaves very small, ovate and pointed, on short petioles, the margins revolute. The lower surface of the leaves and the branches clothed with rusty bristles. Berries bright white.—Bogs and cool woods.

4. ARCTOSTAPH'YLOS, Adans. BEARBERRY.

A. Uva-ursi, Spreng. Flowers in terminal racemes. Leaves alternate, obovate or spathulate, entire, smooth. Berry red.—Bare hillsides.

5. EPIGÆ'A, L. GROUND LAUREL. TRAILING ARBUTUS.

E. re'pens, L. (MAYFLOWER.) Flowers in small axillary clusters from scaly bracts. Leaves evergreen, rounded and heart-shaped, alternate, on slender petioles. Flowers very fragrant.—Dry woods in early spring.

6. GAULTHE'RIA, Kalm. AROMATIC WINTERGREEN.

G. procum'bens, L. (TEABERRY. WINTERGREEN). Flowers mostly single in the axils, nodding. Leaves obovate or

oval, obscurely serrate, evergreen. Berry bright red, edible.
—Cool woods, chiefly in the shade of evergreens.

7. CASSAN'DRA, Don. LEATHER-LEAF.

C. calycula'ta, Don. Flowers in 1-sided leafy racemes. Leaves oblong, obtuse, flat.—Bogs.

8. ANDROM'EDA, L. ANDROMEDA.

A. polifo'lia, L. Stem smooth and glaucous, 6–18 inches high. Leaves oblong-linear, with strongly revolute margins, white beneath.—Bogs.

9. CALLU'NA, Salisb. HEATHER.

C. vulga'ris, Salisb. (HEATHER.) A low evergreen shrub with numerous opposite minute leaves, mostly auricled at the base. Flowers axillary or terminating very short shoots, forming close racemes mostly one-sided, rose-coloured or white. Calyx of 4 sepals. Corolla 4-parted, bell-shaped. Calyx and corolla both persistent and becoming dry. Stamens 8. Capsule 4-celled.—Found sparingly in a few places on the coast of Nova Scotia and Newfoundland.

10. KAL'MIA, L. AMERICAN LAUREL.

I. K. glau'ca, Ait. (PALE LAUREL.) A straggling shrub about a foot high, with few-flowered terminal corymbs. Branchlets 2-edged. Leaves opposite, oblong, the margins revolute. Flowers ½ an inch across.—Bogs.

2. K. angustifo'lia, L. (SHEEP LAUREL. LAMBKILL.) Leaves opposite or in threes, oblong, obtuse, petioled. Corymbs *lateral, many-flowered*. Pod depressed. *Pedicels recurved in fruit.*—Bogs and damp barren grounds, abundant eastward.

11. LE'DUM, L. LABRADOR TEA.

L. latifo'lium, Ait. Flowers white, in terminal umbel-like clusters. Leaves elliptical or oblong. Stamens 5, or occasionally 6 or 7.—Bogs.

12. RHODODEN'DRON, L. (ROSE BAY. AZALEA.)

R. Rhodo'ra, Don. Corolla irregular, nearly an inch long, two-lipped; the upper lip 3-lobed; the lower lip of two oblong-linear, curved, nearly or quite distinct petals. Stamens 10, as long as the rose-coloured corolla. Leaves alternate, oblong, somewhat pubescent. Shrub.—Bogs and damp barrens, from Montreal eastward.

13. PY'ROLA, Tourn. WINTERGREEN. SHIN-LEAF.

* *Style straight, narrower than the 5-rayed stigma.*

1. **P. mi'nor**, L. Leaves roundish, slightly crenulate, thickish, usually longer than the margined petiole. *Raceme not one-sided.* Flowers white or rose-colour. *Style short and included in corolla.*—Cold woods, Atl. Prov. and northward.

2. **P. secun'da**, L. Easily recognized by the *flowers of the dense raceme being all turned to one side.* Leaves ovate. Style long, protruding.—Rich woods. Var. **pu'mila** has *orbicular* leaves, and is 3-8-flowered.—Peat-bogs and swamps.

* * *Style declined, the apex curved upward. Stigma narrower than the ring-like apex of the style.*

3. **P. rotundifo'lia**, L. *Leaves orbicular, thick, shining, usually shorter than the petiole.* Calyx-lobes lanceolate. Flowers white, or in var. **incarna'ta** rose-purple.—Moist woods.

Var. **asarifo'lia**, Hook., has round-reniform leaves, and mostly rose-coloured petals.

Var. **uligino'sa**, Gray, has broadly ovate calyx-lobes, mostly obovate dull leaves, and flesh-coloured petals.

4. **P. ellip'tica**, Nutt. (SHIN-LEAF.) *Leaves elliptical, thin, dull, usually longer than the margined petiole.* Flowers greenish-white.—Rich woods.

5. **P. chloran'tha**, Swartz, has small roundish dull leaves, converging greenish-white petals, and *the anther-cells contracted below the pore into a distinct neck or horn.*—Open woods.

14. MONE'SES, Salisb. ONE-FLOWERED PYROLA.

M. uniflo'ra, Gr. (*M. grandiflora*, Salisb.) Leaves thin, rounded, veiny, and serrate. Scape 2-4 inches high, bearing a single white or rose-coloured flower.—Deep woods.

15. CHIMAPH'ILA, Pursh. PIPSISSEWA.

1. C. umbella'ta, Nutt. (PRINCE'S PINE.) Leaves wedge-lanceolate, *acute at the base.* Peduncles 4-7-flowered. Corolla rose- or flesh-coloured.—Dry woods.

2. C. macula'ta, Pursh. (SPOTTED WINTERGREEN.) Leaves ovate-lanceolate, *obtuse at the base,* the upper surface *variegated with white.*—Dry woods.

16. MONOT'ROPA, L. INDIAN-PIPE. PINE-SAP.

M. uniflo'ra, L. (INDIAN-PIPE. CORPSE-PLANT.) Smooth, waxy-white, turning black in drying.—Dark rich woods.

17. PTEROS'PORA, Nutt. PINE-DROPS.

P. Andromede'a, Nutt. Calyx 5-parted. Corolla ovate, urn-shaped, 5-toothed, persistent. Stamens 10. Stigma 5-lobed. Pod 5-lobed, 5-celled.—Usually under pines in dry woods.

18. HYPOP'ITYS, Scop. PINE-SAP.

H. lanugino'sa, Nutt. Somewhat pubescent. Sepals bract-like. Stigma ciliate. Style longer than the ovary, hollow. Pod globular or oval.—Oak and pine woods.

ORDER LIV. PLUMBAGINA'CEÆ. (LEADWORT F.)

Maritime herbs with regular pentamerous flowers, a plaited calyx, 5 stamens *opposite* the lobes (*or separate petals*) of the corolla, and a 1-celled and 1-seeded ovary.

STAT'ICE, Tourn.

S. Limo'nium, L. (MARSH-ROSEMARY.) A maritime herb, with a thick, woody, astringent root, and oblong, spathulate or obovate-lanceolate radical leaves, tipped with a deciduous bristle. Flowers lavender-colour, panicled on branching scapes. Calyx funnel-form, membranaceous. Corolla of 5 *nearly or quite distinct* petals, with the 5 stamens severally borne on their bases. Ovary 1-celled and 1-ovuled.—Salt marshes, Atl. Prov.

ORDER LV. **AQUIFOLIA'CEÆ.** (HOLLY FAMILY.)

Shrubs or small trees, with small axillary polygamous or diœcious flowers, the parts mostly in fours or sixes. Calyx very minute, free from the ovary. Stamens alternate with the petals, attached to their base, the corolla being almost polypetalous. Anthers opening lengthwise. Stigma nearly sessile. Fruit a berry-like 4–8-seeded drupe.

1. ILEX, L. HOLLY.

I. verticilla'ta, Gr. (BLACK ALDER. WINTERBERRY.) A shrub with the greenish flowers in sessile clusters, or the fertile ones solitary. Parts of the flowers mostly in sixes. Fruit bright red. Leaves alternate, obovate, oval, or wedge-lanceolate, pointed, veiny, serrate.—Swamps and low grounds.

2. NEMOPAN'THES, Raf. MOUNTAIN HOLLY.

N. Canadensis, DC. A branching shrub, with grey bark, and alternate oblong nearly entire smooth leaves on slender petioles. Flowers on long slender axillary peduncles, mostly solitary. Petals 4–5, oblong-linear, *distinct*. Fruit light red.—Moist woods.

ORDER LVI. **PRIMULA'CEÆ.** (PRIMROSE FAMILY.)

Herbs with regular perfect flowers, well marked by having *a stamen before each petal or lobe of the corolla* and inserted on the tube. Ovary 1-celled, the placenta rising from the base. Style 1; stigma 1.

Synopsis of the Genera.
Stemless. Leaves all in a cluster from the root.

1. **Prim'ula.** Flowers in an umbel at the summit of a simple scape. Corolla salver-shaped or funnel-form, open at the throat. Stamens 5, included.
2. **Andros'ace.** Flowers very small, white, in an umbel at the summit of a scape. Corolla salver-shaped or funnel-form, constricted at the throat. Stamens included.—N.W.
3. **Dodecath'eon.** Flowers showy, rose-coloured or white, in an umbel at the summit of a scape. Corolla reflexed, 5-parted. Stamens exserted, connivent in a slender cone.—N.W.

** *Stems leafy. Corolla rotate, wanting in* Glaux.

4. **Trienta'lis.** Leaves in a whorl at the summit of a slender erect stem. Calyx usually 7-parted, the lobes pointed. Corolla usually 7-parted, spreading, without a tube. Filaments united in a ring below. Flowers usually only one, white and star-shaped.

5. **Lysimach'ia.** Leafy-stemmed. Flowers yellow, axillary or in a terminal raceme. Calyx usually 5-parted. Corolla wheel-shaped, mostly 5-parted, and sometimes polypetalous.

6. **Glaux.** A fleshy herb with purplish and white axillary flowers. *Corolla wanting*, the calyx petal-like.—Atl. Prov.

7. **Anagal'lis.** Low and spreading. Leaves opposite or whorled, entire. Flowers variously coloured, solitary in the axils. Calyx 5-parted. Corolla wheel-shaped, 5-parted. Filaments bearded.

8. **Sam'olus.** Smooth and spreading, 6-10 inches high. Corolla bell-shaped, 5-parted, with 5 sterile filaments in the sinuses. *Calyx partially adherent to the ovary.* Flowers very small, white, racemed. *Leaves alternate.*

1. PRIM'ULA, L. PRIMROSE. COWSLIP.

1. **P. farino'sa,** L. (BIRD'S-EYE P.) Lower surface of the leaves covered with a white mealiness. Corolla lilac with a yellow centre.—Shores of Lake Huron and northward.

2. **P. Mistassin'ica,** Michx. Leaves not mealy. Corolla flesh-coloured, the lobes obcordate.—Shores of the Upper Lakes, and northward.

2. ANDROS'ACE, Tourn.

A. septentriona'lis, L. Almost glabrous, 2-10 inches high. Leaves lanceolate to oblong-lanceolate, narrowed at the base, toothed. Bracts of the involucre awl-shaped.—N.W.

3. DODECATH'EON, L. AMERICAN COWSLIP.

D. Mead'ia, L. (SHOOTING-STAR.) Smooth. Leaves oblong or spathulate. Flowers nodding on slender pedicels.—N. W.

4. TRIENTA'LIS, L. CHICKWEED-WINTERGREEN.

T. America'na, Pursh. (STAR-FLOWER.) Leaves thin and veiny, lanceolate, tapering towards both ends. Petals pointed.—Moist woods.

5. LYSIMACH'IA, Tourn. LOOSESTRIFE.

1. **L. thyrsiflo'ra,** L. (TUFTED LOOSESTRIFE.) Flowers in spike-like clusters from the axils of a few of the upper leaves. Petals lance-linear, *purplish-dotted*, as many minute teeth between them. Leaves scale-like below, the upper lanceolate, opposite, sessile, dark-dotted.—Wet swamps.

2. **L. stricta,** Ait. Flowers on slender pedicels *in a long terminal raceme.* Petals lance-oblong, streaked with dark lines. *Leaves opposite*, lanceolate, acute at each end, sessile, dark-dotted.—Low grounds.

3. **L. quadrifo'lia,** L. *Flowers on long slender peduncles from the axils of the upper leaves.* Petals streaked. *Leaves in whorls* of 4 or 5, ovate-lanceolate, dark-dotted.—Sandy soil.

4. **L. cilia'ta,** L. (*Steironema ciliatum*, Raf., in Macoun's Catalogue.) Flowers nodding on slender peduncles from the upper axils. *Petals not streaked or dotted.* Leaves opposite, *not dotted*, ovate-lanceolate, pointed, cordate at the base, *on long fringed petioles.*—Low grounds.

5. **L. longifo'lia,** Walt. (*Steironema longifolium*, Gray, in Macoun's Catalogue.) Petals not streaked or dotted. *Stem-leaves sessile, narrowly linear, 2-4 inches long*, the margins sometimes revolute. Stem 4-angled.—Moist soil, western Ontario.

6. **L. lanceola'ta,** Walt. (*Steironema lanceolatum*, Gray.) Stem erect, 1-2 feet high. Leaves lanceolate to oblong and linear, narrowed into a short margined petiole, or the lowest short and broad on long petioles, not dotted. Petals not streaked or dotted.—S. W. Ontario.

7. **L. nummula'ria,** L., (MONEY-WORT), has escaped from gardens in a few places. Stem trailing and creeping, smooth. Leaves roundish, small, opposite. Peduncles axillary, 1-flowered.—Damp places.

6. GLAUX, Tourn.

G. marit'ima, L. (SEA-MILKWORT.) A fleshy herb, with usually opposite, oblong, entire, sessile leaves. Flowers single in the axils, nearly sessile. Calyx bell-shaped, 5-cleft, purplish and white. Corolla wanting. Stamens 5, on the base of the calyx. Capsule 5-valved, few-seeded.—Sea-shore, Atl. Prov.

7. **ANAGAL'LIS,** Tourn. PIMPERNEL.

A. arven'sis, L. (COMMON PIMPERNEL.) Petals obovate, *fringed with minute teeth*, mostly bluish or purplish. Flowers closing at the approach of rain. Leaves ovate, sessile.— Sandy fields and garden soil.

8. **SAM'OLUS,** L. Tourn. WATER-PIMPERNEL. BROOK-WEED.

S. Valeran'di, L., var. **America'nus,** Gray. Stem slender, diffusely branched. The slender pedicels each with a bractlet at the middle.—Wet places, not common.

ORDER LVII. **PLANTAGINA'CEÆ.** (PLANTAIN FAMILY.)

Herbs, with the leaves all radical, and the flowers in a close spike at the summit of a naked scape. Calyx of 4 sepals, persistent. Corolla 4-lobed, thin and membranaceous, spreading. Stamens 4, usually with long filaments, inserted on the corolla. Pod 2-celled, the top coming off like a lid. Leaves ribbed. The principal genus is

PLANTA'GO, L. PLANTAIN. RIB-GRASS.

1. **P. major,** L. (COMMON P.) *Spike long and slender. Leaves 5-7-ribbed,* ovate or slightly heart-shaped, *with channelled petioles. Pod 7-16-seeded.*—Moist ground about dwellings.

2. **P. Kamtschat'ica,** Hook. (*P. Rugelii,* Decaisne, in Macoun's Catalogue.) Resembling small forms of No. 1, but *pod 4-seeded.*

3. **P. lanceola'ta,** L. (RIB-GRASS. ENGLISH PLANTAIN.) *Spike thick and dense, short.* Leaves 3-5-ribbed, lanceolate or lanceolate-oblong. Scape grooved, long and slender.— Dry fields and banks.

4. **P. corda'ta,** Lam. Tall and glabrous. *Bracts round-ovate, fleshy.* Pod 2-4-seeded.—South-western Ontario.

5. **P. marit'ima,** L., var. **juncoi'des,** Gr., with very narrow and slender spike, and linear fleshy leaves, is found on the sea-coast and Lower St. Lawrence.

6. **P. eriop'oda,** Torr. Usually a mass of *yellowish wool* at the base. Leaves thickish, oblanceolate to obovate, with stout short petioles. Pod never more than 4-seeded.—Atl. sea-coast and N.W.

7. **P. Patagon'ica,** Jacq., var. **gnaphalioides,** Gray. *White with silky wool.* Leaves oblong-linear to filiform. Spike very dense, *woolly.*—Dry soil, N.W.

ORDER LVIII. **LENTIBULARIACEÆ.** (BLADDERWORT F.)

Small aquatic or marsh herbs, with a 2-lipped calyx and a personate corolla with a spur or sac underneath. Stamens 2. Ovary as in Primulaceæ. Chiefly represented by the two following genera :—

1. **UTRICULA'RIA,** L. BLADDERWORT.

1. **U. vulga'ris,** L. (GREATER BLADDERWORT.) Immersed leaves crowded, finely dissected into capillary divisions, furnished with small air-bladders. Flowers yellow, several in a raceme on a naked scape. Corolla closed; the spur conical and shorter than the lower lip.—Ponds and slow waters.

2. **U. interme'dia,** Hayne. Immersed leaves 4 or 5 times forked, the divisions linear-awl-shaped, minutely bristle-toothed on the margin, *not bladder-bearing*, the bladders being on leafless branches. Stem 3-6 inches long. Scape very slender, 3-6 inches long, bearing few yellow flowers. Upper lip of the corolla much longer than the palate; *the spur closely pressed to the broad lower lip.*—Shallow waters.

3. **U. cornu'ta,** Michx., with an awl-shaped spur turned downward and outward, and the lower lip of the corolla helmet-shaped, is not uncommon in the northern parts of Ontario. Flowers yellow. Leaves awl-shaped.

4. **U. clandesti'na,** Nutt. Stems and scapes slender. Leaves hair-like, bearing small bladders. Corolla yellow;

lower lip 3-lobed, longer than the thick, blunt spur. Submerged stems bearing cleistogamous flowers.—Ponds, Atl. Prov.

5. **U. gib'ba,** L. Scape only 1-3 inches high, 1-2-flowered, with very slender short branches at the base, bearing capillary root-like leaves and scattered bladders. Corolla yellow, the lower lip with sides reflexed. Spur very thick and blunt, conical, gibbous.—Shallow water, central and S. W. Ontario.

2. PINGUIC'ULA, L. BUTTERWORT.

P. vulga'ris, L. A small and stemless perennial growing on damp rocks. Scapes 1-flowered. Leaves entire, ovate or elliptical, soft-fleshy, clustered at the root. Upper lip of the calyx 3-cleft, the lower 2-cleft. Corolla violet, the lips very unequal, the palate open, and hairy or spotted.—Shore of Lake Huron.

Order LIX. OROBANCHA'CEÆ. (Broomrape F.)

Parasitic herbs, destitute of green foliage. Corolla more or less 2-lipped. Stamens didynamous. Ovary 1-celled with 2 or 4 parietal placentæ, many-seeded.

1. EPIPHE'GUS, Nutt. (Beech-drops.)

E. Virginia'na, Bart. A yellowish-brown branching plant, parasitic on the roots of beech-trees. Flowers racemose or spiked; the upper sterile, with long corolla; the lower fertile, with short corolla.

2. CONOPH'OLIS, Wallroth. Squaw-root.

C. America'na, Wallroth. A chestnut-coloured or yellow plant found in clusters in oak woods in early summer, 3-6 inches high and rather less than an inch in thickness. The stem covered with fleshy scales so as to resemble a cone. Flowers under the upper scales; stamens projecting.

3. APHYL'LON, Mitchell. Naked Broom-rape. Cancer-root.

1. **A. uniflo'rum,** Torr. and Gr. Plant yellowish-brown. Flower solitary at the top of a naked scape. *Stem subterranean or nearly so,* short and scaly. Scapes 3-5 inches

high. Calyx 5-cleft, the divisions lance-awl-shaped. Corolla with a long curved tube and 5-lobed border, and 2 yellow-bearded folds in the throat. Stigma 2-lipped.—Woods, in early summer.

2. **A. fascicula'tum**, Gray. *Scaly stem erect*, and rising 3 or 4 inches above the ground, mostly longer than the crowded peduncles.—N.W.; parasitic on Artemisia, etc.

ORDER LX. **SCROPHULARIA'CEÆ.** (FIGWORT F.)

Herbs distinguished by a 2-lipped or more or less irregular corolla, stamens usually 4 and didynamous, or only 2, (or in Verbascum 5) and a 2-celled and usually many-seeded ovary. Style 1; stigma entire or 2-lobed.

Synopsis of the Genera.

* *Corolla wheel-shaped, and only slightly irregular.*

1. **Verbas'cum.** *Stamens (with anthers) 5.* Flowers in a long terminal spike. Corolla 5-parted, nearly regular. Filaments (or some of them) woolly.
2. **Veron'ica.** *Stamens only 2*; filaments long and slender. Corolla mostly 4-parted, nearly or quite regular. Pod flattish. Flowers solitary in the axils, or forming a terminal raceme or spike.

** *Corolla 2-lipped, or tubular and irregular.*

— Upper lip of the corolla embracing the lower in the bud, except occasionally in **Mimulus.**

3. **Lina'ria.** Corolla *personate* (Fig. 181, Part I.), *with a longer spur beneath.* Stamens 4. Flowers yellow, in a crowded raceme.
4. **Scrophula'ria.** Corolla tubular, somewhat inflated, 5-lobed; *the 4 upper lobes erect*, the lower one spreading. Stamens with anthers 4, the rudiment of a fifth in the form of a scale on the upper lip of the corolla. Flowers small and dingy, forming a narrow terminal panicle. Stem 4-sided.
5. **Collins'ia.** Corolla 2-cleft, the short tube saccate on the upper side; the middle lobe of the lower lip sac-like and enclosing the 4 declined stamens; the upper lip 2-cleft, the lobes partly turned backward. Fifth stamen rudimentary. Leaves opposite. Flowers blue and white, in umbel-like clusters.
6. **Chelo'ne.** Corolla inflated-tubular (Fig. 189, Part I.). Stamens 4, with woolly filaments and anthers, and a fifth filament without an anther. Flowers white, in a close terminal spike.

7. **Penste'mon.** Corolla 2-lipped, gradually widening upwards. Stamens 4, with a fifth sterile filament, *the latter yellow-bearded.* Flowers white or purplish, in a loose panicle.

8. **Mim'ulus.** *Calyx 5-angled* and 5-toothed. Upper lip of the corolla erect or reflexed-spreading, the lower spreading, 3-lobed. Stamens 4, alike; *no rudiment of a fifth.* Stigma 2-lipped. Flowers blue or yellow, solitary on axillary peduncles.

9. **Grati'ola.** Corolla tubular and 2-lipped. *Stamens with anthers only 2,* included. Flowers with a yellowish tube, on axillary peduncles, solitary. Style dilated at the apex.

10. **Ilysan'thes.** Corolla tubular and 2-lipped. *Stamens with anthers only 2,* included; *also a pair of filaments which are two-lobed but without anthers.* Flowers purplish, axillary. Style 2-lipped at the apex.

← ← *Lower lip of the corolla embracing the upper in the bud.*

11. **Gerar'dia.** Corolla funnel-form, swelling above, *the 5 spreading lobes more or less unequal.* Stamens 4, *strongly didynamous,* hairy. Style long, enlarged at the apex. Flowers purple or yellow, solitary on axillary peduncles, or sometimes forming a raceme.

12. **Castille'ia.** Corolla tubular and 2-lipped, *its tube included in the tubular and flattened calyx;* the upper lip long and narrow and flattened laterally, the lower short and 3-lobed. Stamens 4, didynamous. *Floral leaves scarlet* (rarely yellow) in our species. Corolla pale yellow.

13. **Orthocarp'us.** Corolla tubular and 2-lipped. Calyx tubular-campanulate, 4-cleft. Upper lip of the corolla much narrower than the inflated lower one. Flowers golden yellow, in a dense spike. —N.W.

14. **Euphra'sia.** *Calyx 4-cleft.* Upper lip of the corolla erect, *2-lobed, the sides turned back;* the lower spreading. Stamens 4, under the upper lip. Very small herbs, with whitish or bluish spiked flowers. (Chiefly on the sea-coast, and north of Lake Superior).

15. **Bart'sia.** Calyx 4-cleft. Upper lip of the corolla *entire, the sides not turned back.* Flowers small, rose-red, in loose spikes. Small herbs with opposite sessile leaves.—Atl. sea-coast chiefly.

16. **Rhinan'thus.** *Calyx flat, greatly inflated in fruit,* 4-toothed. Upper lip of the corolla arched, flat, with a minute tooth on each side below the apex. Stamens 4. Flowers yellow, solitary in the axils, nearly sessile, the whole forming a crowded 1-sided spike. (Chiefly on the sea-coast, and north of Lake Superior.)

17. **Pedicula'ris.** Calyx split in front, not inflated in fruit. Corolla 2-lipped, the upper lip arched or hooded, incurved, flat, 2-toothed under the apex. Stamens 4. *Pod flat, somewhat sword-shaped.*

18. **Melampy'rum.** *Calyx 4-cleft, the lobes sharp-pointed.* Corolla greenish-yellow; upper lip arched, compressed, the lower 3-lobed at the apex. Stamens 4; anthers hairy. Pod 1-4-seeded, flat, oblique. Upper leaves larger than the lower ones and fringed with bristly teeth at the base.

1. VERBAS'CUM, L. MULLEIN.

1. **V. Thap'sus,** L. (COMMON MULLEIN.) A tall and very woolly herb, with the simple stem winged by the decurrent bases of the leaves. Flowers yellow, forming a dense spike. —Fields and roadsides everywhere.

2. **V. Blatta'ria,** L. (MOTH M.) Stem slender, *nearly smooth*. Lower leaves petioled, doubly serrate; the upper partly clasping. Flowers whitish with a purple tinge, *in a loose raceme. Filaments all violet-bearded.*—Roadsides; not common northward.

2. VERON'ICA, L. SPEEDWELL.

1. **V. America'na,** Schweinitz. (AMERICAN BROOKLIME.) Flowers pale blue, in *opposite* axillary racemes. *Leaves mostly petioled,* thickish, serrate. *Pod swollen.*—A common plant in brooks and ditches.

2. **V. anagal'lis,** L., (WATER SPEEDWELL) is much like No. 1, but *the leaves are sessile,* with a heart-shaped base.

3. **V. scutella'ta,** L. (MARSH S.) Flowers pale blue, in racemes, chiefly from *alternate* axils. *Leaves sessile,* linear, opposite, hardly toothed. Racemes 1 or 2, *slender and zigzag. Flowers few.* Pods very flat, notched at both ends.— Bogs.

Var. **pubescens,** Macoun, has hairy stem and leaves.

4. **V. officina'lis,** L. (COMMON S.) Flowers light blue. Stem prostrate, rooting at the base, *pubescent.* Leaves short-petioled, obovate-elliptical, serrate. *Racemes dense,* chiefly from alternate axils. Pod obovate-triangular, strongly flattened, notched.—Hillsides and open woods.

5. **V. chamæ'drys,** L. Flowers pale blue. Stem pubescent, at least in 2 lines, from a creeping base. Leaves nearly sessile, ovate or cordate, incisely crenate. Racemes

axillary, *loosely* flowered. Pod triangular-obcordate.—Chiefly eastward ; also at Niagara Falls.

6. **V. serpyllifo'lia,** L. (THYME-LEAVED S.) Flowers whitish or pale blue, in a loose *terminal* raceme. Stem nearly smooth, branched at the creeping base. Leaves obscurely crenate, *the lowest petioled*. Pod flat, notched.—Roadsides and fields. Plant only 2 or 3 inches high.

7. **V. peregri'na,** L. (NECKWEED.) Flowers whitish, *solitary in the axils of the upper leaves, short-pedicelled*. Corolla shorter than the calyx. Stem 4-9 inches high, *nearly smooth*. *Pod orbicular*, slightly notched.—Waste places and cultivated grounds.

8. **V. arven'sis,** L. (CORN SPEEDWELL.) Flowers (blue) as in No. 7, but *the stem is hairy*, and *the pod inversely heart-shaped*.—Cultivated soil.

9. **V. agres'tis,** L. (FIELD SPEEDWELL.) Leaves round or ovate, crenate, petioled. Flowers *small*, in the axils of the ordinary leaves, *long-pedicelled*. *Seeds few, cup-shaped.*—Sandy fields, Atl. Prov. (Int. from Eu.)

10. **V. Buxbaum'ii,** Tenore. Leaves as in the last. Flowers (blue) *large*, long-pedicelled, in the axils of the ordinary leaves. Calyx-lobes widely spreading in fruit. Pod obcordate-triangular, broadly notched, *many-seeded*.—Atl. Prov.

3. LINA'RIA, Tourn. TOAD-FLAX.

1. **L. vulga'ris,** Mill. (TOAD-FLAX. BUTTER-AND-EGGS.) Leaves crowded, linear, pale green. Corolla *pale yellow*, with a deeper yellow or orange-coloured palate.—Roadsides.

2. **L. Canaden'sis,** Spreng. (WILD TOAD-FLAX.) A slender herb, with linear, entire, alternate leaves. Flowers *blue*, small, in a naked, terminal raceme. Spur of corolla curved, filiform.—Sandy soil, Atl. Prov.

3. **L. cymbala'ria,** Mill. A delicate little trailing plant, with thickish 3-5-lobed leaves on long petioles, and small yellow and purple flowers.—A garden escape.

4. SCROPHULA'RIA, Tourn. FIGWORT.

S. nodo'sa, L., var. Marilandica, Gr. Stem smooth, 3–4 feet high. Leaves ovate or oblong, the upper lanceolate, serrate.—Damp thickets. (There appear to be two forms, one with broad leaves and greenish-brown flowers, and the other with narrow leaves and greenish-yellow flowers.)

5. COLLINS'IA, Nutt.

1. C. verna, Nutt. Slender, 6–20 inches high. Lower leaves ovate, upper ovate-lanceolate, clasping. Corolla (blue and white) twice as long as the calyx. Peduncles *long*.—Western Ontario.

2. C. parviflo'ra, Dougl. Smaller. Corolla (blue) very small, scarcely longer than the calyx. Peduncles *short*.— Chiefly N.W.

6. CHELO'NE, Tourn. TURTLE-HEAD.

C. glabra, L. Stem smooth, erect and branching. Leaves short-petioled, lance-oblong, serrate, opposite. Bracts of the flowers concave.—Wet places.

7. PENTSTE'MON, Mitchell. BEARD-TONGUE.

1. P. pubes'cens. Stem 1–3 feet high, pubescent; the panicle more or less clammy. Throat of the corolla almost closed. Stem leaves lanceolate, clasping.—Dry soil.

2. P. gra'cilis, Nutt. Stem 1 foot high or less, glabrous or nearly so, viscid-pubescent above. Corolla tubular-funnel-form, the throat open, lilac-purple or whitish.—N.W.

3. P. acumina'tus, Dougl. Stem 6–20 inches high, stout, *glabrous and glaucous*. Leaves thick, the lower obovate or oblong, the upper lanceolate to broadly ovate or clasping. Panicle narrow, leafy below.—N.W.

4. P. crista'tus, Nutt. Pubescent, and viscid-pubescent above. Leaves from linear-lanceolate to oblong. Corolla funnel-form, the lower lip long-villous within. *Sterile filament copiously yellow-bearded.*—N.W.

5. **P. confer'tus,** Dougl., var. **cæruleo-purpureus,** Gray, is at once recognized by the *interrupted spike-like panicle*, which consists of *2 to 5 dense verticillate clusters.* Corolla narrow, the lower lip conspicuously bearded within. —N.W.

8. MIM'ULUS, L. MONKEY-FLOWER.

1. **M. ring'ens,** L. Stem square, 1-2 feet high. *Corolla blue*, an inch long. (A white-flowered variety is sometimes met with.) Leaves oblong or lanceolate, clasping.—Wet places.

2. **M. Jamesii,** Torr. Stem creeping at the base. *Corolla yellow*, small. Leaves roundish or kidney-shaped, nearly sessile. Calyx inflated in fruit.—In cool springs.

9. GRATI'OLA. L. HEDGE-HYSSOP.

1. **G. Virginia'na,** L. Stem 4-6 inches high, clammy with minute pubescence above. *Sterile filaments minute or none.* Leaves lanceolate. Peduncles slender.—Moist places.

2. **G. au'rea,** Muhl. *Nearly glabrous.* Sterile filaments slender, tipped with a little head. *Corolla golden yellow.*

10. ILYSAN'THES, Raf. FALSE PIMPERNEL.

I. **gratiolol'des,** Benth. Stem 4-8 inches high, much branched, diffusely spreading. Leaves ovate, rounded or oblong, the upper partly clasping.—Wet places.

11. GERAR'DIA, L. GERARDIA.

1. **G. purpu'rea,** L. (PURPLE GERARDIA.) Corolla rose-purple. Leaves linear, acute, rough-margined. *Flowers an inch long, on short peduncles.*—Low grounds.

Var. **pauper'cula,** Gray, is smoother, with smaller and lighter-coloured corolla.

2. **G. tenuifo'lia,** Vahl. (SLENDER G.) Corolla rose-purple. Leaves linear, acute. *Flowers about ½ an inch long, on long thread-like peduncles.*—Dry woods.

3. **G. fla'va,** L. (DOWNY G.) Corolla yellow, woolly inside. Stem 3-4 feet high, *finely pubescent.* Leaves oblong or lance-shaped, the upper entire, the lower usually more or less pinnatifid, downy-pubescent.—Woods.

4. **G. quercifo'lia,** Pursh. (Smooth G.) Corolla yellow, woolly inside. Stem 3-6 feet high, *smooth and glaucous.* Lower leaves twice-pinnatifid, the upper pinnatifid or entire, smooth.—Woods.

5. **G. pedicula'ria,** L. (Cut-leaved G.) Nearly smooth. Flowers nearly as in Nos. 3 and 4. Stem 2-3 feet high, very leafy, much branched. Leaves pinnatifid, *the lobes cut and toothed.*—Thickets.

12. CASTILLE'IA, Mutis. Painted-Cup.

1. **C. coccin'ea,** Spreng. (Scarlet Painted-Cup.) Calyx 2-cleft, yellowish. Stem pubescent or hairy, 1-2 feet high. The stem-leaves nearest the flowers 3-cleft, the lobes toothed, *bright scarlet.* (A yellow-bracted form occurs on the shore of Lake Huron.)--Sandy soil.

2. **C. pal'lida,** Kunth., var. **septentrionalis,** Gray. *Calyx equally cleft, divisions 2-cleft. Upper lip of corolla decidedly shorter than the tube.* Lower leaves linear; upper broader, mostly entire; the floral oblong or obovate, *greenish-white, varying to yellowish, purple or red.*—Atl. Prov. and northward.

3. **C. sessiliflo'ra,** Pursh. *Calyx more deeply cleft in front,* the narrow lobes deeply 2-cleft. Leaves mostly 3-5 cleft, the floral ones similar, and *not coloured.*—N.W.

4. **C. minia'ta,**—Dougl. Calyx about equally cleft before and behind, the lobes 2-cleft. Corolla over an inch long, the upper lip linear, longer than the tube. Leaves lanceolate or linear, entire, the floral ones usually bright red.—N.W.

13. ORTHOCARP'US, Nutt.

O. **lu'teus,** Nutt. Stem about 1 foot high, pubescent, sometimes viscid. Leaves linear to lanceolate, sometimes 3-cleft. Bracts not coloured.—N.W.

14. EUPHRA'SIA, Tourn. Eyebright.

E. **officina'lis,** L., is rather common on the Lower St. Lawrence and the sea-coast. Flowers white, with purple veins. Lowest leaves crenate, those next the flowers *bristly-toothed.*

Var. **Tartar'ica,** Benth., has pale purple flowers.

15. BART'SIA, L.

B. Odonti'tes, Huds. Stem branching, scabrous-pubescent. Leaves oblong-lanceolate, coarsely and remotely serrate. - Atl. sea-coast.

16. RHINAN'THUS, L. YELLOW-RATTLE.

R. Crista-galli, L. (COMMON YELLOW-RATTLE.) Localities much the same as those of Euphrasia. Seeds broadly winged, rattling in the inflated calyx when ripe.

17. PEDICULA'RIS, Tourn. LOUSEWORT.

1. **P. Canadensis,** L. (COMMON LOUSEWORT.. WOOD BETONY.) Stems clustered, simple, hairy. Lowest leaves pinnately-parted. Flowers in a short spike. — Copses and banks.

2. **P. lanceola'ta,** Michx., has a nearly simple, smooth, upright stem, and oblong-lanceolate cut-toothed leaves. Calyx 2-lobed, leafy-crested. Pod ovate.—Grassy swamps.

3. **P. Furbish'iæ,** Watson. Leaves pinnately-parted, and the short oblong divisions pinnately cut, or (in the upper) serrate. Calyx-lobes 5. Upper lip of corolla straight and beakless.—River banks, Atl. Prov.

18. MELAMPY'RUM, Tourn. COW-WHEAT.

M. America'num, Michx. Leaves lanceolate, short-petioled; the lower ones entire.—Open woods.

ORDER LXI. **VERBENA'CEÆ.** (VERVAIN FAMILY.)

Herbs (with us), with opposite leaves, didynamous stamens, and corolla either irregularly 5-lobed or 2-lipped. Ovary in Verbena 4-celled (when ripe splitting into 4 nutlets) and in Phryma 1-celled, *but in no case 4-lobed*, thus distinguishing the plants of this Order from those of the next.

Synopsis of the Genera.

1. **Verbe'na.** Flowers in spikes. Calyx tubular, 5-ribbed. Corolla tubular, salver-form, the border rather irregularly 5-cleft. Fruit splitting into 4 nutlets.

ACANTHACEÆ.

2. **Phry'ma.** Flowers in loose slender spikes, *reflexed in fruit.* Calyx cylindrical, 2-lipped, the upper lip of three slender teeth. Corolla 2-lipped. Ovary 1-celled and 1-seeded.

1. VERBE'NA, L. VERVAIN.

1. **V. hasta'ta,** L. (BLUE VERVAIN.) Stem 3-5 feet high. Leaves oblong-lanceolate, taper-pointed, serrate. Spikes of *purple* flowers dense, erect, corymbed, or panicled.—Low meadows and fields.

2. **V. urticifo'lia,** L. (NETTLE-LEAVED V.) Stem tall. Leaves oblong-ovate, acute, coarsely serrate. Spikes of small *white* flowers very slender, loosely panicled.—Fields and roadsides.

3. **V. angustifo'lia,** Michx. Stem *low.* Leaves narrowly lanceolate, tapering at the base, sessile, roughish, slightly toothed. Flowers purple, in a crowded spike.—Dry soil.

4. **V. bracteo'sa,** Michx. Stem spreading or procumbent, hairy. Leaves wedge-lanceolate, cut-pinnatifid or 3-cleft. Spikes single, thick, sessile, leafy-bracted, the bracts longer than the small purple flowers.—S. W. Ontario.

2. PHRY'MA, L. LOPSEED.

P. Leptostach'ya, L. Corolla purplish or pale rose-coloured. Stem slender and branching, 1-2 feet high. Leaves ovate, coarsely-toothed.—Woods and thickets.

ORDER LXII. **ACANTHA'CEÆ.** (ACANTHUS FAMILY.)

Herbs (with us), with opposite leaves, diandrous (or didynamous) stamens inserted on the tube of the 2-lipped corolla, and a 2-celled and several-seeded capsule. Seeds flat, supported by hooked projections of the placentas. Flowers commonly much bracted: Calyx 5-cleft. Represented with us by the single genus

DIANTHE'RA, Gronov. WATER-WILLOW.

D. America'na, L. A perennial herb growing in water or wet places, with entire leaves and purplish flowers in oblong, dense, long-peduncled, axillary spikes. Corolla deeply 2-lipped, the upper erect, notched; the lower spreading, 3-parted. Stamens 2, *the anther-cells separated*. Pod obovate, flat, 4-seeded.

ORDER LXIII. **LABIA'TÆ.** (MINT FAMILY.)

Herbs with square stems, opposite leaves (mostly aromatic), didynamous (or in one or two genera *diandrous*) stamens, a 2-lipped or irregularly 4- or 5-lobed corolla, and a *deeply 4-lobed ovary*, forming in fruit 4 nutlets or achenes. (See Part I., Section 65, for description of a typical plant.)

Synopsis of the Genera.

* *Stamens 4, curved upwards, parallel, exserted from a deep notch on the upper side of the 5-lobed corolla.*

1. **Teu'crium.** Calyx 5-toothed. The four upper lobes of the corolla nearly equal, *with a deep notch between the upper 2;* the lower lobe much larger. Flowers pale purple.
2. **Isan'thus.** Calyx bell-shaped, 5-cleft, almost equalling the small pale-blue corolla. Lobes of the corolla *almost equally spreading*. Stamens only slightly exserted.

** *Stamens 4, the outer or lower pair longer, or only 2 with anthers, straight and not converging in pairs! Anthers 2-celled!*

+ *Corolla almost equally 4-lobed, quite small.*

3. **Men'tha.** Calyx equally 5-toothed. Upper lobe of the corolla rather the broadest, and sometimes notched. *Stamens 4*, of equal length, not convergent. Flowers either in terminal spikes or in head-like whorled clusters, often forming interrupted spikes. Corolla purplish or whitish.
4. **Lyc'opus.** Calyx-teeth 4 or 5. *Stamens 2*, the upper pair, if any, without anthers. Flowers white, in dense axillary clusters.

+ + *Corolla evidently 2-lipped, but the lobes nearly equal in size; the tube not bearded inside. Stamens with anthers 2.*

5. **Hedeo'ma.** Calyx 2-lipped, bulging on the lower side of the base, hairy in the throat; 2 stamens with good anthers, *and 2 sterile filaments with false anthers*. Low odorous plants, with bluish flowers in loose axillary clusters.

+ + + *Corolla 2-lipped, the lower of the 5 lobes much larger than the other 4; the tube with a bearded ring inside. Stamens 2 (occasionally 4), much exserted.*

6. **Collinso'nia.** Calyx ovate, enlarged and turned down in fruit, 2-lipped. Corolla elongated, the lower lip toothed or fringed. Strong-scented plants with *yellowish flowers on slender pedicels* in terminal panicled racemes.

← ← ← ← *Corolla evidently 2-lipped. Stamens with anthers 4.*

7. **Hysso'pus.** Calyx tubular, 15-nerved, equally 5-toothed. Corolla blue-purple, short; upper lip erect, flat, obscurely notched, the lower 3-cleft, with the middle lobe larger and 2-cleft. Stamens exserted, diverging. Flowers in small clusters, crowded in a spike. Branches simple, wand-like.

8. **Pycnan'themum.** Calyx short-tubular, 10-13-nerved, equally 5-toothed. The whitish or purplish flowers in *small dense* heads, forming terminal corymbs. Aromatic plants, with narrow rigid leaves crowded and clustered in the axils.

9. **Sature'ia.** Calyx bell-shaped, not hairy in the throat, equally 5-toothed. Aromatic plants, with narrow leaves and purplish spiked flowers.

* * * *Stamens only 2, parallel; the anthers only 1-celled. Corolla 2-lipped.*

10. **Monar'da.** Calyx tubular, nearly equally 5-toothed, hairy in the throat. Corolla elongated, strongly 2-lipped, the upper lip narrow. Stamens with long protruding filaments, each bearing a linear anther on its apex. Flowers large, in whorled heads surrounded by bracts.

* * * * *Stamens 4, the upper or inner pair longer! Anthers approximate in pairs. Corolla 2-lipped.*

11. **Nep'eta.** Calyx obliquely 5-toothed. Anthers approaching each other in pairs under the inner lip of the corolla, the cells of each anther divergent.

12. **Dracoceph'alum.** Calyx straight, 5-toothed, the upper tooth much the largest. Whorls of bluish flowers with awn-toothed or fringed leafy bracts, in a crowded head or spike.

13. **Lophan'thus.** Calyx obliquely 5-toothed. Stamens exserted, the upper pair declined, the lower ascending, so that the *pairs cross*. Anther-cells *parallel*. Tall herbs with small flowers in interrupted terminal spikes.

* * * * * *Stamens 4, the lower or outer pair longer! Anthers approximate in pairs. Corolla 2-lipped.*

14. **Calamin'tha.** Calyx tubular, *2-lipped*, often bulging below. Corolla 2-lipped, *the upper lip not arched*, the throat inflated. Flowers pale purple, in globular more or less dense clusters which are crowded with linear or awl-shaped hairy bracts.

15. **Melis'sa.** Calyx with the upper lip flattened and 3-toothed, the lower 2-cleft. Corolla nearly white, the tube recurved-ascending. Stamens curved and conniving under the upper lip. Flower-cluster loose, few-flowered, one-sided, with few bracts resembling the leaves.

16. **Physoste'gia.** Calyx not 2-lipped, 5-toothed or lobed, thin and membranaceous, inflated-bell-shaped in fruit. Anther-cells parallel. Flowers large and showy, rose-colour variegated with purple, opposite, in terminal leafless spikes.

17. **Brunel'la.** *Calyx 2-lipped*, flat on the upper side, closed in fruit; *the upper lip 3-toothed, the lower 2-cleft*. Filaments 2-toothed at the apex, the lower tooth bearing the anther. Flowers violet, in a close terminal spike or head, which is very leafy-bracted.

18. **Scutella'ria.** *Calyx 2-lipped*, short, closed in fruit, the lips rounded and entire, *the upper with a projection on the back*. Corolla blue or violet, the tube elongated and somewhat curved. Anthers of the lower stamens 1-celled, of the upper 2-celled. Flowers solitary in the axils of the upper leaves, or in axillary or terminal 1-sided racemes.

19. **Marru'bium.** *Calyx 10-toothed*, the teeth spiny and recurved after flowering. Stamens 4, *included in the corolla tube*. *Whitish woolly plants* with small white flowers in head-like whorls.

20. **Galeop'sis.** Calyx 5-toothed, the teeth spiny. The middle lobe of the lower lip of the corolla inversely heart-shaped, *the palate with 2 teeth at the sinuses*. Stamens 4, *the anthers opening cross-wise*. Flowers purplish, in axillary whorls.

21. **Stach'ys.** Calyx 5-toothed, beset with stiff hairs, the teeth spiny, diverging in fruit. Stamens 4, *the outer pair turned down after discharging their pollen*. Flowers purple, crowded in whorls, these at length forming an interrupted spike.

22. **Leonu'rus.** Calyx 5-toothed, the teeth spiny, and spreading when old. The middle lobe of the lower lip of the corolla narrowly oblong-obovate, entire. Flowers pale purple, in close whorls in the axils of the *cut-lobed leaves*. Nutlets sharply 3-angled.

23. **La'mium.** Calyx tubular bell-shaped, 5-nerved, with 5 nearly equal awl-pointed teeth. Corolla dilated at the throat; upper lip arched and narrowed at the base, the middle lobe of the lower lip notched at the apex and narrowed at the base. Herbs with purple flowers in few or several whorls or heads.

24. **Ballo'ta.** Calyx nearly funnel-form, 10-ribbed, with a spreading 5-toothed border. Whorls of purplish flowers dense. *Anthers exserted*. Plant erect, hairy (but green).

1. TEU'CRIUM, L. GERMANDER.

1. **T. Canadense,** L. (AMERICAN GERMANDER. WOOD SAGE.) Stem 1-3 feet high, downy. Leaves ovate-lanceolate, serrate, short-petioled, hoary beneath. Flowers in a long spike.—Low grounds.

2. **T. occidenta'le,** Gr., is loosely pubescent, and the calyx is *villous with viscid hairs.*—S. W. Ontario.

2. ISAN'THUS, Michx. FALSE PENNYROYAL.

I. **cæru'leus,** Michx. A low, branching, clammy-pubescent annual. Leaves lance-oblong, 3-nerved, nearly entire. Peduncles axillary, 1-3-flowered.—Gravelly soil.

3. MEN'THA, L. MINT.

1. **M. vir'idis,** L. (SPEARMINT.) *Flowers in a narrow terminal spike.* Leaves ovate-lanceolate, wrinkled, veiny, unequally serrate, *sessile.*—Wet places.

2. **M. piperi'ta,** L. (PEPPERMINT.) *Flowers in loose interrupted spikes.* Leaves ovate or ovate-oblong, acute, *petioled.* Plant smooth.—Wet places.

3. **M. Canadensis,** L. (WILD MINT.) *Flowers in axillary whorled clusters,* the uppermost axils without flowers. Stem more or less hairy, with ovate or lanceolate toothed leaves on short petioles. Var. **glabra'ta,** Benth., is smoothish, and has a rather pleasanter odour.—Sandy wet places.

4. **M. sati'va,** L. (WHORLED MINT.) Flowers in globular clusters in the axils of leaves; the uppermost axils not flower-bearing. Leaves petioled, ovate, sharply serrate. Calyx with very slender teeth.—Atl. Prov. (Int. from Eu.)

5. **M. arvensis,** L. (CORN MINT.) Flowers as in M. sativa, but leaves smaller, obtusely-serrate, and teeth of the calyx short and broader.—Atl. Prov. (Int. from Eu.)

4. LYC'OPUS, L. WATER HOREHOUND.

1. **L. Virgin'icus,** L. (BUGLE-WEED.) *Calyx-teeth 4, bluntish.* Stems obtusely 4-angled, 6-18 inches high, producing slender runners from the base. Leaves ovate-lanceolate, toothed,—Moist places.

2. **L. sinua'tus,** Ell. (*L. Europæus,* var. *sinuatus,* Gray.) *Calyx-teeth 5, sharp-pointed.* Stem sharply 4-angled, 1-3 feet high. Leaves varying from cut-toothed to pinnatifid. —Wet places.

3. **L. lu'cidus,** Turcz., var. **America'nus,** Gray. Calyx-teeth 5, very acute. Corolla hardly exceeding the calyx. Stem strict, short, 2-3 feet high. Leaves lanceolate or oblong-lanceolate, sessile or nearly so, sharply and coarsely serrate.—N.W.

5. HEDEO'MA, Pers. Mock Pennyroyal.

1. **H. pulegioi'des,** Pers. (AMERICAN PENNYROYAL). Stem 5-8 inches high, branching, hairy. Leaves oblong-ovate, *petioled*, obscurely serrate. Whorls few-flowered. Plant with a pungent aromatic odour.—Open woods and fields.

2. **H. his'pida,** Pursh., has the leaves *sessile, linear, and entire,* and the calyx *ciliate and hispid.*—Not common.

6. COLLINSO'NIA, L. Horse-Balm.

C. Canaden'sis, L. (RICH-WEED. STONE-ROOT.) Stem smooth or nearly so, 1-3 feet high. Leaves serrate, pointed, petioled, 3-6 inches long.—Rich woods.

7. HYSSO'PUS, Tourn. Hyssop.

H. officina'lis, L. Escaped from gardens in a few localities. Leaves lanceolate or linear, entire.

8. PYCNAN'THEMUM, Michx. Mountain Mint. Basil.

1. **P. lanceola'tum,** Pursh. Stem 2 feet high, smoothish or minutely pubescent. Leaves lanceolate or lance-linear, entire. Heads downy. Calyx-teeth short. Lips of corolla very short.—Dry soil.

2. **P. mu'ticum,** Pers., var. **pilo'sum,** Gr. *Hoary with loose pubescence.* Leaves oblong-lanceolate, *denticulate.* Bracts and calyx-teeth villous-pubescent. Heads larger than in No. 1.—S. W. Ontario.

9. SATURE'IA, L. Savory.

S. horten'sis, L. (SUMMER SAVORY.) Stem pubescent. Clusters few-flowered.—Escaped from gardens in a few localities.

10. MONAR'DA, L. Horse-Mint.

1. **M. did'yma,** L. (OSWEGO TEA.) *Corolla bright red,* very showy. The large outer bracts tinged with red.—Along shaded streams.

2. **M. fistulo'sa**, L. (WILD BERGAMOT.) *Corolla purplish,* the outer bracts somewhat purplish. Dry and rocky banks and woods.

Var. **mollis**, Benth., with flesh-coloured or lilac corolla, is common in the N.W. prairie region.

11. NEP'ETA, L. CAT-MINT.

1. **N. Cata'ria**, L. (CATNIP.) Flowers in cymose clusters. Stem erect, downy, branching. Leaves oblong, crenate, whitish beneath. Corolla dotted with purple.—Roadsides.

2. **N. Glecho'ma**, Benth. (GROUND IVY.) *Creeping and trailing.* Leaves *round-kidney-shaped*, crenate, green both sides. Corolla light blue.—Damp waste grounds.

12. DRACOCEPH'ALUM, L. DRAGON-HEAD.

D. parviflo'rum, Nutt. Stem erect, 8-20 inches high, leafy. Leaves ovate-lanceolate, cut-toothed, petioled. Corolla small and slender.—N.W. Ontario, in partly cleared lands.

13. LOPHAN'THUS, Benth. GIANT HYSSOP.

1. **L. nepetoi'des**, Benth. Smooth or nearly so, coarsely crenate-toothed. Calyx-teeth ovate, rather obtuse. Corolla greenish-yellow.—Borders of woods.

2. **L. scrophulariæfo'lius**, Benth., has lanceolate calyx-teeth and a purplish corolla.—Near Queenston Heights.

3. **L. anisa'tus**, Benth., has the leaves ovate, acute, and glaucous-white underneath.—N.W. plains.

CALAMIN'THA, Mœnch. CALAMINTH.

1. **C. Clinopo'dium**, Benth. (BASIL.) Stem hairy, erect, 1-2 feet high. Flower-clusters *dense*. Leaves ovate, nearly entire, petioled.—Thickets and waste places.

2. **C. Nuttal'lii**, Benth. Smooth, 5-9 inches high. Leaves narrowly oblong. Clusters *few-flowered*, the flowers on slender naked pedicels. Bracts linear or oblong.—Wet limestone rocks, western and south-western Ontario.

15. MELIS'SA, L. BALM.

M. officina'lis, L. (COMMON BALM.) Upright, branching, pubescent. Leaves broadly ovate, crenate-toothed, lemon-scented.—Escaped from gardens in a few places.

16. PHYSOSTE'GIA, Benth. FALSE DRAGON-HEAD.

P. Virginia'na, Benth. Stem smooth, wand-like. Lower leaves oblong-ovate, upper lanceolate. Corolla an inch long, funnel-form, the throat inflated; upper lip slightly arching, the lower 3-parted, spreading, small.—Wet banks, common on Toronto Island.

17. BRUNEL'LA, Tourn. SELF-HEAL.

B. vulga'ris, L. (COMMON HEAL-ALL.) A low plant with oblong-ovate petioled leaves. Clusters 3-flowered, the whole forming a close terminal elongated head.—Woods and fields everywhere.

18. SCUTELLA'RIA, L. SKULL-CAP.

1. **S. galericula'ta,** L. Flowers blue, ¾ of an inch long, solitary in the axils of the upper leaves. Stem nearly smooth, 1-2 feet high.—Wet places.

2. **S. par'vula,** Michx. Flowers blue, ¼ of an inch long, solitary in the upper axils. Stem *minutely downy*, 2-6 inches high. Lowest leaves round-ovate, the upper narrower, all *entire*. Roots necklace-form.—Dry banks.

3. **S. lateriflo'ra,** L. Flowers blue, ⅛ of an inch long, in 1-sided racemes. Stem upright, much branched, 1-2 feet high.—Wet places.

19. MARRU'BIUM, L. HOREHOUND.

M. vulga're, L. Leaves round-ovate, crenate-toothed. Calyx with 5 long and 5 short teeth, recurved.—Escaped from gardens in some places.

20. GALEOP'SIS, L. HEMP-NETTLE.

G. Tetra'hit, L. (COMMON HEMP-NETTLE.) Stem bristly-hairy, swollen below the joints. Leaves ovate, coarsely serrate. Corolla often with a purple spot on the lower lip.—Waste places and fields.

21. STACH'YS, Tourn. HEDGE-NETTLE.

1. **S. palus'tris**, L. Stem 2-3 feet high, 4-angled, the angles beset with stiff reflexed hairs or bristles. Leaves *sessile*, or the lower short-petioled, oblong or ovate-lanceolate, crenately serrate, downy. Calyx hispid. Upper lip of the corolla pubescent.—Wet grounds.

2. **S. as'pera**, Michx. Taller than the last, the leaves nearly all distinctly petioled. Calyx mostly glabrous. Corolla glabrous throughout.—Wet grounds.

22. LEONU'RUS, L. MOTHERWORT.

L. Cardi'aca, L. (COMMON MOTHERWORT.) Stem tall. Leaves long-petioled, the lower palmately lobed, the upper 3-cleft. Upper lip of the corolla bearded.—Near dwellings.

23. LA'MIUM, L. DEAD-NETTLE.

1. **L. amplexicau'le**, L. Leaves rounded, deeply crenate-toothed, the *upper ones clasping*. In gardens in some places.

2. **L. purpu'reum**, L. Leaves rounded or oblong, heart-shaped, crenate-toothed, *all petioled.*—A garden escape.

3. **L. macula'tum**, L. Taller. Leaves ovate, heart-shaped, often with a white spot above. Flowers rather large, purplish, with hairs at the base inside.—A garden escape.

24. BALLO'TA, L. FETID HOREHOUND.

B. nigra, L. (BLACK HOREHOUND.) Leaves ovate, toothed. Calyx-teeth longer than the tube of the corolla.—S. W. Ontario.

ORDER LXIV. **BORRAGINA'CEÆ.** (BORAGE FAMILY.)

Herbs, with a deeply 4-lobed ovary, forming 4 seed-like nutlets, as in the last Order, *but the corolla is regularly 5-lobed, with 5 stamens inserted upon its tube.*

Synopsis of the Genera.

* *Corolla without any scales in the throat.*

1. **E'chium.** Corolla with a funnel-form tube and a spreading border of 5 *somewhat unequal lobes. Stamens exserted, unequal.* Flowers bright blue, with a purplish tinge, in racemed clusters. Plant bristly.

** *Corolla with 5 scales completely closing the throat.*

2. **Sym'phytum.** Corolla tubular-funnel-form with short spreading lobes; *scales awl-shaped. Flowers yellowish-white*, in nodding raceme-like clusters, the latter often in pairs. *Nutlets smooth.* Coarse hairy herbs.

3. **Echinosper'mum.** *Nutlets prickly on the margin.* Corolla salver-shaped, lobes rounded; scales short and blunt. *Flowers blue,* small, in leafy-bracted racemes. Plant rough-hairy.

4. **Cynoglos'sum.** *Nutlets prickly all over.* Corolla funnel-form; scales blunt. Flowers red-purple or pale blue, in racemes which are naked above, but usually leafy-bracted below. Strong-scented coarse herbs.

5. **Lycop'sis.** Corolla funnel-shaped, the scales in the throat *bristly.* Flowers blue, in leafy raceme-like clusters. *Nutlets rough-bristly,* as is the whole plant.

*** *Corolla open, the scales or folds not sufficient to completely close the throat.*

6. **Merten'sia.** Corolla bell-funnel-shaped, mostly with 5 small glandular folds in the open throat. Calyx deeply 5-cleft. Style long and thread-form. *Nutlets smooth or wrinkled.* Flowers purplish-blue or white, in loose and short raceme-like clusters, only the lower one leafy-bracted. Pedicels slender.

7. **Onosmo'dium.** Corolla tubular, *the 5 lobes acute and erect or converging.* Anthers mucronate; filaments very short. Style thread-form, much exserted. Flowers greenish- or yellowish-white. Rather tall stout plants, shaggy with spreading bristly hairs, or rough with short appressed bristles. *Nutlets smooth.*

8. **Lithosper'mum.** Corolla funnel-form or salver-shaped, the 5 lobes of the spreading limb rounded. Anthers almost sessile. Root mostly red. Flowers small and almost white, or large and deep yellow, scattered and spiked and leafy-bracted. *Nutlets smooth or wrinkled.*

9. **Myoso'tis.** Corolla salver-shaped, with a very short tube, *the lobes convolute in the bud;* scales or appendages of the throat blunt and arching. Flowers blue, in (so-called) racemes without bracts. Low plants, mostly soft-hairy. *Nutlets smooth.*

1. **E'CHIUM,** Tourn. Viper's Bugloss.

E. vulga're, L. (Blue-weed.) Stem erect, 2 feet high. Leaves sessile, linear-lanceolate. Flowers showy, in lateral clusters, the whole forming a long narrow raceme.—Roadsides; common in eastern Ontario and rapidly spreading westward.

BORRAGINACEÆ.

2. SYM'PHYTUM, Tourn. COMFREY.

S. officina'le, L. (COMMON COMFREY.) Stem winged above by the decurrent bases of the leaves, branched. Leaves ovate-lanceolate or lanceolate.—Moist soil; escaped from gardens.

2. ECHINOSPER'MUM, Schwartz. STICKSEED.

1. **E. Lap'pula.** Lehm. A very common roadside weed, 1-2 feet high, hispid, branching above. Leaves lanceolate, rough. Nutlets warty on the back, with a double row of prickles on the margin. Pedicels stout, not deflexed. Calyx becoming foliaceous.

2. **E. floribun'dum,** Lehm. Stem strict, 2 feet high or more, not hispid. Leaves oblong- to linear-lanceolate, the lowest tapering into margined petioles. Racemes numerous, rather strict. Pedicels slender, deflexed in fruit. Nutlets rough and margined with a close row of flat awl-shaped prickles.—Chiefly N.W.

3. **E. Virgin'icum,** Lehm. (*Cynoglossum Morisoni*, DC.) (BEGGAR'S LICE.) Stem 2-4 feet high; hispid. Lower leaves round-ovate or cordate, slender-petioled; upper ones tapering at both ends. Racemes widely spreading. Pedicels slender, deflexed in fruit. Nutlets prickly all over.— Open woods and thickets.

4. CYNOGLOS'SUM, Tourn. HOUND'S TONGUE.

1. **C. officina'le,** L. (COMMON HOUND'S TONGUE.) *Flowers red-purple.* Upper leaves lanceolate, sessile. Stem soft-pubescent.—Nutlets rather flat.—A common weed in fields and along roadsides.

2. **C. Virgin'icum,** L. (WILD COMFREY.) *Flowers pale blue.* Stem roughish with spreading hairs. *Leaves few,* lanceolate-oblong, *clasping.* Racemes corymbed, raised on a long, naked peduncle.—Rich woods.

5. LYCOP'SIS, L. BUGLOSS.

L. arven'sis, L. (SMALL BUGLOSS) Very rough-bristly, 1 foot high. Leaves lanceolate.—Dry or sandy fields, chiefly eastward.

6. MERTEN'SIA, Roth. LUNGWORT.

1. **M. panicula'ta,** Don. *Filaments broad and short.* Nutlets wrinkled when dry. Plant more or less hairy, erect, loosely branched.—N.W.

2. **M. marit'ima,** Don. (SEA-LUNGWORT.) Corolla white, trumpet-shaped, conspicuously 5-lobed, throat crested. Leaves fleshy, glaucous, ovate to spathulate. Stems spreading, often decumbent. *Filaments longer and narrower than the anthers.* Nutlets shining.—Atl. sea-coast.

7. ONOSMO'DIUM, Michx. FALSE GROMWELL.

1. **O. Carolinia'num,** DC. Stem stout, 3-4 feet high. Leaves ovate-lanceolate, acute. *Lobes of the corolla ovate-triangular, very hairy outside.*—Banks of streams.

2. **O. Virgina'num,** DC. Stem slender, 1-2 feet high. Leaves narrowly oblong. *Lobes of the corolla lance-awl-shaped, sparingly bearded outside with long bristles.*—Banks and hillsides; not common.

8. LITHOSPER'MUM, Tourn. GROMWELL. PUCCOON.

* *Corolla almost white. Nutlets wrinkled, gray.*

1. **L. arven'se,** L. (CORN GROMWELL.) Stem 6-12 inches high, erect. Leaves lanceolate or linear.—Waste grounds and in wheat-fields.

** *Corolla deep yellow. Nutlets smooth and shining.*

2. **L. hirtum,** Lehm. (HAIRY PUCCOON.) Stem 1-2 feet high, *hispid.* Stem-leaves lanceolate or linear; those of the flowering branches ovate-oblong, ciliate. Flowers *peduncled. Corolla woolly at the base inside.*—Dry woods.

3. **L. canes'cens,** Lehm. (HOARY PUCCOON. ALKANET.) Stem 6-15 inches high, *soft-hairy. Corolla naked at the base inside.* Flowers *sessile.* Limb of the corolla smaller, and the calyx shorter, than in No. 2.—Open woods and plains.

4. **L. angustifo'lium,** Michx. Erect or diffusely branched from the base, somewhat hoary. Leaves all linear. Flowers pedicelled, leafy-bracted, of two sorts: the earlier large and conspicuous, the later small and pale.—N.W.

* * * *Corolla greenish-white or cream-colour. Nutlets smooth and shining, mostly white.*

5. **L. officina'le**, L. (COMMON GROMWELL.) Much branched above. Leaves broadly lanceolate, acute. *Corolla exceeding the calyx.*—Roadsides and fields.

6. **L. latifo'lium**, Michx. Loosely branched above. Leaves ovate and ovate-lanceolate, mostly taper-pointed. *Corolla shorter than the calyx.*—Borders of woods.

9. MYOSO'TIS, L. FORGET-ME-NOT.

1. **M. palus'tris**, Withering. (FORGET-ME-NOT.) Stem ascending from a creeping base, about a foot high, *smoothish*, loosely branched. Leaves rough-pubescent. *Calyx open in fruit*, its lobes much shorter than its tube. Corolla sky-blue, with a yellow eye. Pedicels spreading.—Wet places.

2. **M. laxa**, Lehm. Stem very slender, *decumbent*, from thread-like subterranean shoots. Pubescence all appressed. Calyx open in front, *its lobes as long as its tube.* Corolla paler blue.—Wet places.

3. **M. arven'sis**, Hoffm. Stem erect or ascending, *hirsute.* Calyx closing in fruit. Corolla blue, rarely white. Pedicels spreading in fruit and longer than the 5-cleft equal calyx. *Racemes naked at the base.*—Fields.

4. **M. verna**, Nutt., differs from the last in having a very small *white* corolla, pedicels *erect* in fruit, and the racemes leafy at the base. The calyx, also, is unequally 5-toothed and hispid.—Dry hills.

ORDER LXV. **HYDROPHYLLA'CEÆ.** (WATERLEAF F.)

Herbs, with alternate cut-toothed or lobed leaves, and regular pentamerous and pentandrous flowers very much like those of the last Order, *but having a 1-celled ovary with the seeds on the walls (parietal).* Style 2-cleft. Flowers mostly in 1-sided cymes which uncoil from the apex.

Synopsis of the Genera.

1. **Hydrophyl'lum.** Stamens exserted. Calyx unchanged in fruit.
2. **Ellis'ia.** Stamens included. Calyx enlarged in fruit.—N.W. only.

1. HYDROPHYL'LUM, L. WATERLEAF.

1. **H. Virgin'icum,** L. Corolla bell-shaped, the 5 lobes convolute in the bud; the tube with 5 folds down the inside, one opposite each lobe. *Stamens and style exserted, the filaments bearded below.* Stem smoothish. Leaves *pinnately* cleft into 5-7 divisions, the latter ovate-lanceolate, pointed, cut-toothed. *Calyx-lobes very narrow, bristly-ciliate.* Flowers white or pale blue. Peduncles *longer* than the petioles of the upper leaves. Rootstocks scaly-toothed.—Moist woods.

2. **H. Canaden'se,** L., differs from the last in having the leaves *palmately* 5-7-lobed, and *rounded*; the peduncles *shorter* than the petioles; and the calyx-lobes *nearly smooth.* —Rich woods.

3. **H. apperdicula'tum,** Michx. Stem, pedicels, and calyx *hairy.* Stem-leaves palmately 5-lobed and rounded, the lowest leaves pinnately divided. *Calyx with a small reflexed appendage in each sinus.* Stamens sometimes not exserted.—Rich woods, S. W. Ontario.

2. ELLIS'IA, L.

E. Nycte'lea, L. A delicate branching annual, minutely or sparingly roughish-hairy. Leaves pinnately parted into 7-13 narrow sparingly cut-toothed divisions. Peduncles 1-flowered, solitary in the forks or opposite the leaves. Flowers small, whitish. Capsule pendulous.—N. W.

ORDER LXVII. POLEMONIA'CEÆ. (POLEMONIUM F.)

Herbs with regular pentamerous and pentandrous flowers, *but a 3-celled ovary and 3-lobed style. Lobes of the corolla convolute in the bud.* Calyx persistent.

Synopsis of the Genera.
1. **Phlox.** Corolla salver-form. Leaves opposite, entire.
2. **Gil'ia.** Corolla tubular-funnel-form or salver-shaped, very slender. Leaves alternate, entire.

1. PHLOX, L. PHLOX.

1. **P. divarica'ta,** L. Corolla salver-shaped, with a long tube. Stamens short, *unequally inserted.* Stem ascending

from a prostrate base, somewhat clammy. Leaves *oblong-ovate*. Flowers lilac or bluish, in a spreading loosely-flowered cyme. *Lobes of the corolla mostly obcordate.*—Moist rocky woods.

2. **P. pilo'sa,** L. Leaves *lanceolate or linear*, tapering to a sharp point. Lobes of the pink-purple corolla obovate, *entire.*—Southwestern Ontario.

3. **P. subula'ta,** L., the Moss Pink of the gardens, has escaped from cultivation in some places. Stem creeping and tufted in broad mats. Flowers mostly rose-colour.—Dry grounds.

4. **P. Hood'ii,** Richards, of the North-West, forms broad, dense mats or tufts, 2-4 inches high. Leaves awl-shaped.

2. GIL'IA, Ruiz and Pav.

G. linea'ris, Gray. (*Collo'mia linea'ris*, Nutt.) A branching herb with alternate, linear-lanceolate or oblong, sessile and entire leaves. Corolla salver-form, with stamens unequally inserted in its narrow tube, lilac-purple to nearly white. Ovules solitary. Found on the sands at the mouth of Eel River, Restigouche Co., N.B.

ORDER LXVIII. CONVOLVULA'CEÆ. (CONVOLVULUS F.)

Chiefly twining or trailing herbs, with alternate leaves and regular flowers. Sepals 5, imbricated. Corolla 5-plaited or 5-lobed and convolute in the bud. Stamens 5. Ovary 2-celled.

Synopsis of the Genera.

1. **Calyste'gia.** *Calyx enclosed in 2 large leafy bracts.* Corolla funnel-form, the border obscurely lobed. Pod 4-seeded.
2. **Convol'vulus.** *Calyx without bracts.*
3. **Cus'cuta.** *Leafless parasitic slender twiners*, with yellowish or reddish stems, attaching themselves to the bark of other plants. Flowers small, mostly white, clustered. Corolla bell-shaped. *Stamens with a fringed appendage at their base.*

1. CALYSTE'GIA, R. Br. BRACTED BINDWEED.

1. **C. se'pium,** R. Br. (*Convolvulus sepium*, L., in Macoun's Catalogue.) (HEDGE BINDWEED.) *Stem mostly*

twining. Leaves halberd-shaped. Peduncles 4-angled. Corolla commonly rose-coloured.—Moist banks.

2. **C. spithamæ'a,** Pursh. (*Convolvulus spithamœus,* L., in Macoun's Catalogue.) Stem low and simple, upright or ascending, *not twining,* 6-12 inches high. Leaves oblong, more or less heart-shaped at the base. Corolla white.—Dry soil.

2. CONVOL'VULUS, L. BINDWEED.

C. arven'sis, L. (BINDWEED.) Stem twining or procumbent and low. Leaves ovate-oblong, sagittate, the lobes acute. Corolla white, or tinged with red.

3. CUS'CUTA, Tourn. DODDER.

1. **C. Grono'vii,** Willd. Stems resembling coarse threads, spreading themselves over herbs and low bushes. Corolla-lobes obtuse, spreading. Capsule globose, *abruptly pointed.* Flowers in loose panicled cymes.—Wet shady places.

2. **C. arven'sis,** Beyrich. Stems pale and slender, low. Flowers in dense clusters. Capsule *depressed-globose.* Corolla-lobes acute, with inflexed points. Stamen-scales deeply fringed.—Dry ridges, N.W.

3. **C. Epil'inum,** Weihe. (FLAX DODDER.) Stems very slender, low. Flowers sessile in dense scattered heads. Corolla short-cylindrical, hardly exceeding the calyx, persistent round the capsule. Stamen-scales short and broad. Capsule globose, circumcissile.—Atl. Prov.; introduced.

ORDER LXVIII. **SOLANA'CEÆ.** (NIGHTSHADE FAMILY.)

Rank-scented herbs (or one species shrubby), with colourless bitter juice, alternate leaves, and regular pentamerous and pentandrous flowers, *but a 2-celled* (in Nicandra 3-5-celled) *ovary, with the placentæ in the axis.* Fruit a many-seeded berry or pod.

Synopsis of the Genera.

1. **Sola'num.** Corolla wheel-shaped, 5-lobed, the margins turned inward in the bud. *Anthers conniving around the style,* the cells opening by pores at the apex; filaments very short. The larger leaves often with an accompanying smaller one. **Fruit a** *berry.*

2 **Phys'alis.** Calyx 5-cleft, enlarging after flowering, *becoming at length much inflated, and enclosing the berry.* Corolla between wheel-shaped and funnel-form. *Anthers separate,* opening lengthwise. Plant clammy-pubescent.

3. **Nican'dra.** Calyx 5-parted, 5-angled, the divisions rather arrow-shaped, enlarged and bladder-like in fruit, enclosing the 3–5-celled globular dry berry. A smooth herb, 2–3 feet high, with pale blue flowers.

4. **Ly'cium.** Corolla funnel-form or tubular. Fruit a small *berry*, the calyx persistent but *not inflated.* A shrubby plant with long drooping branches and greenish-purple flowers on slender peduncles fascicled in the axils.

5. **Hyoscy'amus.** *Fruit a pod, the top coming off like a lid.* Calyx urn-shaped, 5-lobed, persistent. Corolla funnel-form, *oblique*, the limb 5-lobed, dull-coloured and veiny. Plant clammy-pubescent.

6. **Datu'ra.** Fruit a large prickly naked *pod.* Calyx long, *5-angled*, not persistent. Corolla very large, funnel-form, *strongly plaited* in the bud, with 5 pointed lobes. Stigma 2-lipped. Rank-scented weeds, with the showy flowers in the forks of the branching stems.

7. **Nicotia'na.** Fruit a *pod*, enclosed in the calyx. Calyx tubular-bell-shaped, 5-cleft. Corolla dull greenish-yellow, funnel-form, plaited in the bud. Leaves large. Flowers racemed or panicled.

1. SOLA'NUM, Tourn. NIGHTSHADE.

1. **S. Dulcama'ra,** L. (BITTERSWEET.) Stem somewhat shrubby and *climbing.* Leaves ovate and heart-shaped, *the upper halberd-shaped, or with 2 ear-like lobes at the base.* Flowers violet-purple, in small cymes. *Berries red.*—Near dwellings and in moist grounds.

2. **S. nigrum,** L. (COMMON NIGHTSHADE.) *Stem low and spreading,* branched. *Leaves ovate, wavy-toothed.* Flowers small, white, drooping in umbel-like lateral clusters. *Berries black.*—Fields and damp grounds.

3. **S. rostra'tum.** Dunal, is a *prickly* herb with large yellow flowers and *sharp* anthers.—Ottawa.

2. PHYS'ALIS, L. GROUND CHERRY.

1. **P. visco'sa,** L. (*P. Virginiana*, Mill, in Macoun's Catalogue:) Corolla *greenish-yellow*, brownish in the centre. Anthers yellow. Leaves ovate or heart-shaped, mostly toothed. Berry orange, sticky.—Sandy soil.

2. **P. grandiflo'ra**, Hook. Corolla *white*, large, with a woolly ring in the throat. Anthers tinged with blue or violet.

3. NICAN'DRA, Adans. APPLE OF PERU.

N. physalol'des, Gaertn. Leaves ovate, sinuate-toothed or angled. Flowers solitary on axillary and terminal peduncles.—Escaped from gardens in some places.

4. LY'CIUM, L. MATRIMONY-VINE.

L. vulga're, Dunal. Common about dwellings. Berry oval, orange-red.

5. HYOSCY'AMUS, Tourn. HENBANE.

H. niger, L. (BLACK HENBANE.) Escaped from gardens in some localities. Corolla dull yellowish, netted with purple veins. Leaves clasping, sinuate-toothed. A strong-scented and poisonous herb.

6. DATU'RA, L. STRAMONIUM. THORN-APPLE.

1. **D. Stramo'nium**, L. (COMMON THORN-APPLE.) Stem *green*. Corolla *white*, 3 inches long. Leaves ovate, sinuate-toothed.—Roadsides.

2. **D. Tat'ula**, L. (PURPLE T.) Stem *purple*. Corolla *pale violet-purple*.

6. NICOTIA'NA, L. TOBACCO.

N. rus'tica, L. (WILD TOBACCO.) Old fields and in gardens.

ORDER LXIX. **GENTIANA'CEÆ.** (GENTIAN FAMILY.)

Smooth herbs, *distinguished by having a 1-celled ovary with seeds on the walls, either in lines or on the whole inner surface.* Leaves mostly opposite, simple, and sessile, but in one Genus alternate and compound. Stamens as many as the lobes of the regular corolla and alternate with them. Stigmas 2. Calyx persistent. Juice colourless and bitter.

Synopsis of the Genera.

1. **Fra'sera.** Corolla wheel-shaped, 4-parted; a fringed glandular spot on each lobe. Flowers light greenish-yellow, with small purple-brown spots.

GENTIANACEÆ. 177

2. **Hale'nia.** Corolla 4-lobed, *the lobes all spurred at the base.* Flowers yellowish or purplish, somewhat cymose.
3. **Gentia'na.** Corolla not spurred, 4-5-lobed, mostly funnel-form or bell-shaped, generally with teeth or folds in the sinuses of the lobes. Stigmas 2, persistent. Pod oblong. Seeds innumerable. Flowers showy, in late summer and autumn.
4. **Menyan'thes.** A bog-plant. Corolla short, funnel-form, 5-lobed, *densely white-bearded on the upper face. Leaves alternate, compound, of 3 oval leaflets. The flowers in a raceme at the summit of a naked scape*, white or tinged with pink.
5. **Limnan'themum.** An aquatic, with simple round-heart-shaped floating leaves on long petioles. Corolla white, wheel-shaped, 5-parted, bearded at the base only. Flowers in an umbel borne on the petiole.

1. FRA'SERA, Walt. AMERICAN COLUMBO.

F. Carolinien'sis, Walt. Tall and showy. Leaves whorled, mostly in fours. Root thick. Flowers numerous in a pyramidal panicle.—Dry soil.

2. HALE'NIA, Bork. SPURRED GENTIAN.

H. deflex'a, Griseb. Stem erect, 9-18 inches high. Leaves 3-5-nerved, those at the base of the stem oblong-spathulate, petioled; the upper acute and sessile or nearly so. Spurs of the corolla curved.—Not common in Ontario; common on the Lower St. Lawrence.

3. GENTIA'NA, L. GENTIAN.

1. G. crini'ta, Frœl. (FRINGED GENTIAN.) Corolla funnel-form, *4-lobed, the lobes fringed on the margins;* no plaited folds in the sinuses. *Flowers sky-blue, solitary, on long naked stalks,* terminating the stem or simple branches. *Ovary lanceolate.* Leaves lance-shaped or ovate-lanceolate. —Low grounds.

2. G. deton'sa, Fries., (*G. serrata*, Gunner.) (SMALLER FRINGED G.) is distinguished from No. 1 by the shorter or almost inconspicuous fringe of the corolla, the linear or lance-linear leaves, and the broader ovary.—Moist grounds, chiefly in the Niagara District.

3. G. quinqueflo'ra, Lam. (FIVE-FLOWERED G.) Corolla tubular-funnel-form, pale-blue, no folds in the sinuses.

Calyx 5-cleft, the lobes awl-shaped. Lobes of the corolla triangular-ovate, bristle-pointed. Anthers separate. Stem slender and branching, a foot high, the branches racemed or panicled, about 5-flowered at the summit.—Dry hill-sides.

4. **G. puber'ula,** Michx. Stems erect or ascending, 8-16 inches high, *minutely rough above*. Leaves rigid, lanceolate to linear-lanceolate, 1-2 inches long. Flowers mostly clustered. Calyx-lobes *lanceolate*, much shorter than the *bell-funnel-form* open bright blue corolla.—High Park, Toronto.

5. **G. alba,** Muhl. (WHITISH G.) Corolla inflated-club-shaped, *at length open*, 5-lobed, the lobes about twice as long as the *toothed appendages in the sinuses*. Flowers *greenish-white or yellowish*, sessile, crowded in a terminal cluster. Anthers usually cohering. Leaves lance-ovate, with a clasping heart-shaped base.—Low grounds.

6. **G. Andrews'ii,** Griseb. (CLOSED G.) Corolla inflated-club-shaped, *closed at the mouth*, the apparent lobes being really the large fringed-toothed appendages. *Flowers blue*, in a close sessile terminal cluster. Anthers cohering. Leaves ovate-lanceolate from a narrower base. — Low grounds; common northward, flowering later than No. 3.

7. **G. Amarel'la,** L., var. **acu'ta,** Hook. Corolla somewhat funnel-form, mostly blue, its lobes entire, acute, with a fringed crown at their base. Calyx-lobes (4-5) lanceolate or linear, foliaceous.—Atl. Prov. chiefly.

Var. **stricta,** Watson, has stem and branches *strict*, a *whitish* corolla, and a less deeply cleft calyx.—N.W.

8. **G. linea'ris,** Frœl., var. **latifolia,** Gray. Flowers in a terminal cluster with a leafy involucre. Corolla blue, narrow funnel-form, with roundish-ovate lobes, and broad appendages. Leaves sessile, oblong-linear to ovate-lanceolate, smooth. Seeds winged.—Boggy places, Atl. Prov. and northward.

4. MENYAN'THES, Tourn. BUCKBEAN.

G. trifolia'ta, L. A common plant in bogs and wet places, northward. The bases of the long petioles sheathe

the lower part of the scape, or thick rootstock, from which they spring. Plant about a foot high.

5. LIMNAN'THEMUM, Gmelin. FLOATING HEART.

L. lacunos'um, Griseb. In shallow waters, northern Ontario.

ORDER LXX. **APOCYNA'CEÆ.** (DOGBANE FAMILY.)

Herbs or slightly shrubby plants, with milky juice, opposite simple entire leaves, and regular pentamerous and pentandrous flowers with the lobes of the corolla convolute in the bud. *Distinguished by having 2 separate ovaries,* but the 2 stigmas united. Calyx free from the ovaries. Anthers converging round the stigmas. Seeds with a tuft of down on the apex. Represented with us only by the Genus

APO'CYNUM, Tourn. DOGBANE.

1. **A. androsæmifo'lium,** L. (SPREADING DOGBANE.) The corolla bell-shaped, 5-cleft, *pale rose-coloured, the lobes turned back. Branches of the stem widely forking.* Flowers in loose rather spreading cymes. Leaves ovate, petioled. Fruit 2 long and slender diverging pods.—Banks and thickets.

2. **A. cannab'inum,** L. (INDIAN HEMP.) Lobes of the *greenish-white* corolla not turned back. *Branches erect.* Cymes closer than in No. 1, and the flowers much smaller. —Along streams, very variable.

ORDER LXXI. **ASCLEPIADA'CEÆ.** (MILKWEED F.)

Herbs with milky juice and opposite or whorled (rarely scattered) simple entire leaves. Pods, seeds, and anthers as in the last Order, *but the anthers are more closely connected with the stigma, the (reflexed) lobes of the corolla are valvate in the bud, the pollen is in waxy masses, and the (monadelphous) short filaments bear 5 curious hooded bodies behind the anthers.* Flowers in umbels.

Synopsis of the Genera.

1. **Ascle'pias.** Corolla reflexed, deeply 5-parted. A crown of 5 hooded fleshy bodies with an incurved horn rising from the cavity of each hood. Leaves mostly opposite or whorled.

2. **Acera'tes.** Corolla reflexed or merely spreading. Crown as in No. 1, *but no incurved horn.* Leaves mostly alternate.

1. ASCLE'PIAS, L. MILKWEED.

* *Flowers greenish, yellowish or white, or merely purplish-tinged.*

1. **A. Cornu'ti,** Decaisne. (COMMON MILKWEED.) Stem tall and stout. Leaves oval or oblong, short-petioled, pale green, 4-8 inches long. *Flowers dull greenish-purple. Pods ovate, soft-spiny, woolly.*—Mostly in dry soil; very common.

2. **A. specio'sa,** Torr. Finely white-woolly or becoming glabrous, *the many-flowered umbel and calyx densely woolly.* Leaves oval to oblong, slightly cordate. Corolla-lobes purplish. Hoods with a long lanceolate appendage at the summit.—N.W.

3. **A. phytolaccoi'des,** Pursh. (POKE MILKWEED.) Stem tall and smooth. Leaves broadly ovate, acute at both ends, short-petioled. *Pedicels loose and nodding, very long and slender.* Corolla greenish, *with the hooded appendage white.* Pods minutely downy, *but not warty.*—Moist thickets.

4. **A. ovalifo'lia,** Decaisne. Low, soft-downy. Leaves ovate to lance-oblong, acute, short-petioled, soft-pubescent beneath. Umbels loosely 10-18-flowered. Pedicel slender. Corolla-lobes greenish-white, slightly tinged with purple outside.—Hoods yellowish, with a small horn, obtuse, entire. —N.W.

5. **A. quadrifo'lia,** L. Stem slender, 1-2 feet high, mostly leafless below, *with one or two whorls of four* in the middle, and one or two pairs of ovate or ovate-lanceolate taper-pointed leaves. Corolla lobes pale pink; hoods white. —Not common, but abundant near Toronto.

6. **A. verticilla'ta,** L., has slender stems and *filiform-linear leaves with revolute margins,* 3-6 in a whorl. Corolla-lobes greenish-white.—S.W. Ontario, and N.W.

* * *Flowers red.*

7. **A. incarna'ta,** L. (SWAMP M.) Stem tall, leafy, branching, and smooth. Leaves oblong-lanceolate, acute, obscurely heart-shaped at the base. *Flowers rose-purple. Pods very smooth and glabrous.*—Swamps and low grounds.

*** *Flowers orange.*

8. **A. tubero'sa,** L. (Butterfly-weed. Pleurisy-root.) Stem very leafy, branching above, rough-hairy. Leaves linear or oblong-lanceolate, chiefly scattered. Corolla greenish-orange, *with the hoods bright orange-red.* Pods hoary. Dry hill-sides and fields; almost destitute of milky juice.

2. ACERA'TES, Ell. Green Milkweed.

1. **A. viridiflo'ra,** Ell. Stems ascending, 1-2 feet high, minutely soft-downy, becoming smoothish. Leaves oval to linear. The compact umbels of greenish flowers nearly sessile, lateral, many-flowered.—Dry soil, from Niagara Falls westward.

Var. **lanceola'ta,** Gray, has lanceolate leaves.—S.W. Ontario and N.W.

Var. **linea'ris,** Gray, has elongated-linear leaves, and low stems. Umbels often solitary.—N.W.

Order LXXII. OLEA'CEÆ. (Olive Family.)

The only common representative Genus of this Order in Canada is Fraxinus (Ash). The species of this Genus are trees with pinnate leaves, and polygamous or dioecious flowers without petals, and mostly also without a calyx; stamens only 2, with large oblong anthers. Fruit a 1-2-seeded samara. Flowers insignificant, from the axils of the previous year's leaves.

FRAX'INUS, Tourn. Ash.

* *Leaflets with petioles.*

1. **F. America'na,** L. (White Ash.) *Fruit winged from the apex only, the base cylindrical.* Branchlets and petioles *smooth and glabrous.* Calyx very minute, persistent. Leaflets 7-9, *stalked.*—Rich woods.

2. **F. pubes'cens,** Lam., (Red Ash) has the *branchlets and petioles softly pubescent,* and the fruit acute at the base, 2-edged, and gradually expanding into the long wing above. —Same localities as No. 1.

3. **F. vir'idis,** Michx. f. (GREEN ASH.) *Glabrous throughout.* Fruit as in No. 2. Leaflets 5-9, *bright green both sides.*—Western Ontario and N.W., along streams.

4. **F. quadrangula'ta,** Michx. (BLUE ASH.) *Branchlets often square,* smooth. Leaflets sharply serrate, green both sides. Fruit narrowly oblong, of the same width at both ends, often notched at the apex, wing-margined.—Lake Erie coast.

* * *Leaflets sessile.*

5. **F. sambucifo'lia,** Lam. (BLACK or WATER ASH.) Branchlets and petioles smooth. Leaflets 7-9, *sessile,* serrate. *Fruit winged all round.* Calyx wanting, and the flowers consequently naked.—Swamps.

III. APET'ALOUS DIVISION.

Flowers destitute of corolla, and sometimes also of calyx.

ORDER LXXIII. **ARISTOLOCHIA'CEÆ.** (BIRTHWORT F.)

Herbs with perfect flowers, *the tube of the 3-lobed calyx adherent to the 6-celled many-seeded ovary.* Leaves heart-shaped or kidney-shaped, on long petioles from a thick rootstock. Stamens 12 or 6. Flowers solitary. Calyx dull-coloured, the lobes valvate in the bud.

AS'ARUM, Tourn. WILD GINGER.

A. Canaden'se, L. Radiating stigmas 6. Leaves only a single pair, kidney-shaped, and rather velvety, the peduncle in the fork between the petioles, close to the ground. Rootstock aromatic. Calyx brown-purple inside, the spreading lobes pointed.—Rich woods.

ORDER LXXIV. **PIPERACEÆ.** (PEPPER FAMILY.)

A small family having, with us, but a single representative :—

SAURU'RUS, L. LIZARD'S TAIL.

S. cer'nuus, L. A swamp herb, with jointed branching stem, 2 feet high. Leaves petioled, heart-shaped, with converging ribs. Flowers white, in a dense terminal spike, nodding at the end, each flower with a lanceolate bract. Flowers perfect, but entirely destitute of calyx and corolla. Stamens usually 6 or 7, with long slender white filaments. Carpels 3 or 4, slightly united at the base.

ORDER LXXV. PHYTOLACCA'CEÆ. (POKEWEED F.)

Herbs with alternate leaves and perfect flowers, resembling in most respects the plants of the next Order, *but the ovary is composed of several carpels in a ring, forming a berry in fruit.* Only one Genus and one Species.

PHYTOLAC'CA, Tourn. POKEWEED.

P. decan'dra, L. (COMMON POKE.) Calyx of 5 rounded white sepals. Ovary green, of ten 1-seeded carpels united in a ring. Styles 10, short and separate. Stamens 10. Fruit a crimson or purple 10-seeded berry. Stem very tall and stout, smooth. Flowers in long racemes opposite the leaves. —Sandy soil.

ORDER LXXV. ILLECEBRA'CEÆ. (KNOTWORT FAMILY.)

Small diffuse or tufted herbs, with mostly opposite and entire leaves, scarious stipules, and a 4-5-toothed or parted coriaceous persistent calyx. Petals wanting. Stamens perigynous, as many as the lobes of the calyx and opposite them, or fewer. Style 2-cleft at the apex. Fruit a 1-seeded utricle.

PARONYCH'IA, Tourn. WHITLOW-WORT.

P. sessiliflo'ra, Nutt. Flowers terminal, solitary, sessile. Stems densely matted or tufted, from a woody root. The dry, silvery stipules 2-cleft. Sepals oblong-linear, concave, awned at the apex.—N. W. prairies.

ORDER LXXVI. **CHENOPODIA'CEÆ.** (GOOSEFOOT F.)

Homely herbs, with more or less succulent leaves (chiefly alternate), and small greenish flowers mostly in interrupted spikes. Stamens usually as many as the lobes of the calyx and opposite them. Ovary 1-celled and 1-ovuled, forming an achene or utricle in fruit. Stigmas mostly 2.

Synopsis of the Genera.

1. **Chenopo'dium.** Weeds with (usually) mealy leaves, and very small perfect greenish sessile flowers in small panicled spiked clusters. Calyx 5-cleft, more or less enveloping the fruit, and sometimes becoming fleshy and berry-like. Stamens mostly 5, filaments slender.
2. **Monol'epis.** A low annual, glabrous or somewhat mealy, with small fleshy leaves. *Sepal only 1*, bract-like. *Stamen 1.* Styles 2. Seed vertical, flattened.—N. W. prairies only.
3. **Cyclolo'ma.** A much-branched coarse herb, *with very small scattered sessile flowers in open panicles.* Flowers perfect or pistillate. Calyx 5-cleft, the concave lobes strongly keeled, *at length with a broad and continuous horizontal wing.* Stamens 5; styles 3. Seed horizontal, flat.
4. **At'riplex.** Flowers *monœcious or diœcious*, the staminate with a regular calyx, in spiked clusters; the pistillate without a calyx, but with a pair of appressed bracts.
5. **Corisper'mum.** Flowers all *perfect, single*, and sessile in the axils of the upper leaves, usually forming a spike. *Calyx of a single delicate sepal.* Low herbs, with linear 1-nerved leaves
6. **Salicor'nia.** Low saline plants with fleshy leafless jointed stems and opposite branches. Flowers perfect, densely spiked, *3 together sunk in each hollow of the fleshy axis of the spike.* Calyx small and bladder-like, with a toothed or torn margin, at length spongy and narrowly wing-bordered. Stamens 1 or 2. Styles 2, united at the base.
7. **Suæda.** Fleshy saline plants, with alternate terete linear leaves. Flowers perfect, sessile in the axils of leafy bracts. Calyx 5-parted, *very fleshy.* Stamens 5, the anthers exserted. Stigmas 2 or 3. Seed horizontal.
8. **Sal'sola.** A diffusely branching annual, with alternate awl-shaped *prickly-pointed* leaves. Flowers perfect, sessile, with 2 bractlets. Calyx 5-parted, persistent, its divisions at length horizontally winged on the back. Seed horizontal.

CHENOPODIACEÆ.

1. CHENOPO'DIUM, L. GOOSEFOOT. PIGWEED.

* Fruiting calyx dry.

1. **C. album**, L. (LAMB'S QUARTERS.) Stem upright, 1-3 feet high. Leaves varying from rhombic-ovate to lanceolate, more or less toothed, *mealy, as are also the dense flower-clusters.*—Extremely common in cultivated soil.

2. **C. ur'bicum**, L. Rather pale and only slightly mealy, 1-3 feet high, branches *erect.* Leaves triangular, acute, *coarsely and sharply many-toothed.* Spikes *erect,* crowded in a long and narrow racemose panicle.—Waste places in towns.

3. **C. hy'bridum**, L. (MAPLE-LEAVED GOOSEFOOT.) *Bright green.* Stem widely branching, 2-4 feet high. Leaves thin, large, triangular, heart-shaped, sinuate-angled, the angles extended into pointed teeth. Panicles loose, leafless. Plant with a rank unpleasant odour.—Waste places.

4. **C. Bot'rys**, L. (JERUSALEM OAK.) Not mealy, but *sticky;* low, spreading, sweet-scented. Leaves deeply sinuate, *slender-petioled.* Racemes in *loose* divergent corymbs.—Roadsides; escaped from gardens.

5. **C. ambrosioi'des**, L. (MEXICAN TEA.) Not mealy, but sticky. Leaves *slightly* petioled, wavy-toothed or nearly entire. Spikes *densely* flowered.—Streets of towns.

6. **C. glau'cum**, L. (OAK-LEAVED GOOSEFOOT.) Somewhat glaucous-mealy, 5-12 inches high, spreading. Leaves *sinuately pinnatifid-toothed,* oblong, obtuse. Clusters small, in axillary spikes. Seed vertical, exserted, with *sharp* edges.

7. **C. Bonus Henri'cus**, L. (*Blitum Bonus Henricus,* Reichenbach.) (GOOD-KING-HENRY.) Stout, erect, 1-2 feet high. Leaves broadly triangular-hastate, slightly sinuate or entire. Flowers somewhat densely paniculate-spiked. Seed vertical, exserted, with *blunt* edges. Not common.

** Fruiting calyx fleshy and often coloured.

8. **C. capita'tum**, Watson. (*Blitum capitatum,* L.) (STRAWBERRY BLITE.) Stem ascending, branching. Leaves

smooth. The axillary head-like clusters *bright red* in fruit, and resembling strawberries.—Dry soil, margins of woods, etc.

9. **C. ru'brum,** L. (*Blitum maritimum,* Nutt.) (COAST BLITE.) Stem angled, much-branched. Leaves thickish, acuminate, the upper linear-lanceolate. Flower-clusters scattered in axillary *leafy* spikes. Stamen 1.—N.W., in saline soil.

2. MONOL'EPIS, Schrad.

M. chenopodioi'des, Moq. Branched from the base. Leaves lanceolate-hastate or sometimes narrowly spathulate, entire or sparingly sinuate-toothed. Flower-clusters often reddish.—N.W.

3. CYCLOLO'MA, Moquin. WINGED PIGWEED.

C. platyphyl'lum, Moq. Diffuse, 6-15 inches high, light-green or sometimes purple.—S.W. Ontario.

4. AT'RIPLEX, Tourn. ORACHE.

1. **A. pat'ulum,** L. Erect or diffuse, scurfy, green or rather hoary. Leaves varying from triangular or halberd-shaped to lance-linear, petioled.

Var. **hasta'tum,** Gray, has at least the lower leaves broadly triangular-hastate, often toothed.—Atl. Prov. and N.W.

Var. **littora'le,** Gray, is slender, with leaves linear-lanceolate to linear.—Waste places.

2. **A. Nuttal'lii,** Watson. A shrubby densely-appressed-scurfy perennial, with oblong-spathulate to narrowly oblanceolate entire leaves.—N.W. only.

5. CORISPER'MUM, Ant. Juss. BUG-SEED.

C. hyssopifo'lium, L. Somewhat hairy when young, pale. Stamens 1 or 2. Styles 2. Fruit oval, flat.—Sandy beaches, western and south-western Ontario, and N.W.

6. SALICOR'NIA, Tourn. GLASSWORT. SAMPHIRE.

1. **S. herba'cea,** L. (SAMPHIRE.) Flowers perfect, in threes, embedded in hollows on the thickened upper joints,

forming an elongated narrow spike. Calyx small and bladder-like, its margin toothed. Stamens 1 or 2.—Salt marshes, Atl. Prov. and N.W.

2. **S. mucrona'ta,** Bigel., has thick spikes and mucronate-pointed scales. Stem *turning red* when old.—Atl. seacoast.

SUÆ'DA, Forskal. SEA BLITE.

S. linea'ris, Moq. (*Suæda maritima*, Gray.) (SEA BLITE.) A branching fleshy herb, with alternate, roundish, linear leaves. Flowers perfect, sessile in the axils of leafy bracts on slender branchlets. Sepals very thick. Stamens 6, with anthers exserted.—Atl. Prov.

8. SAL'SOLA, L. SALTWORT.

S. Ka'li, L. (SALTWORT.) Flowers perfect, sessile, with two bractlets, single in axils of leaves. Calyx 5-parted, enclosing the depressed fruit. Stamens 5. A branching plant with alternate, awl-shaped, prickly-pointed leaves.— Sandy sea-shore, and rapidly spreading westward.

ORDER LXXVII. **AMARANTA'CEÆ.** (AMARANTH F.)

Homely weeds, a good deal like the plants of the last Order, *but the flower-clusters are interspersed with dry and chaff-like (sometimes coloured) persistent bracts, usually 3 to each flower.*

Synopsis of the Genera.

1. **Amaran'tus.** Flowers *monœcious or polygamous*, all with a calyx of 3 or 5 distinct erect sepals.
2. **Monte'lia.** Flowers *diœcious;* calyx none in the pistillate flowers.

1. AMARAN'TUS, Tourn. AMARANTH.

1. **A. panicula'tus,** L. *Reddish flowers* in terminal and axillary *slender* spikes, the bracts awn-pointed.—In the neighbourhood of gardens.

2. **A. retroflex'us,** L. (PIGWEED.) *Flowers greenish,* in spikes, forming a stiff panicle. Leaves a dull green, long-petioled, ovate, wavy-margined. *Stem erect.*—Common in cultivated soil.

8. **A. albus,** L. Flowers greenish, in small close axillary clusters. Stem low and spreading.—Roadsides.

2. MONTE'LIA, Moquin.

M. tamaris'cina, Gray. (*Acnida ruscocarpa*, Gray.) A tall smooth herb, with lanceolate or oblong-ovate alternate leaves on long petioles, and small clusters of greenish flowers in interrupted spikes.—Wet places.

Var. **concatena'ta,** Gray, (*Acnida tuberculata*, Moq.) has the flowers in the lower part of the stem in close clusters in the axils of the leaves.

ORDER LXXVIII. **POLYGONA'CEÆ.** (BUCKWHEAT F.)

Herbs, *well marked by the stipules of the alternate leaves being in the form of membranous sheaths above the usually swollen joints of the stem* (these obsolete in one Genus). Flowers usually perfect. Calyx 4–6-parted. Stamens 4–9, inserted on the base of the calyx. Stigmas 2 or 3. Ovary 1-celled, with a single ovule rising from the base, forming a little nutlet.

Synopsis of the Genera.

* *Flowers involucrate. Stamens 9. Stipules obsolete.*

1. **Eriog'onum.** Involucre 4–8-toothed, the flowers exserted. Calyx 6-parted, coloured (yellow in ours), enclosing the achene.—N.W. only.

* * *Flowers not involucrate. Stamens 4–8.*

2. **Polyg'onum.** Sepals 5 (occasionally 4), often coloured and petal-like, *persistent*, embracing the 3-angled (or sometimes flattish) nutlet or achene. Flowers in racemes or spikes, or sometimes in the axils.

3. **Ru'mex.** Sepals 6, *the 3 outer ones herbaceous and spreading in fruit*, the 3 inner (called *valves*) somewhat petal-like and, after flowering, convergent over the 3-angled achene, *often with a grain-like projection on the back*. Stamens 6. Styles 3. Flowers asually in crowded whorls, the latter in panicled racemes.

4. **Fagopy'rum.** Calyx 5-parted, petal-like. Stamens 8, *with 8 yellow glands between them*. Styles 3. Achenes 3-angled. Flowers white, in panicles. Leaves triangular heart-shaped or halberd-shaped.

POLYGONACEÆ.

1. ERIOG'ONUM, Michx.

1. **E. fla'vum**, Nutt. Woolly throughout, a few inches high. Leaves oblanceolate. Umbel of 3-9 rays, on a naked peduncle. Flowers yellow, silky.—N.W.

2. POLYG'ONUM, L. KNOTWEED.

* *Flowers along the stem, inconspicuous, greenish-white, nearly sessile in the axils of the small leaves. Sheaths cut-fringed or torn.*

1. **P. marit'imum**, L. (COAST KNOTGRASS.) Prostrate with stout stems, glaucous. Leaves thick, oval to narrowly oblong. Flowers in the axils of leaves, clustered. Stipules very conspicuous. Stamens 8. Achenes smooth and shining, projecting above the calyx.—Sea-coast.

2. **P. avicula're**, L. (KNOTGRASS. GOOSEGRASS.) A weed everywhere in yards and waste places. *Stem prostrate and spreading.* Stamens chiefly 5. Achene 3-sided, *dull.* Stigmas 3. Leaves sessile, lanceolate or oblong. Var. **erectum**, Roth. (*P. erectum*, L.) is upright and larger, with broader leaves.

3. **P. ramosis'simum**, Michx. Erect or ascending, *yellowish-green.* Leaves lanceolate to linear, acute, very small above. Sepals mostly 6; *stamens 3-6.* Achene smooth and shining.—Chiefly westward; sandy places.

4. **P. ten'ue**, Michx. Stem *slender*, *upright*, sparingly branched, *angled*. Leaves sessile, *narrowly linear*, very acute, *3-nerved*. Stamens 8. Achene dull black.—Dry soil and rocky places.

* * *Flowers in terminal spikes or racemes, mostly rose-coloured or pinkish, occasionally greenish.*

— *Leaves not heart-shaped or arrow-shaped.*

5. **P. lapathifo'lium**, L. *Sheaths not fringed, stem nearly smooth*, 3-6 feet high. Leaves long, tapering from near the base to a narrow point, rough on the midrib and margins. *Spikes oblong to linear and erect or nearly so. Stamens 6. Styles 2. Achene flat or hollow-sided.*—In muddy places along streams and ponds.

Var. incarna'tum, Watson, has the spikes more slender and elongated, nodding.

Var. inca'num, Koch, is 6-12 inches high, with small leaves more or less *flocculent-hoary* underneath, and short spikes. Chiefly eastward and N. W.

6. **P. Pennsylvan'icum,** L. Sheaths not fringed. Stem 1-3 feet high, *the upper branches and the peduncles bristly with stalked glands. Spikes thick, erect. Stamens 8.* Achene flat.—Low open grounds.

7. **P. Persica'ria,** L. (LADY'S THUMB.) Sheaths with a somewhat ciliate border. *Stem nearly smooth,* a foot or more in height. *Leaves with a dark blotch on the middle of the upper surface.* Spikes dense, erect, on naked peduncles. Stamens 6. Achene flat or 3-angled, according as the stigmas are 2 or 3.—Very common near dwellings in moist ground.

8. **P. amphib'ium,** L. (WATER PERSICARIA.) Spike of flowers dense, oblong, *showy, rose-red. Stem floating in shallow water* or rooting in soft mud, *nearly glabrous.* Leaves long-petioled, *often floating.* Sheaths not bristly-fringed. *Stamens 5.* Stigmas 2.—In shallow water, mostly northward.

9. **P. Muhlenberg'ii,** Watson, differs from the last in being *rough with appressed hairs* all over.—Ditches.

10. **P. Hartwright'ii,** Gray, is distinguished from P. amphibium by its *foliaceous and ciliate sheaths.*—Muddy margins of ponds and lakes.

11. **P. hydropiperoi'des,** Michx. (MILD WATER-PEPPER.) Stem slender, 1-3 feet high, *in shallow water.* Leaves narrow, roughish. *Sheaths hairy and fringed with long bristles.* Spikes slender, erect, pale rose-coloured or whitish. Stamens 8. Stigmas 3. Achene 3-angled.—In shallow water.

12. **P. acre,** H. B. K. (WATER SMARTWEED.) Sheaths fringed with bristles. *Leaves transparent-dotted.* Stem rooting at the decumbent base, 2-4 feet high, *in shallow water or muddy soil.* Leaves narrow, taper-pointed. Spikes

slender, *erect*, pale rose-coloured. *Sepals glandular-dotted. Stamens 8.* Achene 3-angled, *shining.*—Muddy soil or shallow water.

13. **P. Hydrop'iper,** L. (COMMON SMARTWEED or WATER-PEPPER.) Sheaths and leaves as in the last, the leaves, however, larger. *Spikes slender, nodding,* greenish. *Sepals glandular-dotted. Stamens 6.* Achene *dull.*—Wet places.

14. **P. Virginia'num,** L. Calyx greenish, *unequally 4-parted.* Stamens 5. Styles 2, persistent on the flat achene. Flowers in long and slender naked spikes. Stem upright, nearly smooth. Leaves ovate or ovate-lanceolate, taper-pointed, rough-ciliate. Sheaths hairy and fringed.—Thickets, in rich soil.

← ← *Leaves heart-shaped or sagittate. Sheaths much longer on one side than on the other.*

15. **P. arifo'lium,** L., (HALBERD-LEAVED TEAR-THUMB) with grooved stem, halberd-shaped long-petioled leaves, flowers in short loose racemes, 6 stamens, and a flattish achene, is not uncommon on the Lower St. Lawrence; rare in Ontario.

16. **P. sagitta'tum,** L. (ARROW-LEAVED TEAR-THUMB.) *Stem 4-angled, the angles beset with reflexed minute prickles,* by which the plant is enabled to climb. Leaves arrow-shaped. Stamens 8. Achenes 3-angled.—Common in low grounds, especially beaver-meadows.

17. **P. Convol'vulus,** L. (BLACK BINDWEED.) *Stem twining,* not prickly but *roughish; the joints naked.* Flowers in loose panicled racemes, 3 of the calyx-lobes rigid in fruit. Leaves heart-shaped and partly halberd-shaped. Not climbing so high as the next.—Cultivated grounds and waste places.

18. **P. dumeto'rum,** L., var. **scandens,** Gray. (CLIMBING FALSE BUCKWHEAT.) Stem twining high, *smooth; sheaths naked,* 3 of the calyx-lobes *winged in fruit.*—Moist thickets.

19. **P. cilino'de,** Michx. Stem twining, *minutely downy. Sheaths fringed at the base with reflexed bristles.*—Sandy pine woods and rocky hills.

3. RUMEX, L. Dock. Sorrel.

* *Herbage not sour, nor the leaves halberd-shaped.*

1. **R. orbicula'tus**, Gray. (GREAT WATER DOCK.) Growing in marshes. Stem erect, stout, 5-6 feet high. Leaves lanceolate; *not wavy-margined or heart-shaped*, often over a foot long. Flowers nodding *on thread-like pedicels*. Valves nearly orbicular, finely net-veined, each with a grain on the back.—Wet places.

2. **R. salicifo'lius**, Weinmann, (WHITE DOCK) may be looked for in marshes on the sea-coast and far northward. The whorls of flowers are dense and form a very conspicuous spike, owing to the great size of the grains on the back of the valves.

3. **R. verticilla'tus**, L. (SWAMP DOCK.) Leaves lanceolate or oblong-lanceolate, *not wavy*, the *lowest* often heart-shaped. Stem tall. Fruit-bearing pedicels slender, club-shaped, abrubtly reflexed, several times longer than the fruiting calyx. Valves dilated-rhomboid, *strongly wrinkled*, each bearing a very large grain.—Swamps, common.

4. **R. crispus**, L. (CURLED DOCK.) *Leaves with strongly wavy or curly margins*, lanceolate. Whorls of flowers in long wand-like racemes. Valves grain-bearing.—Cultivated soil and waste places.

5. **R. obtusifo'lius**, L. (BITTER DOCK.) Lowest leaves oblong heart-shaped, obtuse, only slightly wavy-margined; the upper oblong-lanceolate, acute. Whorls loose, *distinct*. Valves somewhat halberd-shaped, *deeply toothed at the base*, usually *one only* grain-bearing.—Waste grounds.

6. **R. sanguin'eus**, L. *Leaves oblong-lanceolate*, often fiddle-shaped, wavy-margined. *Whorls distant*, in long slender leafless spikes. Valves narrowly oblong, obtuse, entire. *Veins of the leaf red or green.*—Atl. Prov. chiefly.

7. **R. marit'imus**, L. (GOLDEN DOCK.) Low, slightly-pubescent, much branched. Leaves linear-lanceolate, wavy-margined, the lower auricled or heart-shaped at base. Flowers in whorls forming leafy spikes. Valves oblong,

lance-pointed, each bearing 2-3 long bristles on each side, and a large grain on the back.—Sea-shore, Atl. Prov.

* * *Herbage sour; leaves halberd-shaped.*

8. **R. Acetosel'la,** L. (FIELD or SHEEP SORREL.) Stem 6-12 inches high. *Flowers diœcious,* in a terminal naked panicle.—A very common weed in poor soil.

3. FAGOPY'RUM, Tourn. BUCKWHEAT.

1. **F. esculen'tum,** Mœnch. (BUCKWHEAT.) Old fields and copses, remaining after cultivation. Achene smooth and shining.

2. **F. Tartar'icum,** Gærtn., has very small flowers, and a *dull roughish* achene.—Escaped from cultivation in a few places.

ORDER LXXIX. LAURA'CEÆ. (LAUREL FAMILY.)

Trees or shrubs with spicy-aromatic bark and leaves, the latter simple (often lobed), alternate and marked with small transparent dots (visible under a lens). Sepals 6, petal-like. Flowers diœcious or polygamo-diœcious. Stamens in sterile flowers 9, inserted at the base of the calyx. Anthers opening by uplifting valves. Ovary in fertile flowers free from the calyx, 1-celled, with a single ovule hanging from the top of the cell. Style and stigma 1. Fruit a 1-seeded drupe.

1. SAS'SAFRAS, Nees. SASSAFRAS.

S. officina'le, Nees. A small or moderate-sized tree with yellowish or greenish-yellow twigs and ovate or 3-lobed entire leaves. Flowers greenish-yellow, in naked corymbs, appearing with the leaves in the axils of the latter. Drupe *blue,* on a reddish pedicel. The 9 stamens in 3 rows, the 3 inner each with a pair of yellow glands at the base of the filament. *Anthers 4-celled, 4-valved.*—Rich woods, in southern and western Ontario.

2. LIN'DERA, Thunberg. WILD ALLSPICE. FEVER-BUSH.

L. Benzo'in, Meisner. (SPICE-BUSH.) A nearly smooth shrub with oblong-obovate leaves, pale beneath. Flowers honey-yellow in lateral umbel-like clusters, before the leaves. Stamens very much as in Sassafras, but the anthers are *2-celled and 2-valved.* Pistillate flowers with 15-18 rudiments of stamens. *Drupe red.*—Damp woods, in early spring.

Order LXXX. THYMELEA'CEÆ. (Mezereum F.)

Shrubs with tough leather-like bark and entire leaves. Flowers perfect. Calyx tubular, resembling a corolla, pale yellow. Stamens twice as many as the lobes of the calyx (in our species 8). Style thread-like. Stigma capitate. Ovary 1-celled, 1-ovuled, free from the calyx. ' Fruit a berry-like drupe. Only one Species in Canada.

DIRCA, L. Leatherwood. Moose-wood.

1. **D. palustris**, L. A branching shrub, 2-5 feet high, with curious *jointed branchlets* and nearly oval leaves on short petioles. Flowers in clusters of 3 or 4, preceding the leaves. Filaments exserted, half of them longer than the others.— Damp woods.

2. **Daphne Meze'reum**, L., has escaped from cultivation in a few places. A low shrub with purple, rose-coloured or whitish flowers, preceding the leaves in early spring.

Order LXXXI. ELÆAGNA'CEÆ. (Oleaster F.)

Shrubs with perfect or diœcious flowers, *and leaves which are scurfy on the under surface.* The calyx-tube in the fertile flowers *becomes fleshy and encloses the ovary, forming a berry-like fruit.* Otherwise the plants of this Order are not greatly different from those of the last.

Synopsis of the Genera.

1. **Elæag'nus.** Flowers perfect. Stamens 4. Leaves alternate.
2. **Shepherd'ia.** Flowers diœcious. Stamens 8. Leaves opposite.

1. ELÆAG'NUS, Tourn.

E. argen'tea, Pursh. (Silver-Berry.) Shrub 6-12 feet high, the young branches covered with rusty scales. Leaves elliptical to lanceolate, silvery-scurfy. Flowers many, deflexed, silvery outside, pale-yellow within, fragrant.—N.W.

2. SHEPHERD'IA, Nutt. SHEPHERDIA.

S. Canadensis, Nutt. Calyx in sterile flowers 4-parted. Stamens 8. Calyx in fertile flowers urn-shaped, 4-parted. Berries yellow. Branchlets brown-scurfy. Leaves opposite, entire, ovate, green above, silvery-scurfy beneath, the small flowers in their axils.—Gravelly banks of streams and lakes.

Order LXXXII. SANTALA'CEÆ. (Sandalwood F.)

Low herbaceous or partly woody plants (with us) with perfect flowers, these *greenish-white*, in terminal or axillary corymbose clusters. Calyx bell-shaped or urn-shaped, 4–5-cleft, adherent to the 1-celled ovary, lined with a 5-lobed disk, the stamens on the edge of the latter between its lobes and *opposite the lobes of the calyx, to which the anthers are attached by a tuft of fine hairs.* Fruit nut-like, crowned with the persistent calyx-lobes.

COMAN'DRA, Nutt. BASTARD TOAD-FLAX.

1. **C. umbella'ta,** Nutt. Stem 8–10 inches high, leafy. Leaves oblong, pale-green, an inch long. Flower-clusters at the summit of the stem. Calyx-tube prolonged and forming a neck to the fruit. Style slender.—Dry soil.

2. **C. liv'ida,** Richardson. Peduncles axillary, slender, several-flowered. Leaves oval, alternate, almost sessile. Fruit pulpy when ripe, red.—Boggy barrens near the Atl. coast, and N.W.

3. **C. pal'lida,** A. DC. Leaves glaucous, linear to narrowly lanceolate, acute. Fruit ovoid, sessile or on short stout pedicels.—N.W.

Order LXXXIII. EUPHORBIA'CEÆ. (Spurge F.)

Plants with milky juice and monœcious flowers, represented in Canada chiefly by the two following genera:—

1. EUPHOR'BIA, L. SPURGE.

Flowers monœcious, the sterile and fertile ones both destitute of calyx and corolla, *but both contained in the same 4-5-lobed cup-shaped involucre which resembles a calyx*, and therefore the whole will probably at first sight be taken for a single flower. Sterile flowers numerous, *each of a single naked stamen from the axil of a minute bract.* Fertile flower only 1 in each involucre; ovary 3-lobed, *soon protruded on a long pedicel*; styles 3, each 2-cleft. Peduncles terminal, often umbellate.

* *Leaves all similar and opposite, short-petioled, green or blotched with brown above, furnished with scale-like or fringed stipules. Stems spreading or prostrate, much forked. Involucres in terminal or lateral clusters, or one involucre in each fork, the involucre invariably with 4 (mostly petal-like) glands in the sinuses.*

1. **E. polygonifo'lia,** L. Leaves *entire*, oblong-linear, mucronate, *very smooth*. Stipules bristly-fringed. Peduncles in the forks. *Glands of the involucre very small, not petal-like.* Pods obtusely angled.—Shores of the Great Lakes, in sandy or gravelly places.

2. **E. serpens,** H. B. K. Leaves *entire*, round-ovate, very small, smooth. Stipules membranaceous, triangular. Peduncles longer than the petioles, in loose clusters. Glands of the small involucre with minute crenulate appendages. Stems thread-like, prostrate. Pods acutely angled. *Seeds smooth.*—London and westward, not common.

3. **E. glyptosper'ma,** Engel. Leaves *serrulate* towards the apex, linear-oblong, *very unequal at the base*. Stipules lanceolate, cut into bristles. Peduncles as long as the petioles, in dense lateral clusters. Glands in the small involucre with crenulate appendages. Stems erect-spreading. Pods sharply angled. *Seeds sharply 4-angled, with 5 or 6 transverse wrinkles.*—Gravelly soil.

* * *Only the uppermost or floral leaves whorled or opposite. Stems erect. Stipules none. Involucres 5-lobed; inflorescence umbelliform, in the forks of the branches, and terminal.*

4. **E. macula'ta,** L. Leaves *serrulate*, oblong-linear, *somewhat pubescent, with a brownish blotch in the centre*, **very**

oblique at the base. *Peduncles in dense lateral clusters.* Glands of the involucre with reddish petal-like attachments. *Pods sharply angled.*—Roadsides.

5. **E. hypericifo'lia,** L. Stem ascending. Leaves *serrate*, often with a red spot or with red margins, oblique at the base, ovate-oblong or oblong-linear. *Peduncles in cymes at the ends of the branches.* Glands of the involucre with *white* or occasionally reddish petal-like attachments. Pod smooth, obtusely angled.—Cultivated soil.

6. **E. corolla'ta,** L. Conspicuous for the *5 bright-white false lobes of the involucre*, resembling petals; the true lobes very small.—Gravelly or sandy soil.

*** *Involucres chiefly in terminal umbels, and their glands always without petal-like attachments. Leaves without stipules or blotches, those of the stem alternate or scattered, the floral ones usually of a different shape, and whorled or opposite.*

7. **E. platyphyl'la,** L. Umbel 5-rayed. Stem erect, 8–18 inches high. Upper stem-leaves lance-oblong, acute, serrulate, the uppermost heart-shaped, the floral ones triangular-ovate and cordate. *Pod warty.*—Shores of the Great Lakes.

8. **E. Heliosco'pia,** L. Umbel first 5-rayed, then with 3, and finally merely forked. Stem ascending, 6–12 inches high. Leaves all obovate, rounded or notched at the apex. serrate. *Pods smooth.*—Along the Great Lakes.

9. **E. Cyparis'sias,** L., with densely clustered stems, and crowded linear stem-leaves (the floral ones round heart-shaped), and a many-rayed umbel, has escaped from gardens in some localities.

10. **E. Peplus,** L., has the umbel 3-rayed, then forking. Glands long-horned. Seeds ash-coloured, 2-grooved on the inner face, and pitted on the back.—Rather rare, waste places.

2. ACALY'PHA, L. THREE-SEEDED MERCURY.

A. Virgin'ica, L. Flowers monœcious, both kinds having a calyx, the staminate 4-parted, the pistillate 3–5-parted; no involucre. Staminate flowers very small, in spikes, with 1–3 pistillate flowers at the base, in the axil of a large leaf-like 5-9-lobed bract. Stamens 8-16, monadelphous at the base, the anther-cells hanging from the apex of the filament. Styles 3, the stigmas cut-fringed, usually red. Pod separating into 3 globular carpels. A nettle-like weed, with ovate, sparsely serrate, alternate, long-petioled leaves.—Fields and open places.

Order LXXXIV. URTICA'CEÆ. (Nettle F.)

Herbs, shrubs, or trees, with monœcious or diœcious (or, in the Elms, sometimes perfect) flowers, with a regular calyx free from the 1-2-celled ovary which becomes a 1-seeded fruit. Stamens opposite the lobes of the calyx. This Order is divided into four well-marked Suborders.

Suborder I. ULMA'CEÆ. (Elm Family.)

Trees, with alternate simple leaves, and deciduous small stipules. Flowers often perfect. Styles 2. *Fruit a samara winged all round, or a drupe.*

* Fruit a samara; anthers extrorse.

1. **Ulmus.** Flowers in lateral clusters, earlier than the leaves, purplish or greenish-yellow. Calyx bell-shaped, 4-cleft. Stamens 4-9; the filaments long and slender. Ovary 2-celled, but the samara only 1-seeded. Stigmas 2.

* * Fruit a drupe; anthers introrse.

2. **Celtis.** Flowers greenish, polygamous, the pistillate solitary or in pairs, appearing with the leaves. Calyx 5-6-parted, persistent. Stamens 5-6. Stigmas 2, long and pointed and recurved. Ovary 1-ovuled.

Suborder II. ARTOCAR'PEÆ. (Bread-fruit and Fig F.)

Flowers monœcious or diœcious, crowded in catkin-like spikes or heads the whole pistillate catkin becoming an aggregate fruit from the enlargement of the calyx in the several flowers. Calyx 4-parted. Stamens 4. Ovary 2-celled, 1 cell eventually disappearing. Styles 2.

3. **Morus.** Pistillate and staminate flowers in separate catkins. Trees with milky juice and rounded leaves. Staminate spikes slender.

URTICACEÆ. 199

SUBORDER III. **URTI'CEÆ.** (NETTLE FAMILY.)

Herbs with watery juice and opposite or alternate leaves, often beset with stinging hairs. Flowers monœcious or diœcious, in spikes or racemes. Stamens as many as the sepals. Style only 1. Ovary 1-celled. Fruit an achene.

4. **Urti'ca.** *Leaves opposite.* Plant beset with stinging hairs. Sepals 4 in both sterile and fertile flowers. Stamens 4. Stigma a small sessile tuft. Achene flat, enclosed between the 2 larger sepals. Flowers greenish.
5. **Laport'ea.** *Leaves alternate.* Plant beset with stinging hairs Sepals 5 in the sterile flowers, 4 in the fertile, 2 of them much smaller than the other 2. Stigma awl-shaped. Achene flat, *very oblique, reflexed on its winged pedicel.*
6. **Pil'ea.** Leaves opposite. *Whole plant very smooth and semi-transparent.* Sepals and stamens 3–4. Stigma a sessile tuft.
7. **Bœhme'ria.** Leaves mostly opposite. No stinging hairs. Sepals and stamens 4 in the sterile flowers. Calyx tubular or urn-shaped in the fertile ones, and enclosing the achene. Stigma long and thread-like.
8. **Parieta'ria.** Leaves alternate, entire, 3-ribbed. No stinging hairs. Flowers polygamous, in involucrate-bracted cymose axillary clusters. Calyx of the pistillate flowers tubular or bell-shaped, 4-lobed. Stigma tufted. Staminate flowers nearly as in the last.

SUBORDER IV. **CANNABIN'EÆ.** (HEMP FAMILY.)

Rough herbs with watery juice and tough bark. Leaves opposite and *palmately divided or lobed.* Flowers diœcious. Sterile ones in compound racemes; stamens 5; sepals 5. Fertile ones in crowded clusters; sepal only 1, embracing the achene. Stigmas 2.

9. **Can'nabis.** A rather tall rough plant with palmately compound leaves of 5–7 linear-lanceolate serrate leaflets. Fertile flowers spiked-clustered.
16. **Hu'mulus.** Leaves 3–5-lobed. *Plant climbing.* Fertile flowers in a short spike, forming a membranaceous catkin in fruit.

1. **ULMUS**, L. ELM.

1. **U. fulva,** Michx. (RED or SLIPPERY ELM.) Flowers nearly sessile. *Leaves very rough above, taper-pointed. Buds downy with rusty hairs.* A medium-sized tree, with mucilaginous inner **bark.**

2. **U. America'na, L.** (AMERICAN or WHITE ELM.) *Leaves not rough above, abruptly pointed.* Flowers in drooping pedicels. *Buds glabrous.* A large ornamental tree, with drooping branchlets.—Moist woods.

3. **U. racemo'sa, Thomas.** (CORKY WHITE ELM.) Resembling the last, but the *bud-scales are downy-ciliate, the branches corky, and the flowers racemed.*—Chiefly along roadsides and borders of fields.

2. CEL'TIS, L. NETTLE-TREE. HACKBERRY.

C. occidenta'lis, L. (SUGARBERRY.) A small tree of Elm-like aspect. *Leaves reticulated,* ovate, taper-pointed, serrate, more or less oblique at the base. Fruit as large as a pea, dark-purple when ripe, the flesh thin.—Low grounds; a few trees here and there through Ontario.

3. MORUS, Tourn. MULBERRY.

1. **M. ru'bra, L.** (RED MULBERRY.) Leaves heart-ovate, rough above, downy beneath, pointed. Fruit red, turning dark-purple, long.—Niagara district, and south-westward.

2. **M. alba, L.** (WHITE M.) Leaves smooth and shining. Fruit *whitish.*—S. W. Ontario.

4. URTI'CA, Tourn. NETTLE.

1. **U. gra'cilis, Ait.** Stem slender, 2-6 feet high. Leaves ovate-lanceolate, pointed, serrate, 3-5-nerved from the base, nearly smooth, the long petioles with a few bristles. Flower-clusters in slender spikes.—Moist ground and along fences.

2. **U. dioi'ca, L.** (STINGING NETTLE.) Plant bristly with very stinging hairs. Leaves ovate, cordate, *very deeply serrate.* Spikes branching.—Waste places.

3. **U. u'rens, L.** Leaves elliptical or ovate, coarsely and deeply serrate with spreading teeth, petioled. Flower-clusters 2 in each axil, composed of both staminate and pistillate flowers.—Waste grounds, Atl. Prov.

5. LAPORT'EA, Gaudichaud. WOOD-NETTLE.

L. Canadensis, Gaudichaud. Stem 2-3 feet high. Leaves large, ovate, long-petioled, *a single 2-cleft stipule in the axil*—Moist woods.

6. PIL'EA, Lindl. RICHWEED. CLEARWEED.

P. pu'mila, Gray. Stem 3-18 inches high. Leaves ovate, coarsely-toothed, 3-ribbed.—Cool moist places.

7. BŒHME'RIA, Jacq. FALSE NETTLE.

B. cylin'drica, Willd. Stem 1-3 feet high, smoothish. Leaves ovate-oblong or ovate-lanceolate, serrate, 3-nerved, long-petioled. Stipules separate.—Moist shady places.

8. PARIETA'RIA, Tourn. PELLITORY.

P. Pennsylvan'ica, Muhl. A low annual, simple or sparingly branched, minutely downy. Leaves oblong-lanceolate, thin, veiny, roughish with opaque dots.—Usually in crevices of limestone rocks; not very common.

9. CAN'NABIS, Tourn. HEMP.

C. sati'va, L. (HEMP.) Common everywhere along roadsides and in waste places.

10. HU'MULUS, L. HOP.

H. Lu'pulus, L. (COMMON HOP.) A twining perennial. Leaves heart-shaped, mostly 3-5-lobed, petioled. Calyx of fertile flower a single sepal. In fruit the calyx, achene, etc., sprinkled with yellow resinous grains, which give the hop its taste and smell.

ORDER LXXXV. **PLATANA'CEÆ.** (PLANE-TREE F.)

Represented only by the Genus

PLAT'ANUS, L. PLANE-TREE. BUTTONWOOD.

P. occidenta'lis, L. (AMERICAN PLANE-TREE or SYCAMORE.) A fine large tree found in south-western Ontario. Leaves alternate, rather scurfy when young, palmately-lobed or angled, the lobes sharp-pointed: *stipules sheathing.* Flowers monœcious, both sterile and fertile ones in catkin-like heads, without calyx or corolla, but with small scales intermixed. Ovaries in the fertile flowers club-shaped, tipped with the thread-like simple style, and downy at the base. Fertile heads solitary, on slender peduncles. The white bark separates into thin plates.

ORDER LXXXVI. **JUGLANDA'CEÆ.** (WALNUT F.)

Trees with alternate pinnate leaves and no stipules. Flowers monœcious. Sterile flowers in catkins. Fertile flowers solitary or in small clusters, with a regular 3–4-lobed calyx adherent to the ovary. Fruit a sort of drupe, the fleshy outer layers at length becoming dry and forming a husk, the inner layer hard and bony and forming a nut-shell. Seed solitary in the fruit, very large and 4-lobed. This Order comprises the Walnuts, Butternuts, and Hickories.

Synopsis of the Genera.

1. **Jug'lans.** Sterile flowers in *solitary* catkins from the previous year's wood. Filaments of the numerous stamens very short. Fertile flowers on peduncles at the ends of the branches. Calyx 4-toothed, *with small petals at the sinuses.* Styles and stigmas 2, the latter fringed. *Exocarp or husk drying without splitting. Shell of the nut very rough and irregularly furrowed.*

2. **Car'ya.** Sterile flowers in slender *clustered* catkins. Stamens 3–10, with very short filaments. Fertile flowers in small clusters at the ends of the branches. Calyx 4-toothed; *no petals.* Stigmas 2 or 4, large. *Exocarp 4-valved, drying and splitting away from the very smooth and bony nut-shell.*

1. JUG'LANS, L. WALNUT.

1. **J. ciner'ea,** L. (BUTTERNUT.) Leaflets oblong-lanceolate, pointed, serrate. *Petioles and branchlets clammy. Fruit oblong, clammy.*—Rich woods.

2. **J. ni'gra,** L. (BLACK WALNUT.) Leaflets ovate-lanceolate, taper-pointed, serrate. Petioles downy *but not clammy.* Fruit spherical. Wood a darker brown than in the Butternut.—Rich woods; rare northward.

2. CAR'YA, Nutt. HICKORY.

1. **C. alba,** Nutt, (SHELL-BARK HICKORY.) *Leaflets 5,* the lower pair much smaller than the others. Husk of the fruit splitting *completely* into 4 valves. Nut flattish-globular, mucronate. Bark of the trunk rough, scaling off in rough strips.—Rich woods.

2. **C. tomento'sa,** Nutt. (WHITE-HEART HICKORY.) Sparingly found in the Niagara district and south-westward

Leaflets 7-9. Bark close but not shaggy, and not scaling off on the old trunks. Husk as in the last. Catkins, shoots, and lower surface of the leaves *tomentose* when young. Nut *globular*.

3. **C. ama'ra,** Nutt. (SWAMP HICKORY or BITTERNUT.) *Leaflets 7-11*. Husk of the fruit splitting *half way* down. Nut *spherical*, short-pointed. Bark smooth, not scaling off. —Moist ground.

4. **C. porci'na,** Nutt. (PIG-NUT. BROOM-HICKORY.) Leaflets 5-7. Shoots, etc., *glabrous*. Husk as in the last. Nut *oblong or oval*.—Niagara district, and south-westward.

LXXXVII. MYRICA'CEÆ. (SWEET-GALE F.)

Shrubs with monœcious or diœcious flowers, both sterile and fertile ones collected in short catkins or heads. *Leaves with resinous dots, usually fragrant.* Fruit a 1-seeded dry drupe or little nut, usually coated with little waxy grains.

Synopsis of the Genera.

1. **Myri'ca.** Flowers chiefly diœcious, catkins lateral, each bract with a pair of bractlets underneath. Stamens in the sterile flowers 2-8. Ovary solitary in the fertile flowers, 1-celled, tipped with 2 thread-like stigmas, and surrounded by 2-4 small scales at the base. In our species the 2 scales form wings at the base of the nut.—A shrub, 3-5 feet high.
2. **Compto'nia.** A low shrub, a foot or more in height, with fern-like very sweet-scented leaves. Flowers monœcious. Sterile catkins cylindrical. Fertile ones spherical, the ovary surrounded by 8 awl-shaped persistent scales, so that the catkin resembles a bur.

I. MYRI'CA, L. BAYBERRY. WAX-MYRTLE.

1. **M. Ga'le,** L. (SWEET GALE.) Leaves wedge-lanceolate, serrate towards the apex, pale. The small nuts in crowded heads, and winged by the 2 scales.—Bogs.

2. **M. cerif'era,** L. (BAYBERRY. WAX-MYRTLE.) Leaves oblong-lanceolate, entire, or wavy-toothed towards the apex, shining and sprinkled with resinous dots on both sides, fragrant. Sterile catkins scattered. Nuts naked

bony, and covered with white wax.—Sandy soil near the coast, Atl. Prov.

2. COMPTO'NIA, Solander. SWEET-FERN.

C. asplenifo'lia, Ait. (*Myrica asplenifolia*, Endl., in Macoun's Catalogue.) Leaves linear-lanceolate in outline, deeply pinnatifid, the lobes numerous and rounded.—Dry soil; especially in Pine barrens.

ORDER LXXXVIII. CUPULIF'ERÆ. (OAK FAMILY.)

Shrubs or trees, with alternate simple leaves, deciduous stipules, and monœcious flowers. Sterile flowers in catkins (but in Beech in small heads); the fertile ones solitary or clustered, and furnished with an involucre which forms a scaly cup or a bur surrounding the nut.

Synopsis of the Genera.

1. **Quer'cus.** Sterile flowers *with a calyx* including few or several stamens with slender filaments. Fertile flowers scattered or somewhat clustered, each in a scaly involucre or cupule. Nut (acorn) rounded, the base enclosed by the cupule. (Part I., sec. 71.)

2. **Castu'nea.** Sterile flowers in long slender catkins. *Calyx 6-parted.* Fertile flowers usually 3 in each involucre, the latter prickly, forming a bur. Calyx 6-lobed. Stigmas bristle-shaped. Nuts enclosed (mostly 2 or 3 together) in the prickly 4-valved involucre, flattened when there are more than one.

3. **Fagus.** Sterile flowers in a small head on drooping peduncles. *Calyx bell-shaped.* Fertile flowers in pairs in the involucre, which consists of awl-shaped bractlets grown together at the bases. Calyx-lobes awl-shaped. *Nuts 3-angled, generally in pairs in the bur-like 4-valved cupule.* Bark close, smooth and light gray.

4. **Cor'ylus.** Sterile flowers in drooping catkins. *No calyx.* Stamens 8 (with 1-celled anthers), *and 2 small bractlets* under each bract. Fertile flowers in a small scaly head; 1 ovary, surmounted by 2 long red stigmas, under each scale, and accompanied by a pair of bractlets which, in fruit, enlarge and form a *leaf-like or tubular fringed or toothed involucre closely enveloping each nut.* Sterile catkins from the axils of the previous year. Fertile flowers terminating the new shoots.

5. **Os'trya.** Sterile flowers in drooping catkins. *Calyx wanting.* Stamens several under each bract, *but not accompanied by bractlets.* Fertile flowers in short catkins, 2 under each bract, each

ovary tipped with 2 long stigmas, and *surrounded by a tubular bractlet which, in fruit, becomes a greenish-white inflated bag*, having the small nut in the bottom.

6. **Carpi'nus.** Sterile flowers in drooping catkins. *Calyx wanting. Stamens several under each bract; no bractlets.* Fertile flowers much as in Ostrya, *but the bractlets surrounding the ovaries are not tubular but open, and in fruit become leaf-like, one on each side of the small nut.*

1. QUERCUS, L. OAK.

* *Acorns ripening the first year, and therefore borne on the new shoots. Lobes or teeth of the leaves not bristle-pointed.*

1. **Q. alba,** L. (WHITE OAK.) A large tree. Leaves (when mature) smooth, bright green above, whitish beneath, obliquely cut into few or several oblong entire *lobes.* The oblong nut much larger than the saucer-shaped rough cupule.—Rich woods.

2. **Q. macrocar'pa,** Michx. (BUR OAK. MOSSY-CUP WHITE OAK.) A medium-sized tree. Leaves *deeply lobed*, smooth above, pale or downy beneath. Acorn broadly ovoid, *half or altogether covered by the deep cup, the upper scales of which taper into bristly points, making a fringed border.* Cup varying greatly in size, often very large.—Rich soil.

3. **Q. bi'color,** Willd. (SWAMP WHITE OAK.) A tall tree. Leaves *sinuate-toothed*, but hardly lobed, *wedge-shaped at the base, downy or hoary* beneath, the main veins 6-8 pairs. Cup nearly hemispherical, about half as long as the oblong-ovoid acorn, sometimes with a fringed border. *Peduncles in fruit longer than the petiole.*—Low grounds.

4. **Q. Pri'nus,** L. (CHESTNUT OAK.) A small tree. Leaves minutely downy beneath, *the main veins 10-16 pairs*, sinuate-toothed, acute or obtuse at the base. *Peduncle shorter than the petiole.* Cup hemispherical; acorn as in the last. —Lake Erie coast.

Var. **hu'milis,** Marsh, (*Q. prinoides*, Willd., in Macoun's Catalogue) is much more abundant with us than the species itself. It has the characters of the species, but is a shrub, 2-4 feet high. Fruit sessile or nearly so.

* *Acorns ripening the second year, and therefore borne on the previous year's wood, below the leaves of the season. Lobes or teeth of the leaves bristle-pointed.*

5. **Q. coccin'ea,** Wang. (SCARLET OAK.) A large tree. Leaves bright green, shining above, turning red in autumn, rounded at the base, deeply pinnatifid, the lobes divergent and sparingly cut-toothed. Bark gray outside, *reddish* inside. *Cup top-shaped or hemispherical, with a more or less conical base*, covering half or more of the rather small acorn.

Var. **tincto'ria,** Gray. (*Q. tinctoria,* Bartram, in Macoun's Catalogue.) (QUERCITRON. YELLOW-BARKED or BLACK OAK.) Leaves usually less deeply pinnatifid, slender-petioled, rather rounded at the base, rusty-downy when young, smooth and shining above when mature, often slightly pubescent beneath, turning brownish, orange, or dull red in autumn, Cup as in the species, but the bark darker and rougher and *yellow or orange* inside.—Western Ontario; mostly in dry soil, but occasionally in moist places.

Var. **ambig'ua,** Gray. Leaves closely resembling those of Q. rubra, but the fruit is that of Q. coccinea.—Belleisle Bay, King's Co., N.B.

6. **Q. rubra,** L. (RED OAK.) A large tree. Leaves moderately. pinnatifid, turning dark-red in the autumn. *Cup saucer-shaped*, sessile or nearly so, very much shorter than the oblong-ovoid acorn.—Rich and poor soil.

7. **Q. palustris,** Du Roi. (PIN OAK.) A medium-sized tree. Cup flat-saucer-shaped, very much shorter than the *ovoid-globose* acorn, which is about half an inch long. Leaves deeply pinnatifid, with divergent lobes and rounded sinuses.—Niagara district and south-westward.

2. CASTA'NEA, Tourn. CHESTNUT.

C. vesca, L., var. **America'na,** Michx. (*C. vulgaris,* var. *Americana,* A. DC., in Macoun's Catalogue.) (CHESTNUT.) A large tree. Leaves oblong-lanceolate, pointed, coarsely and sharply serrate, acute at the base. Nuts 2 or 3 in each bur.—South-western Ontario.

3. FAGUS, Tourn. BEECH.

F. ferrugin'ea, Ait. (AMERICAN BEECH.) A very common tree in rich woods, the branches horizontal. Leaves oblong-ovate, taper-pointed, toothed, the very straight veins terminating in the teeth.

4. CORYLUS, Tourn. HAZEL-NUT. FILBERT.

1. **C. America'na,** Walt. (WILD HAZEL-NUT.) Leaves roundish heart-shaped. *Involucre spreading out above, leaf-like and cut-toothed.*—Chiefly in south-western Ontario; in thickets.

2. **C. rostra'ta,** Ait. (BEAKED HAZEL-NUT.) A rather common shrub, easily distinguished from No. *1 by the involucre, which is prolonged into a narrow tube much beyond the nut, and is densely bristly-hairy.*

5. OSTRYA, Micheli. HOP-HORNBEAM. IRONWOOD.

O. Virgin'ica, Willd. (IRON-WOOD.) A slender tree with brownish furrowed bark. Leaves oblong-ovate, taper-pointed, sharply doubly serrate. Fertile catkin like a hop in appearance. Wood very hard and close.—Rich woods.

6. CARPINUS, L. HORNBEAM.

C. America'na, Michx. (BLUE or WATER BEECH.) Small trees with *furrowed trunks* and close smooth gray bark. Leaves ovate-oblong, pointed, doubly serrate. — Along streams. Resembling a beech in general aspect, but with inflorescence like that of Iron-wood.

ORDER LXXXIX. BETULA'CEÆ. (BIRCH FAMILY.)

Trees or shrubs with monœcious flowers, both sorts in catkins, 2 or 3 flowers under each scale or bract of the catkin. Ovary 2-celled and 2-ovuled, but in fruit only 1-celled and 1-seeded. Fruit a small nut. Stigmas 2, long and slender. Twigs and leaves often aromatic.

Synopsis of the Genera.

1. **Bet'ula.** Sterile catkins long and pendulous, formed during summer and expanding the following spring; each flower consisting of

one small scale to which is attached 4 short filaments; 3 flowers under each scale of the catkin. Fertile catkins stout, oblong, the scales or bracts 5-lobed and with 2 or 3 flowers under each; each flower a naked ovary, becoming a winged nutlet in fruit. Bark easily coming off in sheets.

2. **Alnus.** Catkins much as in Betula, but each fertile and sterile flower has a distinct 3-5-parted calyx. Catkins solitary or clustered at the ends of leafless branchlets or peduncles. Nutlets wingless or nearly so.

(These two genera are included in Cupuliferæ in Macoun's Catalogue.)

1. BETULA, Tourn. BIRCH.

1. **B. lenta,** L. (CHERRY-BIRCH. SWEET or BLACK BIRCH.) Bark of the trunk dark brown, close, aromatic; that of the twigs bronze-coloured. Wood rose-coloured. Leaves ovate, with somewhat heart-shaped base, doubly serrate, pointed, short-petioled. Fruiting catkins sessile, thick, oblong-cylindrical.—Moist woods.

2. **B. lu'tea,** Michx. (YELLOW or GRAY BIRCH.) Bark of the trunk yellowish-gray, somewhat silvery, scaling off in thin layers. Leaves hardly at all heart-shaped. Fruiting catkins thicker and shorter than in No. 1.—Moist woods.

3. **B. populifo'lia,** Ait. (AMERICAN WHITE BIRCH. GRAY BIRCH.) Leaves very tremulous on slender petioles, triangular, very taper-pointed, nearly truncate at the base, smooth and shining except when young. Bark of trunk white, less separable than in Canoe Birch.—Poor soil, Atl. Prov.

4. **B. papyra'cea,** Ait. (*B. papyrifera*, Michx., in Macoun's Catalogue.) (PAPER or CANOE BIRCH.) Bark of the trunk white, easily separating in sheets. Leaves ovate, taper-pointed, heart-shaped, long-petioled. Fruiting catkins cylindrical, usually hanging on slender peduncles.—Woods.

5. **B. pu'mila,** L. (LOW BIRCH.) A *shrub* with brownish bark, *not glandular*. Leaves ovate or roundish, *pale beneath*; veinlets on both surfaces finely reticulated. Catkins mostly erect, on short peduncles.—Bogs and low grounds, northward.

SALICACEÆ. 209

2. ALNUS, Tourn. ALDER.

1. A. inca'na, Willd. (SPECKLED or HOARY ALDER.) A shrub or small tree, growing in thickets in low grounds along streams. Leaves oval or ovate, rounded at the base, serrate, whitish beneath. Flowers preceding the leaves in early spring, from clustered catkins formed the previous summer and remaining naked over winter. *Fruit wingless.*

2. A. vir'idis, DC. (GREEN or MOUNTAIN ALDER.) A shrub 3-8 feet high, along mountain streams. Flowers appearing with the leaves, the *staminate* catkins having remained naked during the winter, *the pistillate enclosed in a scaly bud. Fruit with a thin wing.*—Northward.

ORDER XC. SALICA'CEÆ. (WILLOW FAMILY.)

Trees or shrubs with diœcious flowers, both sorts in catkins, one under each scale of the catkin. No calyx. Fruit 1-celled, *many-seeded*, the seeds furnished with tufts of down. (Part I., section 74, for description of typical flowers.) This Order comprises the Willows and Poplars.

Synopsis of the Genera.

1. **Salix.** Trees with mostly *long and pointed leaves* and slender branches. Bracts or scales of the catkins *not toothed*. Stamens mostly 2 under each bract, but in one or two species as many as 5 or 6. Stigmas short. Catkins appearing before or with the leaves.

2. **Pop'ulus.** Trees with *broad* and more or less heart-shaped leaves. Bracts of the catkins *toothed or cut at the apex*. Stamens 8-30, or even more, under each scale. Stigmas long. Catkins long and drooping, preceding the leaves.

1. SALIX, Tourn. WILLOW.

* *Catkins borne on the ends of the short lateral leafy branchlets. Scales yellowish, deciduous. Filaments hairy below. Trees or large shrubs, with taper-pointed leaves.*

1. S. nigra, Marshall. (BLACK WILLOW.) A tree with a roughish black bark, growing along streams. *Leaves narrowly lanceolate*, tapering at each end, serrate, smooth, green on both sides. Stamens 3-6. Ovary short-pedicelled. Sterile catkins long and narrow.

2. **S. amygdaloi'des,** Anders. A *tree* with lanceolate or ovate-lanceolate leaves, *pale or glaucous beneath*, and with long slender petioles. Fertile catkins becoming *very loose* from the lengthening of the pedicels. Stamens 3-6.—N.W.

3. **S. lu'cida,** Muhl. (SHINING WILLOW.) A shrub or small bushy tree, growing along streams. *Leaves ovate-oblong or narrower*, with a long tapering point, *shining on both sides*, serrate. Stamens most 5. Scales of catkins dentate, hairy at the base. Sterile catkins densely-flowered, showy.

4. **S. frag'ilis,** L. (CRACK WILLOW.) A tall and handsome tree. Leaves lanceolate, long-pointed, pale or glaucous beneath, 3-6 inches long. *Stamens mostly 2*, rarely 3-4. Capsule short-pedicelled.—Atl. Prov.

5. **S. alba,** L., var. **cæru'lea,** Smith. *Leaves ashy-gray or white both sides*, except when old, lanceolate, long and slender-pointed. Stamens mostly 2. Pods sessile or nearly so. Old leaves smooth, glaucous beneath, dull bluish-green. – Cultivated in many places.

6. **S. longifo'lia,** Muhl. (LONG - LEAVED WILLOW.) A shrub or small tree, varying greatly in size, growing along streams in sandy or gravelly places, and often forming dense clumps. Leaves linear-lanceolate, very long, tapering towards both ends, nearly sessile, *serrate with a few spreading teeth*, grayish-hairy when young. Stamens 2.

** *Catkins lateral or terminal. Scales coloured at the tip, persistent. Stamens 2, the filaments not hairy. Shrubs or small trees.*

← *Ovaries woolly.*

7. **S. dis'color,** Muhl. (GLAUCOUS WILLOW.) A shrub or small tree, 8-15 feet high, growing in low grounds and along streams. Leaves lanceolate or ovate-lanceolate, irregularly toothed *in the middle of the margin*, entire at each end, smooth and bright green above, white-glaucous beneath, when young. Stipules moon-shaped, toothed. Catkins sessile, very early in spring before the leaves. Scales very dark and hairy.

Var. **erioceph'ala,** Anders., has densely-flowered and very silky catkins, and the leaves somewhat pubescent even when old.

8. **S. liv'ida,** Wahl. Var. **occidenta'lis,** Gray. (*S. rostra'ta,* Rich., in Macoun's Catalogue.) (LIVID WILLOW.) A good-sized shrub, chiefly in moist situations. *Leaves oblong or obovate-lanceolate, barely toothed, downy above, very veiny, hairy and glaucous beneath.* Stipules semi-lunar, toothed. Ovary at length raised *on a very slender stalk.* Catkins appearing with the leaves.

9. **S. hu'milis,** Marshall. (PRAIRIE WILLOW.) A grayish shrub, 3–8 feet high, growing usually in dry or barren places. Leaves oblanceolate, pointed, the lowest obovate, slightly downy above, thickly so beneath. Stipules semi-ovate or moon-shaped, with a few teeth, shorter than the petioles. Catkins ovoid, sessile, before the leaves, naked at the base. Scales dark red or brownish.

10. **S. petiola'ris,** Smith. (PETIOLED WILLOW.) A low shrub on sandy river banks. Leaves narrowly lanceolate, finely and evenly serrate, silky-gray or glaucous beneath, smooth above. Catkins with a few small leaf-like bracts at the base. Scales of the fertile catkins acute, very hairy. Ovary tapering, silky, stalked.

11. **S. can'dida,** Willd. (HOARY WILLOW.) A shrub not more than 3 or 4 feet high, growing in bogs and wet places; the twigs and leaves clothed with a web-like wool, giving the whole plant a whitish aspect. Leaves lanceolate, narrow, with somewhat revolute margins. Stipules small, lanceolate, toothed. Catkins cylindrical. Anthers red.

← ← *Ovaries glabrous.*

12. **S. corda'ta,** Muhl. (HEART-LEAVED WILLOW.) A shrub or small tree, growing in wet grounds. Leaves lanceolate, not always heart-shaped, sharply serrate, smooth, green both sides. Catkins cylindrical, rather slender, *leafy-bracted at the base,* the sterile ones silky. Var. **angusta'ta** has long narrow leaves.

13. **S. balsamif'era,** Barratt. A small much-branched shrub. Young twigs shining-chestnut on the sunny side. Leaves ovate-lanceolate, usually slightly cordate at base, at first very thin and of a reddish colour, at length rigid, dark-green above, and paler and conspicuously reticulate-veined beneath, slightly serrate, with slender petioles. Sterile catkins very silky, with a few bracts at the base; fertile catkins leafy-peduncled, becoming very loose in fruit. Capsules long-pedicelled.—Swamps, Atl. Prov. and northward.

14. **S. myrtilloi'des,** L. Low shrub, 1-3 feet high. Leaves elliptic-obovate, an inch long, entire, smooth, somewhat coriaceous when mature, revolute, reticulated, pale or glaucous beneath. Fertile catkins loosely few-flowered, on long leafy peduncles. Capsules glabrous, on slender pedicels.—Peat-bogs.

2. POP'ULUS, Tourn. POPLAR.

1. **P. tremuloi'des,** Michx. (AMERICAN ASPEN.) A tree with a greenish-white bark, and *roundish heart-shaped leaves*, continually in a state of agitation, due to the lateral compression of the petiole, and the consequent susceptibility of the leaf to the least motion of the air. *Teeth of the leaves small.*

2. **P. grandidenta'ta,** Michx., (LARGE-TOOTHED ASPEN) has *roundish ovate leaves with large irregular sinuate teeth.*

3. **P. balsamif'era,** L. (BALSAM POPLAR.) A tall tree, growing in swamps and along streams; *the large buds varnished with resinous matter.* Leaves *ovate,* tapering, finely serrate, whitish beneath. Stamens very numerous.

Var. **can'dicans,** Gray, (BALM OF GILEAD) has broader and *more or less heart-shaped* leaves.

4. **P. monilif'era,** Ait. (COTTONWOOD.) A tree with broad deltoid leaves, slightly heart-shaped, serrate with incurved teeth. Young branches *slightly angled,* at length round. Fertile catkins very long, the scales *cut-fringed, not hairy.*—Along the main line of the Grand Trunk Railway.

ORDER XCI. **EMPETRA'CEÆ.** (CROWBERRY FAMILY).

Low shrubby evergreens, resembling heaths as to leaves and general aspect. Flowers diœcious or polygamous. Calyx somewhat petal-like or none. Ovary 3-9-celled, berry-like in fruit.

Synopsis of the Genera.

1. **Em'petrum.** Flowers scattered and solitary in the axils. Sepals 3, somewhat petal-like.
2. **Core'ma.** Flowers in terminal heads. Calyx none.

1. EM'PETRUM, Tourn. CROWBERRY.

E. ni'grum, L. (BLACK CROWBERRY.) A slender procumbent shrub with the foliage and aspect of a heath. Flowers polygamous, inconspicuous in the axils of the leaves. Calyx 3 petal-like sepals. Corolla wanting. Stamens 3. Fruit a blackberry-like drupe.—Atl. Prov. and northward.

2. CORE'MA, Don. BROOM-CROWBERRY.

C. Conrad'ii, Torr. (BROOM-CROWBERRY.) Closely resembling the preceding. Flowers diœcious or polygamous, collected in terminal heads, each in the axil of a scaly bract, having no true calyx, but with 5 or 6 thin, dry bractlets under each. Stamens 3 or 4 with slender filaments. Drupes small, juiceless when ripe.—Atl. Prov.

ORDER XCII. **CERATOPHYLLA'CEÆ.** (HORNWORT F.)

Represented, with us, by a single species.

CERATOPHYL'LUM, L. HORNWORT.

C. demer'sum, L. An aquatic herb, with whorled finely dissected leaves, and minute axillary sessile monœcious flowers, without calyx or corolla, but with an 8-12-cleft involucre. Staminate flowers of 12-24 stamens with large sessile anthers. Pistillate flowers of a single 1-celled ovary, forming an achene, beaked with the slender style. *Embryo with 4 cotyledons.*—Under water in ponds and slow streams.

Subclass II. GYM'NOSPERMS.

Ovules and seeds naked (not enclosed in a pericarp), and fertilized by the direct application of the pollen. Represented in Canada by a single Order.

Order XCIII. CONIF'ERÆ. (Pine Family.)

Trees or shrubs with resinous juice and mostly monœcious flowers, these in catkins, except the last Genus (Taxus), in which the fertile flower is solitary and the fruit berry-like. Leaves awl-shaped or needle-shaped. (See Part I., Cap. xvi., for descriptions of typical plants.)—The Order comprises three well-marked Suborders.

Suborder I. ABIETIN'EÆ. (Pine Family Proper.)

Fruit a true *cone*, the imbricated scales *in the axils of bracts*. Ovules 2 on the inside of each scale at the base, in fruit coming off with a wing attached to each. Leaves scattered or fascicled. (Part I., Figs. 197, 198.)

Cones not ripening till the second year.

1. **Pinus.** Leaves needle-shaped, *2-5 in a cluster, evergreen*, in the axil of a thin scale. Sterile catkins in spikes at the bases of the new shoots, consisting of many almost sessile anthers spirally inserted on the axis. Cones more or less woody, the scales widely spreading when ripe. Cotyledons of the embryo several.

** Cones ripening the first year.

← *Cones pendulous, bracts smaller than the scales.*

2. **Pi'cea.** Leaves sessile, keeled on both sides, scattered.
3. **Tsu'ga.** Leaves petioled, flat, scattered, whitened beneath.

← ← *Cones erect, the bracts longer than the scales.*

4. **A'bies.** Leaves linear or needle-shaped, *scattered uniformly along the new shoots, evergreen*. Sterile catkins in the axils of last year's leaves. Cones with thin scales.
5. **Larix.** Leaves needle-shaped, *clustered or fascicled on lateral spurs of last year's wood, many in each bundle, falling off in the autumn;* those on the new shoots scattered, but *deciduous like the rest.*

SUBORDER II. **CUPRESSIN'EÆ.** (CYPRESS F.)

Fertile flowers of only a few scales, these *not in the axils of bracts*, forming in fruit either a very small loose and dry cone, or a sort of false berry owing to the thickening of the scales.

  * *Flowers monœcious. Fruit a small loose cone.*
6. **Thuja.** Leaves some *awl-shaped*, others *scale-like*, closely imbricated on the *flat branches*. Catkins ovoid, terminal.
  * * *Flowers mostly diœcious. Fruit berry-like, black with a bloom.*
7. **Juniperus.** Leaves awl-shaped or scale-like, sometimes of both shapes, evergreen, prickly-pointed, glaucous-white on the upper surface, and in whorls of 3, or opposite.

SUBORDER III. **TAXIN'EÆ.** (YEW FAMILY.)

Fertile flower solitary, consisting of a naked ovule surrounded by a disk which becomes pulpy and berry-like in fruit, enclosing the nut-like seed. *Berry red.*

8. **Taxus.** Flowers chiefly diœcious. Leaves evergreen, *mucronate*, rigid, scattered.—A low straggling bush, usually in the shade of other evergreens.

1. PINUS, Tourn. PINE.

1. **P. resino'sa,** Ait. (RED PINE.) *Leaves in twos*, slender. Bark rather smooth, *reddish*.—Common northward.

2. **P. Banksia'na,** Lambert. (GRAY or NORTHERN SCRUB PINE.) *Leaves in twos, about 1 inch long.* Cones conical, usually curved, smooth and hard, about *one and one-half inches long*.—Barren soil, eastward and northward.

3. **P. rig'ida,** Mill. (PITCH PINE.) *Leaves in threes.* Scales of the cones tipped with a short stout recurved prickle. —Atl. Prov.

4. **P. stro'bus,** L. (WHITE PINE.) *Leaves in fives,* slender. Bark smooth except on old trees, not reddish.—Common.

2. PI'CEA, Link. SPRUCE.

1. **P. nigra,** Link. (*Abies nigra,* Poir.) (BLACK SPRUCE.) *Leaves needle-shaped and 4-sided, pointing in all directions.* Cones hanging, *persistent*, scales with thin edges.—Swamps and cold woods.

2. **P. alba,** Link. (*Abies alba,* Michx.) (WHITE SPRUCE.)

Leaves as in No. 1. Cones hanging, *deciduous*, the scales with thickish edges.—Swamps and cold woods.

3. TSU'GA, Carrière. HEMLOCK.

T. Canadensis, Carr. (*Abies Canadensis*, Michx.) (HEMLOCK SPRUCE.) *Leaves flat*, lighter beneath, *pointing out in two directions*, *i.e.*, right and left on each side of the branch, obtuse. Cones hanging, persistent.—Hilly or rocky woods.

4. A'BIES, Link. FIR.

A. balsa'mea, Miller. (BALSAM FIR.) *Leaves flat*, the lower surface whitish and the midrib prominent, crowded, pointing mostly right and left on the branches. *Cones erect* on the upper sides of the branches, violet-coloured, *the scales slender-pointed*.—Damp woods and swamps.

5. LARIX, Tourn. LARCH.

L. America'na, Michx. (AMERICAN LARCH. TAMARAC.) A slender and very graceful tree with soft leaves in fascicles, falling off in autumn.—Swamps.

6. THUJA, Tourn. ARBOR VITÆ.

T. occidenta'lis, L. (AMERICAN ARBOR VITÆ.) The well-known cedar of cedar-swamps.—Common.

7. JUNIP'ERUS, L. JUNIPER.

1. J. commu'nis, L. (COMMON JUNIPER.) A spreading shrub with ascending stems, growing on dry hill-sides. *Leaves in whorls of 3*, whitish above, prickly-pointed.

2. J. Virginia'na, L. (RED CEDAR.) A shrub or small tree with *mostly opposite leaves* of two forms, viz.: awl-shaped and loose, and scale-shaped and appressed. Fruit small, *erect*. Wood red and odorous.—Dry sterile soil.

3. J. Sabi'na, L., var. procumbens, Pursh. A procumbent or creeping shrub with two sorts of leaves, awl-shaped and scale-shaped, the latter acute. Fruit on short *recurved peduncles*.—Rocky banks and margins of swamps.

8. TAXUS, Tourn. YEW.

T. bacca'ta, L., var. Canadensis, Gray. (AMERICAN YEW. GROUND HEMLOCK.) A low straggling shrub. Leaves green on both sides. *Berry globular*, red.

Class II. MONOCOTYLE'DONS.

For characters of the Class see Part I., chap. xv.

I. SPADIC'EOUS DIVISION.

Flowers aggregated on a *spadix* (Part I., sec. 94), with or without a *spathe*, or sheathing bract.

Order XCIV. ARA'CEÆ. (Arum Family.)

Herbs with pungent juice and simple or compound leaves, *these sometimes net-veined* and hence suggesting that the plants may be Dicotyledons. Spadix usually accompanied by a spathe. Flowers either without a perianth of any kind, or with 4-6 sepals. Fruit usually a berry.

Synopsis of the Genera.

* *Leaves not linear. Flowers without perianth of any sort. Spadix accompanied by a spathe.*

1. **Arisæ'ma.** Flowers mostly diœcious, collected on the lower part of the spadix only. Spathe (in our common species) arched over the spadix. Scape from a solid bulb. Leaves compound, net-veined, sheathing the scape below with their petioles. Berries bright red.

2. **Peltan'dra.** Flowers monœcious, covering the whole spadix; anthers above, ovaries below. Spathe convolute throughout, wavy on the margin, mostly green. Leaves arrow-shaped. Scapes from a root of thick fibres. Fruit a fleshy green berry, 1-3 seeded.

3. **Calla.** Flowers (at least the lower ones) *perfect*, covering the whole spadix. Spathe open and spreading, with a white upper surface, tipped with an abrupt point. Scape from a creeping rootstock. Leaves not net-veined, simple, heart-shaped.

** *Leaves not linear. Flowers with a perianth of 4 sepals. Spadix surrounded by a spathe.*

4. **Symplocar'pus.** Leaves all radical, very large and veiny, appearing after the spathes, which are close to the ground and are produced very early in spring. Flowers perfect, their ovaries immersed in the spadix, the latter globular and surrounded by the shell-shaped spathe. Sepals hooded. Stamens 4. Fruit consisting of the soft enlarged spadix in which the seeds are sunk.

*** *Leaves linear, sword-shaped. Spadix on the side of the scape. Flowers with a perianth of 6 sepals. No spathe.*

5. **Ac'orus.** Scape 2-edged, resembling the leaves, the cylindrical spadix borne on one edge. Sepals hollowed. Stamens 6.

1. ARISÆ'MA, Martins. INDIAN TURNIP.

1. **A. triphyl'lum**, Torr. (INDIAN TURNIP.) For full description and engraving of this plant see Part I., sections 94–97.

2. **A. Dracon'tium**, Schott., (GREEN DRAGON) is reported from low grounds near London, Ont. Leaf usually solitary, pedately divided into 7–11 oblong-lanceolate pointed leaflets. Spathe convolute, pointed; the slender point of the spadix extending beyond it.

2. PELTAN'DRA, Raf. ARROW ARUM.

P. undula'ta, Raf. (*P. Virginica* of most authors.) Root of thick tufted fibres. Scape 12–18 inches high. Staminate part of the spadix much longer than the pistillate.—Shallow water; apparently rare.

3. CALLA, L. WATER ARUM.

C. palustris, L. (MARSH CALLA.) This plant is fully described and illustrated in Part I., section 98.

4. SYMPLOCAR'PUS, Salisb. SKUNK CABBAGE.

S. fœ'tidus, Salisb. Leaves 1–2 feet long, ovate or heart-shaped, short-petioled. Spathe purplish and yellowish. incurved. Plant with skunk-like odour.—Bogs and wet places; not common northward.

5. AC'ORUS, L. SWEET FLAG. CALAMUS.

A. Cal'amus, L. Scape much prolonged beyond the spadix.—Swamps and wet places.

ORDER XCV. LEMNA'CEÆ. (DUCKWEED FAMILY.)

Very small plants floating about freely on the surface of ponds and ditches, consisting merely of a little frond, commonly with a single root or a tuft of roots from the lower surface, and producing minute monœcious flowers from a cleft in the edge of the frond, or from the upper surface. Fruit a 1-7-seeded utricle. The flowers are rarely to be seen.

Synopsis of the Genera.

1. **Lem'na.** Flowers from a cleft in the edge of the frond, usually 2 of a single stamen each, and 1 of a simple pistil, the three surrounded by a spathe. Roots several or only one.
2. **Wolff'ia.** Flowers from the centre of the upper surface of the frond, only 2; one of a single stamen, the other of a simple pistil. *Roots none.* The smallest of flowering plants.

1. LEMNA, L. DUCKWEED.

1. **L. polyrrhi'za,** L. (*Spirodela polyrrhiza,* Schleiden.) Fronds round-obovate, green above, purplish beneath, mostly 7-nerved. *Rootlets several.*—Common in ponds and pools.

2. **L. trisul'ca,** L. Fronds oblong to oblong-lanceolate, half an inch or more long, narrowed at the base into a slender stalk, denticulate at the tip, obscurely 3-nerved. *Rootlet single,* often wanting.—Ponds and springy places.

3. **L. minor,** L. Fronds round to elliptic-obovate, very small. *Rootlet single.*—Stagnant waters.

2. WOLFF'IA, Horkel.

1. **W. Columbia'na,** Karsten. Globular or nearly so, light green all over, *not dotted.*—Stagnant waters, usually rather below the surface.

2. **W. Brasilien'sis,** Weddell. Oblong, deep green above, pale below, *dotted all over with brown.*—With the last, but floating.

Order XCVI. TYPHA'CEÆ. (Cat-tail Family.)

Aquatic or marsh herbs with linear sword-shaped leaves, erect or floating, and monœcious flowers, either in separate heads or on different parts of the same spike or spadix, but without a spathe, and destitute of true floral envelopes. Fruit an achene, 1-seeded.

Synopsis of the Genera.

1. **Typha.** Flowers in a very dense and long cylindrical terminal spike, the upper ones staminate, the lower pistillate, the ovaries

long-stalked and surrounded by copious bristles forming the down of the fruit. Leaves sword-shaped, erect, sheathing the stem below.

2. **Sparga'nium.** Flowers in separate globular heads along the upper part of the stem, the higher ones staminate, the lower ones pistillate, each ovary sessile and surrounded by a few scales not unlike a calyx. Both kinds of heads leafy-bracted. Leaves flat or triangular, sheathing the stem with their bases.

1. TYPHA, Tourn. CAT-TAIL FLAG.

1. **T. latifolia**, L. (COMMON CAT-TAIL.) Stem 5–8 feet high. Leaves *flat*. No space between the staminate and pistillate parts of the spike.—Marshy places.

2. **S. angustifo'lia**, L. (NARROW-LEAVED or SMALL CAT-TAIL.) Leaves *channelled* towards the base, narrowly linear. The two parts of the spike usually with an interval between them.—Central and eastern Ontario.

2. SPARGA'NIUM, Tourn. BUR-REED.

1. **S. eurycar'pum**, Engelm. Stem erect, *stout*, 2–4 feet high. Leaves mostly flat on the upper side, keeled and hollow-sided on the lower. Heads several, panicled-spiked, the pistillate *an inch across in fruit*. Nutlets or achenes with a broad abruptly-pointed top.—Borders of slow waters and ponds.

2. **S. simplex**, Huds. *Stem slender*, erect, 6–24 inches high. *Leaves more or less channelled and 3-angled*, about one-third of an inch wide. Fertile heads about *half an inch broad in fruit*. Stigma linear.—Borders of ponds, etc.

Var. **androc'ladum**, Engelm., (*S. androcladum*, Morong.) is stouter and taller, with usually broader leaves, and *branching inflorescence*. Fruiting heads also somewhat larger.

Var. **angustifo'lium**, Engelm., (*S. affine*, Schnitzlein.) is very slender, with leaves floating, long and narrow and flat. Inflorescence simple, and fruiting heads smaller.

3. **S. min'imum**, Fries. Usually floating, with very slender stems, and thin flat narrow leaves. Fertile heads only 1 or 2. Stigma oval. Fruit oblong-obovate, pointed, somewhat triangular.

Order XCVII. NAIADA'CEÆ. (Pondweed Family.)

Immersed aquatic herbs, with jointed stems and sheathing stipules. Leaves flat, immersed or floating. Flowers perfect or imperfect, inconspicuous, naked, or with a free scale-like calyx. Ovaries 1-celled, 1-seeded, solitary or 2–4, and distinct.

Synopsis of the Genera.

* *Flowers perfect, in spikes or clusters.*

1. **Potamoge'ton.** Flowers with sepals, stamens, and sessile ovaries each 4.

* * *Flowers monœcious or diœcious, axillary, naked, monandrous.*

2. **Nai'as.** Flowers diœcious. *Ovaries solitary* and naked. Stigmas 2 or 4, awl-shaped. Staminate flower (of 1 stamen) enclosed in a membranous spathe. Stems floating. Leaves linear, opposite, somewhat crowded into whorls, sharp-toothed, sessile and spreading at the base.

3. **Zannichell'ia.** Flowers monœcious, usually both kinds in the same axil. *Ovaries 2-5*, from a cup-shaped involucre. Leaves linear-thread-form, entire.

4. **Zoste'ra.** Flowers monœcious, the two kinds naked and sessile, arranged alternately in 2 vertical rows on the inner side of a leaf-like enclosed spadix. Sterile flowers of single sessile 1-celled anthers; the fertile of single ovate-oblong ovaries. Stigmas 2, bristle-form, deciduous.

1. POTAMOGE'TON, Tourn. Pondweed.

* *Leaves of 2 sorts, the floating ones with a dilated petioled blade, different in form from the submerged ones.*

1. **P. natans.** L. *Submerged leaves grass-like or capillary;* upper stipules *very long, acute.* Spikes cylindrical, all out of the water. Stem hardly branched. Floating leaves *long-petioled*, elliptical, with a somewhat heart-shaped base, with a blunt apex, 21–29-nerved.

2. **P. Clayto'nii,** Tuckerman. *Stem compressed.* Submerged leaves *linear*, 2–5 inches long, 2-ranked, 5-nerved; stipules *obtuse.* Floating leaves *short-petioled*, chiefly opposite, oblong, 11–17-nerved. Spikes all above water.

3. **P. Spiril'lus,** Tuckerman. Stems very slender. Floating leaves when present oval to lanceolate, about as long as

the petiole; submerged leaves *narrowly linear*, or the upper ones broad-linear, or lance-oblong. Emersed flowers in many-flowered spikes; submersed flowers usually solitary. Fruit either winged and 4-5-toothed, or wingless and entire.

4. **P. rufes'cens**, Schrad. Stem simple. Submersed leaves almost sessile, *lanceolate and lance-oblong*; floating leaves (often wanting) *wedge-oblanceolate*, narrowed into a short petiole, 11-17-nerved. Spike dense, many-flowered. Fruit obovate, lenticular, with acute margin, and pointed with the long style.

5. **P. flu'itans**, Roth. (*P. lonchites*, Tuckerm.) Stem often branching below. Submersed leaves very long, lanceolate and lance-linear, 7-15-nerved; floating leaves lance-oblong, or oblong-elliptical, long-petioled, 17-23-nerved. Spike dense. Fruit obliquely obovate, 3-keeled when dry, the middle one winged above.—In streams mostly.

6. **P. amplifo'lius**, Tuckerman. Submersed leaves large, *lanceolate or oval*, acute at each end, recurved, wavy; *stipules long and tapering*. Floating leaves large, oblong or lance-ovate, or slightly cordate, long-petioled, 30-50-nerved.

7. **P. heterophyl'lus**, Schreb. (*P. gramineus*, Fries.) Stem slender, *very much branched below*. Submersed leaves lanceolate or linear-lanceolate, acuminate or cuspidate, 3-7-nerved, upper ones petioled, lower sessile; floating leaves variable, with short blunt points, 9-15-nerved. Stipules *obtuse*.

8. **P. Ziz'ii**, Mert. and Koch. (*P. lucens*, var. *minor*, Nolte.) Much branched at the base. Upper leaves long-petioled and sometimes emersed, the others nearly sessile, all usually numerous, wavy and *shining*. Resembling the next species, but smaller.

** *Leaves all submersed and similar, mostly sessile, membranaceous and dilated. Stipules obtuse, becoming loose.*

9. **P. lucens**, L. Stem thick, branching. Leaves petioled oval or lanceolate, *mucronate*, frequently shining. Fruit roundish, compressed, with blunt edges, slightly keeled.

10. **P. prælon′gus,** Wulfen. Stem very long and branching. Leaves all submersed and similar, lanceolate, half-clasping, with a boat-shaped cavity at the end. Spikes loose-flowered with very long peduncles. Fruit sharply keeled when dry. Stem white.—Ponds and large rivers.

11. **P. perfolia′tus,** L. Leaves all submersed, varying in width from *orbicular to lanceolate, clasping by a heart-shaped base.* Stem branching.

Var. **lanceola′tus,** Robbins, has long-lanceolate acuminate leaves. Peduncles thickened upward.

* * * *Leaves all submersed and similar, mostly sessile, linear or bristle-like.*

12. **P. compressus,** Fries. (*P. zosteræfolius,* Schum., in Macoun's Catalogue.) Leaves linear, grass-like, sessile, abruptly pointed, with *three large nerves and many fine ones.* Stem branching, *wing-flattened.* Stipules free from the sheathing base of the leaf. Spikes cylindrical.

13. **P. pauciflo′rus,** Pursh. *Stem filiform,* flattish and much branched. Leaves narrowly-linear, acute, *obscurely 3-nerved.* Spikes capitate.

14. **P. pusil′lus,** L. Stem slender, somewhat flattish, often much branched. Leaves *narrow- or setaceous-linear,* 1-3-nerved, *with translucent glands on each side at the base.*

15. **P. mucronat′us,** Schrad. Like the last, but the stem less branching, and the leaves *broader* (nearly an inch wide), often 5-nerved.

16. **P. pectina′tus,** L. Leaves bristle-shaped. Stem repeatedly forking, filiform. *Spikes interrupted, on long slender peduncles. Stipules united with the sheathing base of the leaf.*

17. **P. Robbin′sii,** Oakes. Leaves narrowly lanceolate or linear, crowded in 2 ranks, recurved, serrulate, many-nerved. Stems rigid with numerous branches. Fruit keeled with a broadish wing. Stipules united with the sheathing base of the leaf.

2. NAI'AS, L. NAIAD.

N. flex'ilis, Rostk. and Schmidt. Stems very slender. Leaves very narrowly linear, minutely serrulate.

3. ZANNICHEL'LIA, Micheli. HORNED PONDWEED.

Z. palus'tris, L. Fruit obliquely oblong, flattish and somewhat incurved. Style half as long as the nutlet.

4. ZOSTE'RA, L. EEL-GRASS. GRASS-WRACK.

Z. mari'na, L. Leaves obscurely 3-5-nerved.—Atl. sea-coast, in shoal water.

II. PETALOI'DEOUS DIVISION.

Flowers with a perianth coloured like a corolla.

ORDER XCVIII. **ALISMA'CEÆ.** (WATER PLANTAIN F.)

Marsh herbs, with flowers having 3 distinct sepals and 3 distinct petals, pistils either apocarpous or separating at maturity into distinct carpels, and hypogynous stamens 6–many. Flowers on scapes or scape-like stems. Leaves sheathing at the base either rush-like or, when broad, mostly heart-shaped or arrow-shaped.

Synopsis of the Genera.

* *Calyx and corolla both greenish. Carpels more or less united, but spreading at maturity. Leaves rush-like and fleshy, or grass-like.*

1. **Triglo'chin.** Flowers small, in a spike or close raceme, without bracts. Carpels *united to the top;* when ripe, splitting away from a central persistent axis.
2. **Scheuchze'ria.** A low bog-herb, with a creeping jointed rootstock, and grass-like leaves. Stamens 6. Carpels 3, globular, *nearly distinct.*

(These two genera are included in Naiadaceæ in Macoun's Catalogue.)

* * *Calyx green, persistent. Corolla white. Pistil apocarpous. Leaves, with distinct blades and petioles.*

3. **Alis'ma.** *Flowers perfect.* Stamens usually 6. Carpels numerous in a ring. Leaves all radical. *Scapes with whorled panicled branches.*

4. **Sagitta'ria.** *Flowers monœcious, sometimes diœcious.* Stamens numerous. Carpels numerous, in more or less globular heads. Leaves arrow-shaped, but varying greatly. Flowers mostly in whorls of 3 on the scapes, the sterile ones uppermost.

1. TRIGLO'CHIN, L. ARROW-GRASS.

1. **T. palus'tre,** L. A slender rush-like plant, 6-18 inches high, found growing in bogs northward. *Carpels 3*, awl-pointed at the base, splitting away from below upwards. Spike or raceme slender, 3 or 4 inches long.

2. **T. marit'imum,** L. The whole plant is stouter than No. 1, and the carpels are usually 6 in number.—Atl. sea-coast, and saline marshes.

2. SCHEUCHZE'RIA, L. SCHEUCHZERIA.

S. palustris, L. Stem zigzag. Flowers in a loose terminal raceme, with sheathing bracts.—Bogs.

3. ALIS'MA, L. WATER-PLANTAIN.

A. Planta'go, L., var. **America'num,** Gray. Leaves long-petioled, mostly oblong-heart-shaped, but often narrower, 3-9-nerved or ribbed, and with cross veinlets between the ribs. Flowers small, white, in a large and loose compound panicle.—Low and marshy places, often growing in the water.

4. SAGITTA'RIA, L. ARROW-HEAD.

* *Filaments narrow, as long as the anthers.*

1. **S. varia'bilis,** Engelm. Very variable in size and in the shape of the leaves. Scape angled.—Common everywhere in shallow water.

Var. **obtusa,** Engelm., is diœcious, and has large obtuse leaves.

Var. **latifolia,** Engelm., is monœcious, with large broad acute leaves.

Var. **angustifo'lia,** Engelm., has narrow leaves, with long and linear diverging lobes.

* * *Filaments very short, with enlarged mostly glandular base.*

2. **S. heterophyl'la,** Pursh. Scape weak and at length procumbent. Leaves lanceolate or lance-ovate, entire, or with one or two narrow basal sagittate appendages.

3. **S. gramin'ea,** Michx. Scape very slender, erect. Leaves varying from ovate-lanceolate to linear, *scarcely ever sagittate.*

4. **S. calyci'na,** Engelm., var. **spongiosa,** Engelm. Scape weak, and at length usually procumbent. Fertile flowers perfect. Leaves broadly halberd-shaped with wide-spreading lobes. *Submerged leaves without blades.*—Atl. Prov.

ORDER XCIX. **HYDROCHARIDA'CEÆ.** (FROG'S-BIT F.)

Aquatic herbs, with diœcious or polygamo-diœcious flowers on scape-like peduncles from a kind of spathe of one or two leaves, the perianth in the fertile flowers of 6 pieces united below into a tube which is adherent to the ovary. Stigmas 3. Fruit ripening under water.

Synopsis of the Genera.

1. **Elode'a.** Growing under water, the pistillate flowers alone coming to the surface. Stem leafy and branching. Perianth of the fertile flowers with a 6-lobed spreading limb, the tube prolonged to an extraordinary length, thread-like. Leaves crowded, pellucid, 1-nerved, sessile, whorled in threes or fours. Stamens 3-9.

2. **Vallisne'ria.** Nothing but the pistillate flowers above the surface, these on scapes of great length, and after fertilization drawn below the surface by the spiral coiling of the scapes. Tube of the perianth not prolonged. Leaves linear, thin, long and ribbon-like.

(In both genera the staminate flowers break off spontaneously and float on the surface around the pistillate ones, shedding their pollen upon them.)

'ELODE'A, Michx. WATER-WEED.

E. Canadensis, Michx. (*Anach'aris Canadensis,* Planchon.)—Common in slow waters.

2. **VALLISNE'RIA,** L. TAPE-GRASS. EEL-GRASS.

V. spira'lis, L. Leaves 1-2 feet long.—Common in slow waters.

ORDER C. **ORCHIDA'CEÆ.** (ORCHIS FAMILY.)

Herbs, well marked by the peculiar arrangement of the stamens, these being *gynandrous,* that is, borne on or adherent to the stigma or style. There is also usually but a single stamen, of two rather widely separated anthers, but in the last genus of the following list there are 2 distinct stamens, with the rudiment of a third at the back of the stigma. As explained in Part I., sections 90-93, the Orchids as a rule require the aid of insects to convey the *pollinia*, or pollen-masses, to the stigma, but occasionally it happens that when the anther-cells burst open the pollinia fall forward and dangle in front of the viscid stigma beneath, being sooner or later driven against it either by the wind or by the head of some insects in pursuit of honey. In all cases where the student meets with an Orchid in flower, he should, by experiment, endeavour to make himself acquainted with the method of its fertilization.

The Orchis family is a very large one, there being probably as many as 3,000 different species, but the greater number are natives of tropical regions. Many of them are *epiphytes*, or air-plants, deriving their support chiefly from the moisture of the air, through their long aerial roots which never reach the ground. The perianth in many species, and particularly the *labellum*, or lip, assumes the most fantastic shapes, making the plants great favourites for hot-house cultivation. In Canada the representatives of this great Order, though not very numerous, are among the most interesting and beautiful of our wild flowers. They are, as a rule, bog-plants, and will be found in flower in early summer.

Synopsis of the Genera.

* *Anther only one, but of 2 cells, these separated in the first genus.*

+ *Lip with a spur underneath. Anther on the face of the stigma.*

1. **Orchis.** The 3 sepals and 2 of the petals erect and arching over the centre of the flower; the lip turned down. The 2 glands or viscid disks of the base of the pollen-masses enclosed in a little pouch just over the concave stigma. Leaves 2, large. Flowers few, in a spike.

2. **Habena'ria.** The lateral sepals usually spreading. The glands or viscid disks of the pollen-masses not enclosed in a covering. Flowers in spikes.

++ *Lip without a spur. Anther on the back of the column.*
    ↦ *Flowers small, white, in a slender spike.*

3. **Spiran'thes.** Spike (of white or whitish flowers) more or less spirally twisted. Sepals and petals narrow and generally connivent. Lip oblong, the lower part embracing the column, *and with a protuberance on each side of the base.*

4. **Goodye'ra.** Flowers very much as in Spiranthes, *but the lip sac-shaped, and without protuberance at the base.* Leaves white-veiny, in a tuft at the base of the scape.

↦ ↦ *Flowers racemose, varying from greenish-yellow to purple. Stem very leafy.*

5. **Epipac'tis.** Stem 1–2 feet high, stout, leafy. Leaves broadly ovate, the upper narrower. Sepals and petals nearly equal, spreading. Lip deeply concave at the base, constricted and somewhat joined in the middle, dilated and petaloid above. Anther sessile behind the stigma, on a slender-jointed base. Ovaries reflexed at maturity.

+++ *Lip without a spur. Anther on the apex of the style, hinged like a lid.*

↦ *Pollen-masses 2 or 4, powdery or pulpy, without stalk or gland.*

6. **Lis'tera.** Flowers small, greenish or brownish-purple, in a spike or raceme. Stem bearing a pair of opposite sessile roundish leaves near the middle. Lip flat, *mostly drooping, 2-lobed at the apex.*

7. **Calopo'gon.** *Ovary not twisted, the lip consequently turned toward the stem.* Flowers large, pink-purple, 2–6 at the summit of the scape; the lip spreading at the outer end, and beautifully bearded above with coloured hairs. Leaf grass-like, only one. Pollen-masses 4.

8. **Arethu'sa.** Flower solitary, large, rose-purple. Lip dilated, recurved, spreading at the end. Sepals and petals lanceolate, nearly alike, arching over the column. Pollen-masses 4. Scape low, sheathed, from a globular solid bulb, with a single linear-nerved leaf hidden in the sheaths of the scape.

9. **Pogo'nia.** Flower solitary, irregular, large, sweet-scented, pale rose-colour or white. Column club-shaped. Lip crested and fringed. Pollen-masses 2. Stem 6–9 inches high, with a single oval or lance-oblong leaf near the middle, and a smaller one, or bract, near the flower.

↦ ↦ *Pollen-masses 4, smooth and waxy, attached directly to a large gland: no stalks.*

10. **Calyp'so.** Flower solitary, large, showy, variegated with purple, pink, and yellow. Lip large, inflated, sac-shaped, 2-pointed under the apex. Scape short, from a solid bulb, with a single ovate or slightly heart-shaped leaf below.

ORCHIDACEÆ. 229

** ** ** *Pollen-masses 4: no stalks or glands.*

11. **Micros'tylis.** Small herbs from solid bulbs; the scape bearing a single leaf and a raceme of minute greenish flowers. Column very small, terete, with 2 teeth at the top, and the anther between them. Petals thread-like or linear, spreading.

12. **Lip'aris.** Small herbs from solid bulbs; the low scape bearing 2 radical leaves and a raceme of a few greenish flowers. Column elongated, incurved, margined at the apex. Petals thread-like or linear, spreading. Anther lid-like.

13. **Corallorhi'za.** Brownish or yellowish plants, with the small dull flowers in spikes or scapes which are leafless or have mere sheaths instead of leaves. Rootstocks branching and coral-like. *Perianth gibbous or slightly spurred below.* Lip with 2 ridges on the inner part of the face.

14. **Aplec'trum.** Somewhat like the last, but the perianth is *not gibbous below*, and the rootstock, instead of being coral-like, is slender, and produces each year a solid bulb or corm. Lip with 3 ridges on the palate. Scape with 3 greenish sheaths below.

** *Anthers 2, one on each side of the stigma, and a triangular body, which is the rudiment of a third, at the back of the stigma. Pollen loose and powdery or pulpy.*

15. **Cypripe'dium.** *Lip a large inflated sac*, into the mouth of which the style is declined. Sepals and the other petals much alike, the former apparently only 2, two of them being generally united into one under the lip. Leaves large, many-nerved. Flowers solitary or few.

1. ORCHIS, L. ORCHIS.

O. specta'bilis, L. (SHOWY ORCHIS.) Scape 4-angled, 4-7 inches high, bearing a few flowers in a spike. The arching upper lip pink-purple, the *labellum* white; each flower in the axil of a leaf-like bract.

2. HABENA'RIA, Willd., R. Br. REIN-ORCHIS.

1. **H. tridenta'ta,** Hook. *Spike few-flowered*, the flowers very small, greenish-white. *Lip wedge-shaped, truncate and 3-toothed at the apex. Spur slender, longer than the ovary*, curved upwards. Stem less than a foot high, slender, with one oblanceolate leaf below and 2 or 3 much smaller ones above.—Wet woods.

2. **H. vires'cens,** Spreng. Stem 10–20 inches high. Spike of small greenish flowers at first dense, with the bracts longer than the flowers, at length long and loose. Lip

oblong, almost truncate at the tip; a tooth on each side at the base, and a nasal protuberance on the face. Spur slender, club-shaped. Leaves ovate-oblong or oblong-lanceolate, the upper ones gradually narrowing and passing into bracts. —Wet places.

3. **H. vir'idis**, R. Br., var. **bracteata**, Reichenbach. (*H. bracteata*, R. Br., in Macoun's Catalogue.) *Spike many-flowered*, close. Flowers small, *greenish*. Lip oblong-linear, 2-3-lobed at the tip, *much longer than the very short and sac-like spur*. Stem 6-12 inches high, *leafy*, the lower leaves obovate, the upper oblong or lanceolate, gradually reduced to bracts much longer than the flowers.

4. **H. hyperbo'rea**, R. Br. Spike many-flowered, *long and dense*. Flowers small, greenish. Lip lanceolate, *entire, about the same length as the slender incurved spur*. Stem 6-24 inches high, *very leafy, the leaves lanceolate and erect*, and the bracts longer than the flowers.—Bogs and wet woods.

5. **H. dilata'ta**, Gray. Not unlike No. 4, but more slender and with *linear leaves* and *white flowers*.

6. **H. rotundifo'lia**, Richardson. (*Orchis rotundifolia*, Gray, in Macoun's Catalogue.) Spike few-flowered, loose. Flowers rose-purple, *the lip usually white, spotted with purple, 3-lobed*, the middle larger and notched, longer than the slender spur. Stem 5-9 inches high, *naked and scape-like above, bearing a single roundish leaf* at the base.—Bogs and wet woods.

7. **H. obtusa'ta**, Richardson. Stem as in the last, but the leaf is obovate or spathulate-oblong. Spike few-flowered, the flowers *greenish-white*. Upper sepal broad and rounded, the others and the petals lance-oblong. Lip *entire*, deflexed, as long as the tapering and curving spur.—Bogs.

8. **H. Hook'eri**, Torr. Spike many-flowered, *strict*. Flowers *yellowish-green*, the lip lanceolate, pointed, incurved; petals lance-awl-shaped. *Spur slender, acute, nearly an inch long*. Stem scape-like above, *2-leaved at the base, the leaves orbicular*.—Woods.

Var. **oblongifo'lia,** J. A. Paine, has *oblong* leaves.

9. **H. orbicula'ta,** Torr. Spike many-flowered, *loose and spreading.* Flowers *greenish-white.* Lip narrowly linear, obtuse. Spur curved, *more than an inch long, thickened towards the apex.* Scape 2-leaved at the base, the leaves *very large, orbicular and lying flat on the ground,* shining above, silvery beneath.—Rich woods.

10. **H. blephariglot'tis,** Hook. (WHITE FRINGED-ORCHIS.) Spike many-flowered, open. *Flowers white,* very handsome; *the lip fringed, but not lobed,* at the apex. *Spur thread-shaped, three times as long as the lip.* Stem a foot high, leafy; the leaves oblong or lanceolate, the bracts shorter than the ovaries.—Peat-bogs, etc.

11. **H. leucophæ'a,** Gray. (GREENISH FRINGED-ORCHIS.) Spike as in the last, but the flowers *greenish or yellowish-white.* Petals obovate, minutely cut-toothed. *Lip 3-parted above the stalk-like base, the divisions fan-shaped, fringed.* Spur *gradually thickened downward,* longer than the ovary. Stem leafy, 2–4 feet high. Leaves oblong-lanceolate; bracts a little shorter than the flowers.—Wet meadows.

12. **H. la'cera,** R. Br. (RAGGED FRINGED-ORCHIS.) Like the last, but the *petals are oblong-linear and entire.* The divisions of the lip also are *narrow* and the *fringe is less copious.*—Bogs and rich woods.

13. **H. psyco'des,** Gray. (PURPLE FRINGED-ORCHIS.) Spike cylindrical, many-flowered, the *flowers pink-purple,* fragrant. Lip fan-shaped, 3-parted above the stalk-like base, *the divisions fringed.* Spur curved, somewhat thickened downward, very long.—Low grounds.

14. **H. fimbria'ta,** R. Br. Resembling the last, but the flowers 3 or 4 times larger, and the petals toothed down the sides. Divisions of the large lip more fringed. Spike loosely flowered.—Wet meadows, chiefly eastward.

3. SPIRAN'THES, Richard. LADIES' TRESSES.

1. **S. latifo'lia,** Torr. Flowers white, in 3 ranks, forming a narrow spike. Lip oblong, yellowish on the face, not

contracted in the middle, wavy-crisped at the blunt apex. Stem nearly naked. Leaves oblong or lance-oblong.—Moist banks.

2. **S. Romanzovia'na,** Chamisso. Spike dense, oblong or cylindrical. Flowers *pure white, in 3 ranks in the spike.* Lip ovate-oblong, *contracted below the wavy recurved apex. Sepals and petals all connivent above.* Beak of the stigma 2-horned, *short.* Stem 5-15 inches high, leafy below, leafy-bracted above; the leaves oblong-lanceolate or linear.—Cool bogs.

3. **S. cer'nua,** Richard. Resembling the last, but the lower sepals are not upturned or connivent with the upper. Margins of the lower lip much incurved. Gland of the stigma linear, *in a long and slender beak.*—Wet places.

4. **S. grac'ilis,** Bigelow. *Flowers in a single spirally-twisted rank* at the summit of the very slender scape. Leaves with blades all in a cluster at the base, ovate or oblong. Scape 8-18 inches high.—Sandy plains and pine barrens.

4. GOODYE'RA, R. Br. RATTLE-SNAKE PLANTAIN.

1. **G. repens,** R. Br. *Flowers in a loose 1-sided spike.* Lip with a recurved tip. Scape 5-8 inches high. Leaves thickish, petioled, *intersected with whitish veins.*—Woods, usually under evergreens.

2. **G. pubes'cens,** R. Br. *Spike not 1-sided.* Plant rather larger than the last, and the leaves more strongly white-veined.—Rich woods.

3. **G. Menzies'ii,** Lindl., is found in the north-western part of Ontario. The leaves are less white-veiny than the preceding, and the column is continued above the stigma into a *conspicuous awl-shaped beak.* The lip, also, is hardly, if at all, saccate below.

5. EPIPAC'TIS, Haller.

E. Hellebori'ne, Crantz. Near Toronto, on the bank of the Humber, and at Montreal.

6. LIS'TERA, R. Br. TWAYBLADE.

1. **L. corda'ta,** R. Br. Racemes crowded; pedicels not longer than the ovary. Lip linear, 2-cleft. Column very short.—Damp cold woods.

2. **L. convallariol'des,** Nutt. Racemes loose and slender; pedicels longer than the ovary. Lip wedge-oblong, 2-lobed. Column longer than the last.—Damp thickets.

7. CALOPO'GON, R. Br. CALOPOGON.

C. pulchel'lus, R. Br. Leaf linear. Scape a foot high. Flowers an inch across.—Bogs.

8. ARETHU'SA, Gronov. ARETHUSA.

A. bulbo'sa, L. A beautiful little bog-plant, bearing a single large flower (rarely 2), with the lip bearded-crested on the face.

9. POGO'NIA, Juss. POGONIA.

P. ophioglossoi'des, Nutt. A bog-plant. Sepals and petals nearly equal and alike. Root of thick fibres.

10. CALYP'SO, Salisb. CALYPSO.

C. borea'lis, Salisb. A beautiful little plant growing in mossy bogs. The lip woolly inside; the petals and sepals resembling each other, lanceolate, sharp-pointed. Column winged.

11. MICROS'TYLIS, Nutt. ADDER'S MOUTH.

1. **M. monophyl'los,** Lindl. Leaf sheathing the base of the stem, ovate-elliptical. Raceme spiked, long and slender. Lip long-pointed.—Cold swamps.

2. **M. ophioglossoi'des,** Nutt. Leaf near the middle of the stem, ovate, clasping. Raceme *short*. Lip 3-toothed.—Damp woods, not so common as the last.

12. LIP'ARIS, Richard. TWAYBLADE.

L. Lœse'lii, Richard. Lip yellowish-green, mucronate, shorter than the unequal petals and sepals. Leaves elliptical-lanceolate or oblong, keeled.—Bogs.

13. CORALLORHI'ZA, Haller. CORAL-ROOT.

1. **C. inna'ta,** R. Br. Flowers small; the lip whitish or purplish, often crimson-spotted, 3-lobed above the base. Spur very small. Stem slender, brownish-yellow, *with a few-flowered spike.*—Swamps.

2. **C. multiflo'ra,** Nutt. *Spike many-flowered.* Stem purplish, stout. Lip deeply 3-lobed. Spur more prominent than in No. 1.—Dry woods.

3. **C. Macræ'i,** Gray. (*C. striata*, Lindl., in Macoun's Catalogue.) Spike crowded, of numerous large flowers, *all the parts of the perianth strikingly marked with 3 dark lines.* Lip not lobed. Spur none, but the base of the perianth gibbous.—Rich woods; not common.

14. APLEC'TRUM, Nutt. PUTTY-ROOT. ADAM-AND-EVE.

A. hyema'le, Nutt. Scape a foot high. Perianth greenish-brown.—Rich mould in woods.

15. CYPRIPE'DIUM, L. LADY'S SLIPPER. MOCCASIN-FLOWER.

* *The three sepals separate.*

1. **C. arieti'num,** R. Br. (RAM'S-HEAD LADY'S SLIPPER.) The smallest species. Stem slender, 6–10 inches high, leafy. Leaves 3 or 4, elliptical-lanceolate, nearly smooth. Lip only half an inch, red and whitish veiny, prolonged at the apex into a deflexed point.—Swamps; rare.

* * *Two sepals united into one piece under the lip.*

2. **C. parviflo'rum,** Salisb. (SMALLER YELLOW LADY'S SLIPPER.) Stem leafy to the top, 1–3-flowered. Lip yellow, *flattish above*, rather less than an inch long. Sepals and petals wavy-twisted, brownish, pointed, longer than the lip. —Bogs and wet woods.

3. **C. pubes'cens,** Willd. (LARGER YELLOW L.) Lip flattened *laterally*, rounded above, larger than No. 2, but the two species are not sufficiently distinct.

4. **C. specta'bile,** Swartz. (SHOWY L.) *Lip very large, white, pinkish in front.* Sepals and petals *rounded, white,* not longer than the lip.—Bogs.

5. **C. acau'le**, Ait. (STEMLESS L.) *Scape naked*, 2-leaved *at the base, 1-flowered. Lip rose-purple*, split down the whole length in front, veiny. Sepals and petals greenish.—Dry or moist woods, under evergreens.

ORDER CI. **IRIDA'CEÆ.** (IRIS FAMILY.)

Herbs with equitant leaves and perfect flowers. The 6 petal-like divisions of the perianth in 2 (similar or dissimilar) sets of 3 each; the tube adherent to the 3-celled ovary. Stamens 3, distinct or monadelphous, opposite the 3 stigmas, and with anthers extrorse, that is, on the outside of the filaments, facing the divisions of the perianth and opening on that side. Flowers from leafy bracts. (See Part I., sections 88 and 89.)

Synopsis of the Genera.

1. **Iris.** The 3 outer divisions of the perianth redexed, the 3 inner erect and smaller. Stamens distinct, the anther of each concealed under a flat and petal-like arching stigma. The styles below adherent to the tube of the perianth. Pod 3-angled. Flowers blue, large and showy. Leaves sword-shaped or grass-like.
2. **Sisyrin'chium.** The 6 divisions of the perianth alike, spreading. Stamens monadelphous. Stigmas thread-like. Pod globular, 3-angled. Stems 2-edged. Leaves grass-like. Flowers blue, clustered, from 2 leafy bracts. Plants low and slender.

1. IRIS, L. FLOWER-DE-LUCE.

1. **I. versic'olor**, L. (LARGER BLUE FLAG.) Stem stout and leafy, from a thickened rootstock. Leaves sword-shaped, Flowers violet-blue, 2 or 3 inches long. Inner petals much smaller than the outer.—Wet places.

2. **I. prismat'ica**, Pursh. (SLENDER BLUE FLAG.) Distinguished from the last by its very slender stem and narrowly linear leaves. Tube of the flower *very short.*—Atl. sea-coast.

3. **I. lacus'tris**, Nutt. (LAKE DWARF IRIS.) Stem low, 3-6 inches high. Inner petals nearly equal to the outer. Tube of the perianth slender, less than an inch long, dilated upwards, rather shorter than the divisions of the perianth. Leaves lanceolate, 3-5 inches long.—Shore of Lake Huron.

2. SISYRIN'CHIUM, L. BLUE-EYED GRASS.

1. S. angustifo'lium, Mill. (*S. Bermudiana*, var. *mucronatum*, Gray.) A pretty little plant, rather common in moist meadows among grass. The divisions of the delicate blue perianth obovate, notched at the end, and bristle-pointed from the notch. The spathe solitary. Roots fibrous.

2. S. anceps, Cav., (*S. Bermudiana*, var. *anceps*, Gray.) has the scape 6–18 inches high, and usually bearing 2 *or more* peduncled spathes.

ORDER CII. AMARYLLIDA'CEÆ. (AMARYLLIS F.)

Bulbous and scape-bearing herbs, with linear flat root-leaves, and regular and perfect 6-androus flowers, the tube of the petal-like 6-parted perianth adherent to the 3-celled ovary. Lobes of the perianth imbricated in the bud. Style single. Anthers introrse.—Represented with us by one species of the genus

HYPOX'YS, L. STAR-GRASS.

H. erecta, L. A small herb sending up a slender scape from a solid bulb. Leaves linear, grass-like, longer than the umbellately 1–4-flowered scape. Perianth hairy and greenish outside, yellowish within, 6-parted nearly down to the ovary. Stamens 6, sagittate. Pod indehiscent, crowned with the withered perianth.—Meadows and open woods.

ORDER CIII. DIOSCOREA'CEÆ. (YAM FAMILY.)

Represented with us by the genus

DIOSCORE'A, Plumier. YAM.

D. villo'sa, L. (WILD YAM-ROOT.) A slender twiner, with knotted rootstocks, and net-veined, heart-shaped, 9–11-ribbed petioled leaves. Flowers diœcious, small, in axillary racemes. Stamens 6. Pod with three large wings.—Reported only from the warm and sheltered valley lying between Hamilton and Dundas, Ont., the banks of the Thames at London, Ont., and Niagara Falls.

ORDER CIV. **SMILA'CEÆ.** (SMILAX FAMILY.)

Climbing plants, more or less shrubby, with alternate *ribbed and net-veined* petioled leaves, and small diœcious flowers in umbels. Perianth regular, of 6 greenish sepals, free from the ovary. Stamens as many as the sepals, with 1-celled anthers. Ovary 3-celled, surrounded by 3 sessile spreading stigmas. Fruit a small berry. Represented by the single genus

SMILAX, Tourn. GREEN-BRIER. CAT-BRIER.

(*Included in Liliaceæ, in Macoun's Catalogue.*)

1. **S. his'pida,** Muhl. *Stem woody, densely covered below with long weak prickles.* Leaves large, ovate or heart-shaped, pointed, thin, 5-9-nerved. Peduncles of the axillary umbels *much longer than the petioles.* Berry black.—Moist thickets.

2. **S. rotundifo'lia,** L., var. **quadrangularis,** Gray. (*S. quadrangularis,* Pursh, in Macoun's Catalogue.) *Stem woody,* it and the branches armed with *scattered prickles. Branches 4-angular.* Peduncles *not longer than the petioles.* Leaves ovate, broader than long, slightly cordate. Berry blue-black.—South-western Ontario.

3. **S. herba'cea,** L. (CARRION-FLOWER.) *Stem herbaceous, not prickly.* Leaves ovate-oblong and heart-shaped, 7-9-ribbed, long-petioled, mucronate. Flowers carrion-scented. Berry bluish-black.—Meadows and river banks.

ORDER CV. **LILIA'CEÆ.** (LILY FAMILY.)

Herbs, distinguished as a whole by their regular and symmetrical flowers, having a 6-leaved perianth (but 4-leaved in one species of Smilacina) free from the usually 3-celled ovary, and as many stamens as divisions of the perianth (*one before each*) with 2-celled anthers. Fruit a pod or berry, generally 3-celled. The outer and inner divisions of the perianth coloured alike, except in the genus Trillium. (See Part. I., sections 82–87, for description of typical plant of this Order.)

Synopsis of the Genera.

* *Leaves net-veined, all in one or two whorls. The stem otherwise naked, rising from a fleshy rootstock. Styles 3.*

1. **Tril'lium.** Leaves 3, in a whorl at the top of the stem. Divisions of the perianth in 2 sets, the outer green, the inner coloured. (See Part I., sections 85 and 86.)

2. **Mede'ola.** Leaves in 2 whorls, the lower near the middle of the stem, and consisting of 5–9 leaves, the upper of (generally) 3 small leaves, near the summit. Stem tall, covered with loose wool. Flowers small, in an umbel. Divisions of the perianth alike, greenish-yellow, recurved. Anthers turned outwards. Styles thread-shaped. Berry globular or nearly so, dark purple.

* * *Leaves straight-veined, linear, grass-like, alternate. Stem simple or tufted. Styles 3.*

3. **Zygade'nus.** Flowers rather large, perfect or polygamous, greenish-white, in a few-flowered panicle; the divisions of the perianth each with a conspicuous obcordate spot or gland on the inside, near the narrowing base. Stem smooth and glaucous, *from a coated bulb.*

4. **Tofield'ia.** Flowers small, perfect, greenish-white, in a terminal raceme or spike, which, however, develops from above downward; the pedicels in clusters of 3, from little involucres of 3 bracts. Pod triangular. Roots *fibrous.* Stem leafy at the base only, in our species *sticky.* Leaves 2-ranked, equitant.

* * * *Leaves straight-veined, but broad (not grass-like), alternate. Stem from a rootstock of fibrous roots, at all events not from a bulb. Style one, or more or less divided into 3 above.*

— *Perianth of completely separate pieces (polyphyllous.)*

5. **Uvula'ria.** Stem leafy, *forking above.* Flowers yellow, at least an inch long, drooping, lily-like, usually solitary (but occasionally in pairs) at the end of the forks of the stem. Style *deeply* 3-cleft. Pod triangular. *Leaves clasping-perfoliate or sessile.*

6. **Clinto'nia.** *Stemless,* the naked scape sheathed at the base by 2, 3, or 4 large oblong or oval, ciliate leaves. Flowers few, greenish-yellow, in an umbel at the top of the scape. Filaments long and slender. Style long, the stigmas hardly separate. Berry blue.

7. **Prosar'tes.** Downy low herbs, branching above. Flowers greenish, bell-shaped, rather large, solitary or in pairs, drooping on terminal

slender peduncles. Sepals taper-pointed. Stigmas 3. Leaves ovate-oblong, taper-pointed, closely sessile, downy underneath. Berry oblong or obovate, red.

8. **Strep'topus.** Stem leafy and forking. Flowers small, not quite in the axils of the ovate clasping leaves, *on slender peduncles which are bent near the middle*. Anthers arrow-shaped, *2-horned at the apex*.

9. **Vera'trum.** Stem tall and stout, very leafy to the top, from a thickened base, producing coarse fibrous roots. Flowers yellowish green, in a pyramidal panicle of dense spike-like racemes. Perianth somewhat spreading, slightly adherent to the ovary. Filaments of the stamens recurved. Anthers kidney-shaped. Styles 3.

+ + *Perianth of one piece (gamophyllous.)*

10. **Smilaci'na.** Flowers small, white, in a terminal raceme. Perianth 6-parted, *but 4-parted in one species*, spreading. Style short and thick. Stigma obscurely lobed. Filaments slender.

11. **Polygona'tum.** Flowers small, greenish, nodding, *mostly in pairs* in the axils of the nearly sessile leaves. Perianth cylindrical, 6-lobed at the summit, *the 6 stamens inserted on or above the middle of the tube*. Stem simple, from a long and knotted rootstock. Leaves glaucous beneath.

12. **Aspar'agus.** Flowers small, greenish-yellow, axillary, on jointed pedicels. Perianth 6-parted, spreading above, the 6 stamens on its base. Stigma 3-lobed. Berry globular, 3-celled. Stem much branched. Leaves thread-like. Rootstock thick and matted.

* * * * *Leaves straight-veined, not grass-like. Stem usually from a coated or scaly bulb. Style 1, not divided above, but the stigma sometimes 3-lobed. Fruit a pod, splitting open midway between the partitions (loculicidal.)*

13. **Lil'ium.** Stem leafy, from a *scaly* bulb, *the leaves often whorled or crowded*. Anthers at first erect, *at length versatile*. Style long, rather club-shaped. Stigma 3-lobed. Pod oblong. Flowers large and showy, one or more.

14. **Erythro'nium.** For full description see Part I., sections 82 and 83. (Dog's-tooth Violet.)

15. **Allium.** *Scape naked*, from a *coated* bulb. Flowers in an umbel, from a spathe. Style thread-like. *Strong-scented plants.*

16. **Hemerocal'lis.** Scape *not* from a bulb, but from a fleshy-fibrous root. Perianth large, orange-yellow, lily-like, the short tube enclosing the ovary, the spreading limb 6-parted. Stamens 6, on the perianth. Leaves long, linear, keeled, in 2 ranks, at the base of the scape. Flowers several, bracted.

1. TRIL'LIUM, L. WAKE-ROBIN.

1. T. grandiflo'rum, Salisb. (LARGE WHITE TRILLIUM.) *Leaves sessile,* longer than broad. *Peduncle erect.* Petals white (rose-coloured when old), *obovate.*—Rich woods.

2. T. erectum, L. (*T. erectum,* L., var. *atropurpureum,* Hook, in Macoun's Catalogue.) (PURPLE TRILLIUM.) *Leaves sessile,* about as broad as long. *Peduncles erect.* Petals dull purple, *ovate.*—Rich woods. Var. **album,** *with greenish-white petals,* is found along with the purple form. It does not appear to be clearly distinguished from No. 1.

3. T. cer'nuum, L. *Leaves sessile* or nearly so, broadly rhomboid, abruptly pointed. *Peduncle recurved under the leaves.* Petals white, oblong-ovate, acute. Chiefly eastward.

4. T. erythrocar'pum, Michx. (PAINTED TRILLIUM.) *Leaves distinctly petioled,* rounded at the base. Petals pointed, white, *with purple stripes inside at the base.*—Not uncommon northward in damp woods and low grounds.

2. MEDE'OLA, Gronov. INDIAN CUCUMBER-ROOT.

M. Virgin'ica, L. Stem 1-3 feet high.—Rich woods.

3. ZYGADE'NUS, Michx. ZYGADENE.

Z. glaucus, Nutt. (*Z. elegans,* Pursh.) Not uncommon in bogs and beaver-meadows northward. Leaves flat and pale.

4. TOFIELD'IA, Hudson. FALSE ASPHODEL.

T. glutino'sa, Willd. Stem and pedicels very sticky with dark glands. Leaves short.—Lake Huron coast.

5. UVULA'RIA, L. BELLWORT.

1. U. grandiflo'ra, Smith. Leaves clasping-perfoliate. Rootstock short.—Rich woods.

2. U. sessilifo'lia, L. Leaves *sessile* or partly clasping, lance-oblong. Rootstock *creeping.*—Chiefly eastward.

6. CLINTO'NIA, Raf. CLINTONIA.

C. borea'lis, Raf. Umbel 2-7-flowered. Leaves 5-8 inches long. Perianth pubescent outside.—Damp woods, often under evergreens.

LILIACEÆ.   241

7. PROSAR'TES, Don. PROSARTES.

1. **P. lanugino'sa,** Don. (*Disporum lanuginosa*, Don., in Macoun's Catalogue.) Leaves taper-pointed. Fruit oblong, pointed.—Rich woods, western Ontario.

2. **P. trachycar'pa,** Watson, with whitish perianth, ovate to oblong-lanceolate leaves, and broadly obovate fruit, is not uncommon in the N.W.

8. STREP'TOPUS, Michx. TWISTED-STALK.

1. **S. ro'seus,** Michx. Flowers *rose-purple.* Leaves green both sides, finely ciliate. Stigma 3-cleft.—Damp woods.

2. **S. amplexifo'lius,** DC. Flowers *greenish - white.* Leaves very smooth, glaucous beneath. Stigma entire.—Chiefly Atl. Prov. and N.W.

9. VERA'TRUM, L. FALSE HELLEBORE.

V. vir'ide, Ait. (AMERICAN WHITE HELLEBORE. INDIAN POKE.) Leaves broadly oval, pointed, sheath-clasping. Segments of the perianth ciliate-serrulate.—Swamps and low grounds, Atl. Prov. and Eastern Townships, Q.

10. SMILACI'NA, Desf. FALSE SOLOMON'S SEAL.

1. **S. racemo'sa,** Desf. (FALSE SPIKENARD.) *Raceme compound.* Stem pubescent, 2 feet high. Leaves many, oblong, taper-pointed, ciliate. Berries speckled with purple.—Rich woods and thickets.

2. **S. stella'ta,** Desf. *Raceme simple.* Stem nearly smooth, 1-2 feet high. Leaves 7-12, oblong-lanceolate, slightly clasping. Berries black.—Moist woods and copses.

3. **S. trifo'lia,** Desf. *Raceme simple.* Stem low (3-6 inches), glabrous. Leaves usually 3, oblong, the bases sheathing. Berries red.—Bogs.

4. **S. bifo'lia,** Ker., var. **Canadensis,** Gray. (*Maianthemum Canadense*, Desf., in Macoun's Catalogue.) Distinguished at once by the *4-parted perianth and the 4 stamens.* Raceme simple. Stem 3-5 inches high. Leaves usually 2, but sometimes 3.—Moist woods.

11. POLYGONA'TUM, Tourn. SOLOMON'S SEAL.

1. **P. biflo'rum,** Ell. (SMALLER SOLOMON'S SEAL.) Stem slender, 1-3 feet high. Leaves ovate-oblong or lance-oblong. Peduncles mostly 2-flowered. Filaments *hairy.*—Rich woods.

2. **P. gigante'um,** Dietrich, (GREAT S.) is occasionally met with westward and south-westward. The stem is taller and stouter than in the last, the peduncles *several-flowered,* and the filaments are *not hairy.*

12. ASPAR'AGUS, L. ASPARAGUS.

A. **officina'lis,** L. (GARDEN ASPARAGUS.) Escaped from gardens in a few places.

13. LIL'IUM, L. LILY.

1. **L. Philadel'phicum,** L. (WILD ORANGE-RED LILY.) Divisions of the perianth *narrowed into claws* below, not recurved at the top. Flowers *erect,* 1-3, orange, spotted with purple inside. Leaves linear-lanceolate, the upper mostly in whorls of 5-8.—Sandy soil.

2. **L. Canadense,** L. (WILD YELLOW LILY.) Divisions of the perianth *recurved above the middle.* Flowers *nodding,* few, orange, spotted with brown inside. Leaves remotely whorled, 3-ribbed.—Swamps and wet meadows.

3. **L. super'bum,** L. (*L. Carolinianum,* Michx, in Macoun's Catalogue.) (TURK'S-CAP LILY.) Divisions of the perianth *very strongly recurved.* Flowers nodding, *often numerous,* in a pyramidal raceme, bright orange, dark-purple-spotted within. Lower leaves whorled, 3-ribbed or nerved. Stem taller than either of the first two, 3-7 feet. —Rich low grounds, commoner southward and south-westward.

14. ERYTHRO'NIUM, L. DOG'S-TOOTH VIOLET.

1. **E. America'num,** Smith. (YELLOW ADDER'S TONGUE.) Perianth *light yellow,* sometimes spotted at the base.—Copses and rich meadows.

2. **E. al'bidum,** Nutt. (WHITE DOG'S-TOOTH VIOLET.) Leaves less blotched than the last. Perianth *pinkish-white.* —Not common.

15. ALLIUM, L. ONION. LEEK.

* *Bulb cespitose, crowning a rhizome.*

1. **A. tricoc'cum,** Ait. (WILD LEEK.) Leaves 2 or 3, *flat, lance-oblong,* 5-9 inches long, 1-2 inches wide, appearing in early spring and withering before the flowers are developed. Sepals *white.* Pod strongly 3-lobed. Scape 9 inches high.—Rich woods.

2. **A. Schœnop'rasum,** L. (CHIVES.) Leaves linear, hollow. Scape naked, or leafy at the base. Flowers rose-purple, in a globular umbel. Sepals lanceolate, pointed. *Ovary not crested.*—Margin of rivers, Atl. Prov. and northward.

3. **A. cer'nuum,** Roth. (WILD ONION.) Scape naked, angular, nodding at the apex, bearing a loose or drooping umbel of rose-coloured flowers. Leaves linear, flattened, sharply keeled. *Capsule 6-crested.*—N.W.

** *Bulbs mostly solitary. Leaves narrowly linear.*

4. **A. Canadense,** Kalm. (WILD GARLIC.) Leaves *narrowly linear.* Ovary crested with 6 teeth. Umbel few-flowered, *often bearing a head of bulbs instead of flowers.* Sepals pale rose-colour.—Along river-banks; rather rare.

5. **A. stella'tum,** Nutt. Scape terete, 6-18 inches high, slender, bearing an *erect* umbel. *Stamens and style exserted.* Bulb-coats membranous. Capsule 6-crested.—N.W., not very common.

6. **A. reticula'tum,** Fraser. Scape 3-8 inches high. Bulbs densely and coarsely *fibrous coated.* Stamens not exserted. Capsule crested.—N.W.

16. HEMEROCAL'LIS, L. DAY-LILY.

H. fulva, L. (COMMON DAY-LILY.) Inner divisions of the tawny orange perianth wavy and obtuse.—Escaped from gardens occasionally.

ORDER CVI. **JUNCA'CEÆ.** (RUSH FAMILY.)

Grass-like or sedge-like plants, with, however, flowers similar in structure to those of the last Order. Perianth greenish and glumaceous, of 6 divisions in 2 sets of 3 each.

Stamens 6 (occasionally 3). Style 1. Stigmas 3. Pod 3-celled, or 1-celled with 3 placentæ on the walls. The plants of the Order are not of any very great interest to the young student, and the determination of the species is rather difficult. A brief description of a few of the most common is given here, as an easy introduction to the study of the Order with the aid of more advanced text-books.

Synopsis of the Genera.

1. **Lu'zula.** Plant less than 1 foot high. Leaves linear or lance-linear. flat, *usually hairy*. *Pod 1-celled, 3-seeded.* Flowers in umbels or in spikes. Plants usually growing in *dry* ground.
2. **Juncus.** Plants *always smooth*, growing in water or *wet soil*. Flowers small, greenish or brownish, panicled or clustered. Pod 3-celled, *many-seeded.*

1. LU'ZULA, DC. WOOD-RUSH.

1. **L. pilo'sa,** Willd. (*L. vernalis*, DC.) Flowers umbelled, long-peduncled, brown-coloured. Sepals pointed.—Shady banks.

2. **L. campestris,** DC., has the flowers (light brown) in 4–12 spikes, the spikes umbelled. Sepals bristle-pointed.—Fields and woods.

2. JUNCUS, L. RUSH.

* *Scapes single and leafless, but with sheaths at the base. Flowers in sessile panicles, apparently from the side of the scape, owing to the involucral leaf being similar to and continuing the scape.*

1. **J. effu'sus,** L. (COMMON or SOFT RUSH.) Scape 2–4 feet high, soft and pliant, furnished at the base with merely *leafless sheaths*, the inner sheaths awned. Panicle many-flowered. Flowers small, *greenish, only 1 on each pedicel. Stamens 3.* Pod *greenish-brown*, triangular-obovate, *not pointed.*—Marshes.

2. **J. filifor'mis,** L., has a very slender scape (1–2 feet high), fewer flowers than No. 1, and *6 stamens* in each. Pod *greenish*, broadly ovate, and *short-pointed. No leaves.*

3. **J. Bal'ticus,** Dethard, var. **littoralis,** Engelm. Scape *rigid*, 2–3 feet high. *No leaves.* Panicle loose. Flowers brownish. Pod elliptical, somewhat triangular, obtuse but pointed, *deep-brown.*

Var. **monta'nus**, Engelm., has an ovate-pyramidal, angled, and beaked capsule.—N.W.

* * *Stem leafy at the base or throughout; the leaves flat or channelled but never knotted. Panicle terminal.*

4. **J. styg'ius**, L. Scape slender, 1-3-leaved below, naked above. Leaves thread-like, hollow, not knotted. Heads 1 or 2, of 3-4 flowers, about as long as the dry, awl-pointed sheathing bract. Stamens 6.—Peat bogs, Atl. Prov.

5. **J. Greenii**, Oakes and Tuckerm. Stem slender, simple, tufted. Leaves nearly terete, *deeply channelled* on the inner side. Flowers solitary, panicled. The principal leaf of the involucre usually much longer than the panicle. Pod ovoid-oblong, obtuse, longer than the acute sepals. Seeds ribbed, and delicately cross-lined.—Wet sandy places.

6. **J. Va'seyi**, Engelm. Stem slender, rigid, tufted, leafy below, 12-30 inches high, not branching. Leaves nearly terete, *very slightly channelled.* Panicle longer than the involucral leaf. Flowers greenish or light brown, few, often 1-sided. Pod oblong, retuse, greenish-brown.—Wet meadows, N.W.

7. **J. Gerard'i**, Loisel. (BLACK GRASS.) Stems rigid, 1-2 feet high. Panicle contracted, usually longer than the involucral leaf. Flowers brown and green. Sepals incurved, as long as the mucronate capsule. Seeds obovate, delicately ribbed and cross-lined.—Salt marshes, Atl. Prov.

8. **J. longis'tylis**, Torr. and Gr. *Flowers in heads.* Heads few, or rarely single. Flowers greenish with brown lines.—N.W.

9. **J. bufo'nius**, L. Stem *leafy*, slender, 3-9 inches high, branching from the base. Panicle terminal, spreading. Flowers greenish, single on the *pedicels.* Sepals awl-pointed, the outer set much longer than the inner and than the blunt pod. *Stamens 6.*—Ditches along roadsides.

10. **J. ten'uis**, Willd. Stems *leafy below*, wiry, 9-18 inches high, *simple*, tufted. Panicle loose, shorter than the slender involucral leaves. Flowers greenish, single on the pedicels; the sepals longer than the blunt pod. *Seeds white-pointed at both ends.*—Open low grounds.

*** *Stem leafy; the leaves terete or laterally compressed, knotted by internal cross-partitions. Panicle terminal, the flowers mostly in heads.*
← *Stamens 6.*

11. **J. pelocar'pus**, E. Meyer. Stems slender and erect. 6-18 inches high. Leaves few, thread-like, slightly knotted. Flowers greenish with red, *single or in pairs* in the forks and along one side of the branches of the panicle, and often with accompanying tufts of leaves. The 3 inner sepals longer than the outer ones, but shorter than the oblong taper-beaked 1-celled pod. Seeds obovate, short-pointed.

12. **J. articula'tus**, L. Stems 9-15 inches high, tufted from a short creeping rootstock, with 1 or 2 slender leaves, bearing crowded 3-8-flowered heads in short spreading panicles. Capsule deep-brown.—Wet places, Atl. Prov.

13. **J. alpi'nus**, Villars, var. **insignis**, Fries. Stems erect, 9-18 inches high, with 1 or 2 slender leaves. Branches of the meagre panicle erect, bearing *numerous* distant greenish or brownish *heads* of 3-6 flowers each. Outer sepals mucronate or cuspidate, and longer than the rounded inner ones. Pod short-pointed, light-brown. Seeds spindle-shaped. —Lake- and river-margins, common.

14. **J. nodo'sus**, L. Stem erect, 6-15 inches high, slender, from a creeping slender and *tuber-bearing* rootstock, usually with 2 or 3 slender leaves. *Heads few*, 8-20-flowered, and overtopped by the involucral leaf. Flowers brown. Pod slender, taper-pointed, 1-celled. Seeds obovate, mucronate. —Common.

Var. **megacepha'lus**, Torr., has large 30-80-flowered heads.

← ← *Stamens 3. Seeds tailed.*

15. **J. Canadensis**, J. Gay, var. **coarcta'tus**, Engel. Stems slender, 9-18 inches high, tufted, bearing 2 or 3 leaves. Panicle somewhat erect, contracted; the heads 3-5-flowered, deep-brown. Pod prismatic, abruptly pointed, deep-brown. Seeds slender, with short tails.—A very late-flowering species.

Var. **longicauda'tus**, Engelm., is taller, and the heads are many-flowered.

Order CVII. **PONTEDERIA'CEÆ.** (Pickerel-weed F.)

The most common representatives of this Order with us are

1. PONTEDE'RIA, L. Pickerel-weed.

P. corda'ta, L. A stout plant growing in shallow water, sending up a scape bearing a single large arrow-heart-shaped blunt leaf, and *a spike of violet-blue flowers with a spathe-like bract*. Perianth 2-lipped, the 3 upper divisions united, the 3 lower spreading, the whole revolute-coiled after flowering, the fleshy base enclosing the fruit. Stamens 6, 3 of them exserted on long filaments, the rest short.

Var. **angustifolia**, Torr., has narrow scarcely cordate leaves.

2. SCHOL'LERA, Schreber. Water Star-grass.

S. gramin'ea, Willd. (*Heteranthera graminea*, Vahl., in Macoun's Catalogue.) A grass-like herb, wholly under water, only the small yellowish flowers reaching the surface, the latter single, from spathes. Perianth salver-shaped, regular. Stamens 3, anthers sagittate.

Order CVIII. **XYRIDA'CEÆ.** (Yellow-eyed-grass F.)

Rush-like herbs, with equitant leaves sheathing the base of a naked scape, terminated by a head of perfect 3-androus flowers, with glumaceous calyx and coloured corolla. Capsule 3-valved and 1-celled, containing several or many seeds.

XYRIS, Gronov. Yellow-eyed Grass.

X. flexuo'sa, Muhl., var. **pusill'a**, Gray. Scape small and very slender, 2-9 inches high, not bulbous at the base. Flowers in the axils of scale-like bracts densely imbricated in a head. Sepals 3. Petals 3. Fertile stamens 3, with 3 alternating sterile filaments bearded at the apex.—Peat bogs and wet sandy shores, Atl. Prov. and northward.

Order CIX. **ERIOCAULONA'CEÆ.** (Pipewort F.)

Represented with us by the genus

ERIOCAU'LON, L. Pipewort.

E. septangula're, Withering. A slender plant with a

naked scape 2–6 inches high, growing in shallow water in the margins of our northern ponds. Leaves short, awl-shaped, in a tuft at the base. Flowers in a small woolly head at the summit of the scape, monœcious. Perianth double; the outer set or calyx of 2–3 keeled sepals; the corolla tubular in the sterile flowers and of 2–3 separate petals in the fertile ones. Scape 7-angled. The head (except the beard) lead-coloured.

III. GLUMACEOUS DIVISION.

Flowers without a proper perianth, but subtended by thin scales called *glumes*.

This Division includes two very large Orders—Cyperaceæ and Gramineæ—both of which present many difficulties to the beginner. Accordingly no attempt will be made here to enumerate and describe all the commonly occurring species of these Orders. In chapter XIV., Part I., the student will find descriptions and illustrations of several typical Grasses. We shall here, therefore, only describe two or three of the commonest representatives of the Order Cyperaceæ, so as to put the beginner in a position to continue his studies with the aid of Gray's Manual or other advanced work.

Order CX. **CYPERA'CEÆ.** (Sedge Family.)

Grass-like or rush-like herbs, easily distinguished from Grasses by the sheaths of the leaves, which in the Sedges are *closed* round the clum, not split. Flowers in spikes, each flower in the axil of a glume-like bract, either altogether without a perianth or with a few bristles or scales inserted below the ovary. Ovary 1-celled, becoming an achene (2- or 3-angled). Style 2- or 3-cleft. Stamens mostly 3, occasionally 2.

CYPERACEÆ.

We shall describe one species of each of five genera.

1. CYPERUS DIANDRUS.

The plant (Fig. 256) is from 4 to 10 inches in height. The culm is *triangular*, leafy towards the base, but naked above. At the summit there is an umbel the rays of which are unequal in length, and on each ray are clustered several *flat brown-coloured* spikes, the scales of which are imbricated in two distinct rows. At the base of the umbel there are 3 leaves of very unequal length, forming a sort of involucre, and the base of each ray of the umbel is sheathed. In each spike every scale except the lowest one contains a flower in its axil. The flower (Figs. 257 and 258) is entirely destitute of perianth, and consists of 2 *stamens and an ovary surmounted by a 2-cleft style*, being consequently *perfect*. The plant is pretty easily met with in low wet places.

Fig. 257.

Fig. 259.

Fig. 256.

2. ELEOCH'ARIS OBTU'SA.

In this plant, which grows in muddy soil in tufts 8 to 14 inches in height, there is but a single spike at the summit of each slender culm, and the scales of the spikes, instead of being imbricated in 2 rows and thus producing a flat form, are *imbricated all round*. The scales are very thin in texture, with a midrib somewhat thicker, and are usually brownish in colour. Each of them contains a perfect flower in its axil. Instead of a perianth, there are *6 or 8 hypogynous barbed bristles*. The stamens (as is generally the case in this Order) are 3 in number, and the style is usually 3-cleft. Observe that the style is enlarged into a sort of bulb at the base, *this bulbous portion persisting as a flattish tubercle on the apex of the achene.* The culms are without leaves, being merely sheathed at the base.

3. SCIRPUS PUNGENS.

A stout marsh-plant, 2 or 3 feet high, with a sharply triangular hollow-sided culm, and bearing at the base from 1 to 3 channelled or boat-shaped leaves. The rusty-looking spikes vary in number from 1 to 6, and are in a single sessile cluster which appears to spring from the side of the culm, owing to the 1-leaved involucre resembling the culm and seeming to be a prolongation of it. Each scale of the spike is 2-cleft at the apex, and bears a point in the cleft. The flowers are perfect, with 2 to 6 bristles instead of perianth, 3 stamens, and a 2-cleft style, *but there is no tubercle on the apex of the achene.* The culms of this plant spring from stout running rootstocks.

4. ERIOPH'ORUM POLYSTACH'YON.

A common bog-plant in the northern parts of Canada, resembling Scirpus in the details as to spikes, scales, etc., but differing chiefly in this, that the bristles of the flowers are very delicate and become very long after flowering, so that *the spike in fruit looks like a tuft of cotton.* The culm of our plant is triangular, though not manifestly so, and its leaves are hardly, if at all, channelled. The spikes are several in number, and are on nodding peduncles, and the involucre consists of 2 or 3 leaves. Culm 15 or 20 inches high.

5. CAREX INTUMES'CENS.

The species of the genus Carex are exceedingly numerous and difficult of study. The one we have selected (Fig. 259) is one of the commonest and at the same time one of the easiest to examine. In this genus the flowers are monœcious, the separate kinds being either borne in different parts of the same spike or in different spikes. The genus is distinguished from all the others of this Order by the fact of the achene *being enclosed in a bottle-shaped more or less inflated sac*, which is made by the union of the edges of two inner bractlets or scales. To this peculiar sac. (Figs. 260 and 261) which encloses the achene the name *perigynium* is given. The culms are always triangular and the leaves grass-like, usually roughened on the margins and on the keel.

In the species under examination (which may be found in almost any wet meadow) the culm is some 18 inches high. The staminate spike (only one) is separate from and *above* the fertile ones, which are 2 or 3 in number, few- (5 to 8) flowered, and quite near together. The perigynia are very much inflated, that is, very much larger than the achene; they are distinctly marked with many nerves, and taper gradually into a long 2-toothed beak from which protrude the 3 stigmas. The bracts which subtend the spikes are leaf-like, and extend much beyond the top of the culm.

Fig. 259.
Fig. 260.
Fig. 261.

Order CXI. GRAMIN'EÆ. (Grass Family.)

Herbs somewhat resembling those of the last Order, but the culms are hollow except at the joints, and the sheaths of the leaves are split on the opposite side of the culm from the blade.

SERIES II.

FLOWERLESS OR CRYPTOG'AMOUS PLANTS.

PLANTS not producing true flowers, but reproducing themselves by means of *spores* instead of seeds, the spores consisting merely of simple cells, and not containing an embryo.

This series is subdivided into three classes:

1. **Pteridophytes,** embracing Ferns, Horsetails, and Club-Mosses.
2. **Bryophytes,** embracing the true Mosses and Liverworts.
3. **Thallophytes,** embracing Algæ and Fungi.

Types of all of these have already been described and illustrated in Part I. We shall here enumerate the common representatives of the Pteridophytes only.

FERNS.

These beautiful plants are favourites everywhere, and we shall therefore enter into a description of their characteristics with sufficient minuteness to enable the young student to determine with tolerable certainty the names of such representatives of the Family as he is likely to meet with commonly.

In Chapter XXI. of Part I. will be found a full account of the common Polypody, with which it is assumed the student is already familiar.

Fig. 262 shows a portion of the frond of the Common Brake (Pteris aquilina). Here the frond is several times compound. The first or largest divisions to the right and left are called *pinnæ*.

The secondary divisions (or first divisions of the pinnæ) are the *pinnules*. The stem, as in the Polypody, and in fact in all our Ferns which have a stem at all, is a rootstock or rhizome. But here we miss the fruit-dots or sori, so conspicuous in our first example. In this case it will be found that there is a *continuous line of sporangia around the margin* of every one of the pinnules of the frond, and that the edge of the pinnule is reflexed *so as to cover the line of spore-cases*. Fig. 263 is a very much magnified view of one of the lobes of a pinnule,

Fig. 263.

Fig. 262.

with the edge rolled back to show the sporangia. Some of the sporangia are removed to show a line which runs across the ends of the forking veins. To this the sporangia are attached. The veins, it will be seen, do not form a net-work, and so are free, as in Polypody. Observe, then, that in Polypody the sori are not covered, whilst in Pteris the opposite is the case. The covering of the fruit-dots is technically known as the *indusium*. The individual spore-cases are alike in both plants

Fig. 264 shows a frond of one of our commonest Shield-Ferns (Aspidium acrostichoides). It is simply pinnate. The stipe is thickly beset with rusty-looking, chaff-like scales. The veins are free, as before. The *sori* or *fruit-dots* are on the back of the upper pinnæ, but they are neither collected in naked clusters, as in Polypody, nor are they covered by the edge of the frond as in the Brake. Here each cluster has an *indusium* of its own. The indusium is round, and attached to the frond by its depressed centre (peltate). Fig. 265 shows an enlarged portion of a pinna, with the sporangia escaping from beneath the indusium. From one forking vein the sporangia are stripped off to show where they have been attached. The separate sporangia discharge their spores in the manner represented in the account of Polypody.

In some Ferns the fruit-dots are elongated instead of being round, and the indusium is attached to the frond by *one edge* only, being free on the other. Sometimes two long fruit-dots will be found side by side, the free edges of the indusia being towards each other, so that there is the *appearance* of one long fruit-dot with an indusium split down the centre.

Fig. 264.
or Sensitive Fern.

Fig. 265.

Fig. 266 represents a frond of a very common swamp Fern, Onoclea Sensibilis, It is deeply pinnatifid, and on one of the lobes the veining is represented. Here the veins are *not* free, but as they form a net-work they are said to be *reticulated*. You will look in vain on this frond for fruit-dots, but beside it grows

Fig. 266.   Fig. 267.   Fig. 268.
Fig. 269.

another, very different in appearance,—so different that you will hardly believe it to be a frond at all. It is shown in Fig. 267. It is twice pinnate, the pinnules being little globular bodies, one of which, much magnified, is shown in Fig. 268. You may open out one of these little globes, and then you will have something like what is shown in an enlarged form in Fig. 269. It now looks more like a pinnule than when it was rolled up, and it now also displays the fruit-dots on the veins inside. Here, then, we have evidently two kinds of frond. That bearing the fruit-dots we shall call the *fertile* frond, and the other we shall call the *sterile* one. You must not look upon the pinnule in which the sori are wrapped up as an indusium. Sori which are wrapped up in this way have an indusium of their own besides, but in this plant it is so obscure as to be very difficult to observe.

The spore-cases burst open by means of an elastic ring as before.

Fig. 270 represents one of the Moonworts (Botrychium Virginicum), very common in our rich woods everywhere. Here we have a single frond, but made up manifestly of two distinct portions, the lower sterile and the upper fertile. Both portions are thrice-pinnate. The ultimate divisions of the fertile segment are little globular bodies, but you cannot unroll them as in the case of the Onoclea. Fig. 271 shows a couple of them greatly enlarged. There is a slit across the middle of each, and one of the slits is partially open, disclosing the *spores* inside. Each little globe is, in fact, a *spore-case* or *sporangium*, so that here we have something quite different from what we have so far met with. Up to this point we have found the sporangia collected into dots or lines or clusters of some sort. In the Moonwort the sporangia are separate and naked, and instead of bursting through the action of an elastic ring, they open by a horizontal slit and discharge their spores. In other Ferns, as the Osmunda, the sporangia are somewhat similar, but burst open by a *vertical* instead of a horizontal slit.

Observe that the frond of Botrychium is *not circinate* in the bud.

We shall now proceed to describe the commonly occurring representatives of the Fern Family.

ORDER CXII. **FILICES.** (FERN FAMILY.)

Flowerless plants with distinct leaves known as *fronds*, these circinate in the bud, except in one suborder, and bearing on the under surface or margin the clustered or separate sporangia or spore-cases.

Synopsis of the Genera.

SUBORDER I. **POLYPODIA'CEÆ.** (THE TRUE FERNS.)

Sporangia collected into various kinds of clusters called *sori*. Each sporangium pedicelled and encircled by an elastic jointed ring, by the breaking of which the sporangium is burst and the spores discharged. Sori sometimes covered by an *indusium*.

1. **Polypo'dium.** Fruit-dots on the back of the fronds near the ends of the veins. *No indusium.* Veins free. (See Fig. 231, Part I.)
2. **Adian'tum.** Fruit-dots *marginal*, the edge of the frond being reflexed so as to form an indusium. *Midrib of the pinnules close to the lower edge or altogether wanting.* Stipe black and shining. All the pinnules distinct and generally minutely stalked. Veins free.
3. **Pte'ris.** Fruit-dots *marginal.* Indusium formed by the reflexed edge of the frond. Midrib of the pinnules in the centre and prominent. Veins free. Stipe light-coloured. (See Fig. 262.)
4. **Pellæ'a.** Fruit-dots *marginal*, covered by a broad indusium, formed by the reflexed margin of the frond. Small ferns with once- or twice-pinnate fronds, the fertile ones very much like the sterile, but with narrower divisions. Stipe shining, purple or brown.

5. **Asple'nium.** Fruit-dots elongated (rarely curved), on veins on the back of the pinnules, *oblique to the midrib, but only on the upper side of the vein* (rarely attached to both sides). *Indusium attached to the vein by one edge*, the other edge free. Veins free.

6. **Woodwar'dia.** Fruit-dots elongated, on cross-veins *parallel to the midrib*, forming a chain-like row on each side of the latter. Indusium as in the last. *Veins reticulated.*

7. **Scolopen'drium.** Fruit-dots elongated, *occurring in pairs on contiguous veinlets,* the free edges of the two indusia facing each other, so that the sori appear to be single, with an indusium split down the centre. Veins free. *Frond simple, ribbon-shaped,* about an inch broad, generally wavy-margined.

8. **Camptoso'rus.** Fruit elongated, those near the base of the midrib double, as in Scolopendrium; others single, as in Asplenium. *Fronds simple,* ½ to ¾ of an inch wide at the heart-shaped base, and tapering into a long and narrow point; growing in tufts on limestone rocks, and commonly rooting at the tip of the frond, like a runner. Veins reticulated.

9. **Phegop'teris.** Fruit-dots roundish, on the back (not at the apex) of the veinlet, rather small. *Indusium obsolete or none.* Veins free. *Fronds triangular in outline,* in one species twice-pinnatifid, with a winged rhachis, and in the other in three petioled spreading divisions, the divisions once- or twice-pinnate.

10. **Aspid'ium.** Fruit-dots round. Indusium evident, flat, orbicular or kidney-shaped, fixed by the centre, opening all round the margin. Veins free. Generally rather large Ferns, from once- to thrice-pinnate. (See Fig. 264.)

11. **Cystop'teris.** Fruit-dots round. Indusium not depressed in the centre, but rather raised, attached to the frond not by the centre, but by the edge partly under the fruit-dot, and generally breaking away on the side towards the apex of the pinnule, and becoming reflexed as the sporangia ripen. Fronds slender and delicate, twice- or thrice-pinnate.

12. **Struthiop'teris.** Fertile frond much contracted and altogether unlike the sterile ones, the latter very large and growing in a cluster with the shorter fertile one in the centre. Rootstock very thick and scaly. Fertile fronds simply pinnate, the margins of the pinnæ rolled backward so as to form a hollow tube containing the crowded sporangia. Very common in low grounds.

13. **Onocle'a.** Fertile and sterile fronds unlike. (See Figs. 266, 267, 268, 269, and accompanying description.)

14. **Wood'sia.** Small ferns with free veins. Sori round. Indusium very thin and delicate, attached by its base all round *under the sorus,* the top at length bursting into more or less narrow segments.

FILICES. 259

15. **Dickso'nia.** Fruit-dots round, *very small, each on a recurved toothlet* on the upper margin of the lobes of the pinnules, usually one to each lobe. Sporangia on an elevated globular receptacle, and enclosed in a cup-shaped indusium open at the top and partly adherent to the reflexed toothlet of the frond. Fronds minutely globular or hairy, 2-3 feet high, ovate-lanceolate in outline, pale green, very thin, without chaff.

SUBORDER II. **OSMUNDA'CEÆ.**

Sporangia naked, globular, pedicelled, *reticulated*, opening by a vertical slit.

16. **Osmun'da.** Fertile fronds or *fertile portions* of the frond much contracted, bearing naked sporangia, which are globular, short-pedicelled, and opening by a *vertical* slit to discharge the spores. Frond tall and upright, once- or twice-pinnate, from thick rootstocks.

SUBORDER III. **OPHIOGLOSSA'CEÆ.**

Sporangia naked, *not reticulated*, opening by a horizontal slit. *Fronds not circinate in the bud.*

17. **Botrych'ium.** Sporangia in compound spikes, *distinct*, opening by a horizontal slit. Sterile part of the frond compound. Veins free. (See Figs. 270 and 271).
18. **Ophiglos'sum.** Sporangia *coherent* in 2 ranks on the edges of a simple spike. Sterile part of the frond simple. Veins reticulated.

1. **POLYPO'DIUM**, L. POLYPODY.

P. vulga're, L. Fronds evergreen, 4-10 inches long, deeply pinnatifid, the lobes obtuse and obscurely toothed. Sori large.—Common on shady rocks.

2. **ADIAN'TUM**, L. MAIDENHAIR.

A. peda'tum, L. Stipe upright, black and shining. The frond forked at the top of the stipe, the two branches of the fork recurved, and each bearing on its inner side several slender spreading divisions, the latter with numerous thin pinnatifid pinnules which look like the *halves* of pinnules, owing to the midrib being close to the lower edge. Upper margin of the pinnules cleft.—Common in rich woods.

3. PTE'RIS, L. BRAKE. BRACKEN.

P. aquili'na, L. Stipe stout and erect. Frond large and divided into 3 large spreading divisions at the summit of the stipe, the branches twice-pinnate, the pinnules margined all round with the indusium.—Common in thickets and on dry hillsides.

4. PELLÆ'A, Link. CLIFF-BRAKE.

1. **P. gra'cilis,** Hook. Fronds 3-6 inches high, slender, of few pinnæ, the lower ones once- or twice-pinnatifid into 3-5 divisions, those of the fertile fronds narrower than those of the sterile ones. Stipe polished, *brownish*, darker at the base.—Shady limestone rocks; not common.

2. **P. atropurpu'rea,** Link. Larger than the last, 6-15 inches high, the stipe *dark-purple* and shining. Frond pale, once- or (below) twice-pinnate, the divisions broadly linear or oblong (or the sterile sometimes oval), stalked at the base. Stipes clustered.—Dry rocks.

5. ASPLE'NIUM, L. SPLEENWORT.

* *Indusium straight, attached to the upper side of the vein.*

1. **A. Trichom'anes,** L. A very delicate little fern growing in tufts on shaded cliffs. Fronds 3-6 inches long, linear in outline, pinnate, the little pinnæ oval and unequal-sided, about $\frac{1}{4}$ of an inch long. The stipes thread-like, purplish-brown and shining. This species is evergreen.

2. **A. vir'ide,** Hudson. Resembling A. Trichomanes, but less rigid, and the stipe brownish at base, becoming green upwards. Pinnæ roundish-ovate or ovate-rhomboid, short-stalked, crenately toothed.—Shaded cliffs, Atl. Prov. chiefly.

3. **A. ebe'neum,** Ait. Evergreen. Fronds 9-18 inches high, upright, linear, oblanceolate in outline, fertile ones much the taller, simply pinnate. Pinnæ mostly alternate, sessile, finely serrate, the base auricled on the upper or both sides. Stipe and rhachis blackish-purple and shining.—Rocky open woods, Ontario.

4. **A. thelypteroi'des,** Michx. Not evergreen. Fronds 2-3 feet high, *pinnate*, the pinnæ linear-lanceolate in outline,

3-5 inches long, *deeply pinnatifid*, each of the crowded lobes bearing 3-6 pairs of oblong fruit-dots.—Rich woods.

5. **A. angustifo'lium,** Michx. Not evergreen. Fronds simply pinnate, somewhat resembling Aspidium acrostichoides, *but very smooth and thin*, and larger. Pinnæ crenulate, short-stalked. Fruit-dots linear, crowded.—Rich woods; not common.

** *Indusium curved, delicate, often crossing the vein, thus appearing reniform.*

6. **A. Filix - fœ'mina,** Bernh. Fronds 1-3 feet high, broadly lanceolate in outline, *twice-pinnate*, the pinnæ lanceolate in outline, and the pinnules confluent by a narrow margin on the rhachis of the pinna, doubly serrate. *Indusium curved*, often shaped something like a horse-shoe, *owing to its crossing the vein and becoming attached to both sides of it.*—Rich woods.

6. WOODWARD'IA, Smith. CHAIN FERN.

W. Virgin'ica, Smith. Fronds 2-3 feet high, pinnate; pinnæ lanceolate, pinnatifid. Veins forming a single row of meshes next the midrib.—Wet swamps.

7. SCOLOPEN'DRIUM, Smith. HART'S TONGUE.

S. vulga're, Smith. Frond simple, bright green, a foot or more in length, and an inch or more in width.—Shaded ravines and limestone cliffs; not very common.

8. CAMPTOSO'RUS, Link. WALKING-LEAF.

C. rhizophyl'lus, Link. A curious little fern, growing in tufts on shaded limestone rocks. Frond simple, with a very long narrow point.—not very common.

9. PHEGOP'TERIS, Féc. BEECH FERN.

1. **P. polypodioi'des,** Féc. Fronds triangular, longer than broad, 4-6 inches long, hairy on the veins, twice-pinnatifid, *the rhachis winged.* The pinnæ sessile, linear-lanceolate in outline, *the lowest pair deflexed and standing forwards.* Fruit-dots small and *all* near the margin. Stipes rather longer than the fronds, from a slender, creeping

rootstock.—Apparently not common, but growing in rich woods near Barrie, Ont.

2. **P. hexagonop'tera,** Fée. Fronds triangular, generally broader than long, 7-12 inches broad. Pinnæ lanceolate; the lowest very large, their divisions elongated and pinnatifid, the basal divisions decurrent on the main rhachis and forming a many-angled wing. Fruit-dots not exclusively near the margin.—Rich woods.

3. **P. Dryop'teris,** Fée. Fronds broadly triangular in outline, primarily divided into 3 triangular spreading petioled divisions, smooth, the three divisions once- or twice-pinnate. Fronds from 4 to 6 inches wide. Fruit-dots near the margin.—Rich woods; common. Whole plant delicate, and light green in colour.

4. **P. calca'rea,** Fée. Closely resembling P. Dryopteris, but differs in the fronds being *minutely glandular and somewhat rigid*, and in the lowest pinnæ on the lower side of the lateral divisions *proportionally smaller.*—Rocky hillsides, Restigouche River.

10. ASPID'IUM, Swartz. SHIELD FERN. WOOD FERN.
* *Stipes not chaffy.*

1. **A. thelyp'teris,** Swartz. Fronds tall and narrow, lanceolate in outline, pinnate, the pinnæ deeply pinnatifid, nearly at right angles to the rhachis, linear-lanceolate in outline, *the margins of the lobes strongly revolute in fruit.* Stipe over a foot long, and usually longer than the frond.—Common in low, wet places.

2. **A. Noveboracen'se,** Swartz. *Fronds much lighter in colour than the preceding*, tapering towards both ends, pinnate, the pinnæ deeply pinnatifid, much closer together than in No. 1, and not at right angles with the rhachis. Veins simple. Lower pinnæ short and deflexed.—Swamps.

. *Stipes chaffy.*

3. **A. spinulo'sum,** Swartz. Stipes *slightly* chaffy or scaly. Fronds large, ovate-lanceolate in outline, *twice-pinnate*, the pinnules deeply pinnatifid (*nearly pinnate*), and

spiny-toothed. Pinnæ triangular-lanceolate in outline. The variety **intermedium**, D. C. Eaton, which is very common in Canadian woods, has the margin of the indusium denticulate and beset with minute stalked glands, the few scales of the stipes pale brown *with a dark centre*, and *the lower pinnæ unequal-sided.* Var. **Boottii** (*A. Boottii*, Tuckerman) has the scales of the stipe pale brown, the frond elongated-oblong or elongated-lanceolate and pinnules less dissected. Var. **dilata'tum,** Hook., has the scales of the stipe brown with a dark centre, the frond broader, ovate or triangular-ovate in outline, often twice-pinnate, and the indusium smooth and naked.

4. **A. fra'grans,** Swartz. Fronds 4-12 inches high, fragrant, narrowly lanceolate, with narrowly-oblong pinnately-parted pinnæ, their divisions nearly covered beneath by very large thin indusia.—Rocks, Atl. Prov. and northward.

5. **A. crista'tum,** Swartz. Stipes chaffy with broad scales. Fronds large, linear-lanceolate in outline, once-pinnate, the pinnæ deeply pinnatified, the upper ones triangular-lanceolate in outline, the lower considerably broader, the lobes cut-toothed. Fruit-dots large and conspicuous, *half way between the midrib of the lobe and the margin.*—Swamps.

Var. **Clintonia'num,** D. C. Eaton, is distinguished chiefly by its greater size and more numerous pinnæ and segments.

6. **A. Goldia'num,** Hook. A fine fern, the large fronds growing in a circular cluster from a chaffy rootstock. Frond ovate or ovate-oblong in outline, once-pinnate, the pinnæ deeply pinnatifid, 6-9 inches long, *broadest in the middle*, the lobes slightly scythe-shaped, finely serrate. Fruit-dots large, *near the midrib of the lobe.*—Rich moist woods.

7. **A. Fi'lix-mas,** Swartz. (MALE-FERN.) Fronds lanceolate, very chaffy at the base, twice-pinnate except that the upper pinnules run together. Pinnæ linear-lanceolate, tapering from the base to the summit. Pinnules very obtuse, the basal ones incisely lobed. Fruit dots rather closer to the midvein than the margin. Indusium convex, persistent.—Rocky woods, Atl. Prov.

8. **A. margina'le,** Swartz. Stipes very chaffy at the base. Fronds ovate-oblong in outline, twice-pinnate, the pinnæ lanceolate in outline, broadest above the base. Pinnules crenate-margined. *Fruit-dots large, close to the margin.*—Rich woods, mostly on hill-sides.

9. **A. acrostichoi'des,** Swartz. (See Figs. 264 and 265, and accompanying description.)—Rich woods.

10. **A. Lonchi'tis,** Swartz. Not unlike No. 9, but the fronds are *narrower and longer*, more rigid and with hardly any stipe. Pinnæ densely spinulose-toothed.—Apparently not common, but plentiful in rocky woods west of Collingwood, Ont.

11. **A. aculea'tum,** Swartz, var. **Braun'ii,** Koch. Fronds *twice pinnate*, oblong-lanceolate, narrowing gradually toward the base. Pinnules ovate or oblong, truncate and nearly rectangular at the base (the lower short-stalked), beset with long and soft as well as chaffy hairs. Indusium fixed by the centre. Fronds evergreen, very chaffy on the stalk and rhachis.—Ravines and deep woods, Atl. Prov.

11. CYSTOP'TERIS, Bernhardi. BLADDER FERN.

1. **C. bulbif'era,** Bernh. Frond large (1-2 feet), narrow and very delicate, twice-pinnate, the pinnæ nearly at right angles to the rhachis. Rhachis and pinnæ *usually with bulblets beneath*. Pinnules toothed.—Shady, moist ravines.

2. **C. fra'gilis,** Bernh. Frond only 4-8 inches long, with a stipe of the same length, twice- or thrice-pinnate. *Rhachis winged.*—Shady cliffs.

12. STRUTHIOP'TERIS, Willd. OSTRICH FERN.

S. German'ica, Willd. (*Onoclea Struthiopteris*, Hoff.) Sterile fronds with the lower pinnæ gradually much shorter than the upper ones. Pinnæ deeply pinnatifid.—Common in low, wet grounds along streams.

13. ONOCLE'A, L., SENSITIVE FERN.

O. sensib'ilis, L. (See Figs. 266, 267, 268 and 269, and accompanying description.)—Common in wet grounds along streams.

14. WOOD'SIA, R. Brown.

1. **W. Ilven'sis**, R. Brown. Stalks indistinctly jointed at some distance above the base. Fronds oblong-lanceolate, 2-6 inches long, rather smooth and green above, *thickly clothed below with bristly rusty chaff*, pinnate; the pinnæ oblong, blunt, sessile, pinnately-parted with the segments indistinctly crenate. Fruit dots near the margin.—Exposed rocks, Atl. Prov. and northward.

2. **W. hyperbo'rea**, R. Brown. Stalks jointed. Fronds narrowly oblong-lanceolate. *Sparingly hairy beneath with chaffy hairs*, pinnate; the pinnæ triangular-ovate, pinnately lobed, *the lobes few and almost entire.*—Ravines, Atl. Prov. and northward.

3. **W. glabel'la**, R. Brown. Stalks jointed as in the two preceding species. Frond *linear*, very delicate, *smooth*, pinnate. Pinnæ roundish-ovate, the lower somewhat distant, *crenately-lobed.*—Moist rocks, Atl. Prov. and northward.

4. **W. obtu'sa**, Torr. *Stalks not jointed.* Frond broadly lanceolate, beset with small glandular hairs, once or nearly twice pinnate. Pinnæ pinnately parted. Segments of pinnæ crenately toothed.—Cliffs and rocky places; rare.

15. DICKSO'NIA, L'Her. DICKSONIA.

D. punctilo'bula, Kunze. Pleasantly odorous.—Moist shady places.

16. OSMUN'DA, L. FLOWERING FERN.

1. **O. rega'lis**, L. (FLOWERING FERN.) Fronds twice-pinnate, *fertile at the top*, very smooth, pale green. Sterile pinnules oblong-oval, finely serrate towards the apex, 1-2 inches long, either sessile or short-stalked, usually oblique and truncate at the base.—Swamps, along streams and lake-margins.

2. **O. Claytonia'na**, L. Fronds large, once-pinnate, pale green, densely white-woolly when unfolding from the bud, *with fertile pinnæ among the sterile ones.* Pinnæ deeply pinnatifid, the lobes entire.—Low grounds.

O. cinnamo'mea, L. (CINNAMON FERN.) *Fertile fronds [dist]inct from the sterile ones*, contracted, twice-pinnate, [cov]ered with cinnamon-coloured sporangia. Sterile fronds [hair]y-woolly when young, smooth afterwards, once-pinnate, pinnæ deeply pinnatifid. The long, sterile fronds in a [clus]ter, with the fertile ones in the centre.—Low grounds.

17. BOTRYCH'IUM, Swartz. MOONWORT.

* *Sterile portion of the frond not fleshy.*

B. Virginia'num, Swartz. (See Figs. 270 and 271, and [acc]ompanying description.)—Rich woods everywhere.

** *Sterile portion of the frond more or less fleshy.*

+ *Sterile segment sessile or nearly so.*

2. **B. Luna'ria**, Swartz. Sterile segment very fleshy, [ne]arly sessile, oblong, simply pinnate, with 5-15 lunate or [fa]n-shaped fleshy divisions. Fertile segment panicled, 2-3-[p]innate.—Rare.

3. **B. sim'plex**, Hitchcock. Fronds small, seldom 6 inches high, *the sterile segment borne nearer the middle of the plant*, short-petioled, fleshy, simple, and roundish, or pinnately lobed, with roundish lobes decurrent on the broad and flat indeterminate rhachis. *Veins all forking from the base.*—Atl. Prov. chiefly.

4. **B. lanceola'tum**, Angstrœm. Frond 3-10 inches high. *Sterile part closely sessile at the top* of the slender common stalk, very slightly fleshy, triangular, ternately twice-pinnatifid, with *acute*, toothed lobes. Veinlets branching from the continuous midvein. The fertile part twice- or thrice-pinnate.—Atl. Prov.

5. **B. matricariæfo'lium**, Braun. Resembling the preceding, but the sterile segment not quite sessile, somewhat fleshy, pinnate to twice-pinnatifid, with *obtuse* lobes. Midvein broken up into forking veinlets.—Damp woods, Atl. Prov. chiefly.

+ + *Sterile segment long-stalked.*

6. **B. terna'tum**, Swartz. Very fleshy, 4-16 inches high, sparingly hairy. Sterile segment attached near the base of

the plant, broadly triangular, variously decompound, with stalked divisions. Fertile segment erect, 2-4-pinnate.—Not uncommon.

18. OPHIOGLOS'SUM, L. ADDER'S TONGUE.

O. vulga'tum, L. Sterile part of the frond ovate or elliptical-oblong, 2-3 inches long, rather fleshy, sessile, near the middle of the stalk; the latter 6-12 inches high.—Bogs and grassy meadows.

ORDER CXIII. **EQUISETA'CEÆ.**

(HORSETAIL FAMILY.)

The only genus of the Order is

EQUISE'TUM, L. HORSETAIL. SCOURING RUSH.

Fig. 273.

Fig. 274.

Fig. 272 is a view of the fertile stem of **Equise'tum arvense,** the COMMON HORSETAIL, of about the natural size. It may be observed early in spring almost anywhere in moist sandy or gravelly soil. It is of a pale brown colour, and in place of leaves there is at each joint a sheath split into several teeth. At the summit of the stem is a sort of conical catkin, made up of a large number of six-sided bodies, each attached to the stem by a short pedicel. Each of these six-sided bodies turns out on examination to be made up of six or seven sporangia or spore-cases, which open down their inner margins to discharge their spores. Figs. 273 and 274 are enlarged outer and inner views of one of them. The spores themselves are of a similar nature to those of the Ferns, and reproduction is carried on in the same manner; but each spore of the Horsetail is furnished with four minute tentacles which closely envelope it when

Fig. 272.

moist, and uncoil themselves when dry. The use of these tentacles is doubtless to assist in the escape and dispersion of the spores.

The fertile stems will have almost withered away by the time the sterile ones appear. The latter are of the same thickness as the fertile ones, but they are very much taller and are green in colour. Observe, also, the grooving of the sterile stem, and the whorls of 4-angled branches produced at the nodes.

* Annual-stemmed, not surviving the winter.

+ Fruiting in spring, the fertile stems pale or brownish, and earlier than the herbaceous sterile stems or branches.

1. E. arven'se, L. (COMMON HORSETAIL.) *Fertile stems not branching*, 4-10 inches high. Sterile stems slender, taller, with many 4-angular green branches.—Moist sandy or gravelly soil, common.

2. E. praten'se, Ehrh. Stems more slender, with *3-sided simple branches* shorter than in E. arvense. Stem-sheaths with short, ovate-lanceolate teeth (those of the branches 3-toothed.) The *fertile stems produce branches*, when older, except at the top, which perishes after fructification.—Low meadows, Atl. Prov. and northward.

3. E. sylvat'icum, L. Fertile stems branching. *Branches compound, with loose sheaths*, those of the stem having 8-14 somewhat blunt teeth, while those of the branches have 4-5 (of the branchlets 3) lance-pointed diverging teeth. Top withering away after fructification.—Damp shady places, Atl. Prov. and northward.

+ + Fruiting in summer. Stems all similar, or the fertile contemporaneous with the sterile ones.

4. E. palus'tre, L. Stems 10-18 inches high, slender, *very deeply 5-9-grooved*, the lance-awl-shaped teeth whitish-margined.—Atl. Prov. and northward.

5. E. limo'sum, L. Stem 2-5 feet high, *slightly many-furrowed*. Sheaths appressed, with 10-22 dark-brown acute short rigid teeth.—In shallow water

* * *Stems evergreen, all similar, not branching, or only slightly so.*

6. **E. hyema'le,** L. (Scouring Rush.) Stems stout and tall. Sheaths elongated, with a black girdle above the base, and about 20 narrow linear teeth, 1-keeled at the base, and with awl-shaped deciduous points.—Wet banks.

7. **E. læviga'tum,** Braun. Stems 1-4 feet high, rather slender, pale green, 14-30-grooved, the ridges almost smooth. Sheath slightly enlarged upward, with a black girdle at the base of the *white-margined teeth.*—N.W.

8. **E. variega'tum,** Schleicher. Stem *slender*, in tufts, with 5-10 grooves, ascending, 6-18 inches high. Sheaths green, variegated with black above, 5-10 toothed.—Shores and river-banks.

9 **E. scirpoi'des,** Michx. Stem *slender, very numerous in a tuft*, filiform, 3-6 inches high, curving, mostly 6-grooved. *Sheaths 3-toothed.*—Wooded hill-sides.

Order CXIV. **LYCOPODIA'CEÆ.** (Club-Moss F.)

Chiefly moss-like plants; often with long running and branching stems, the sporangia solitary in the axils of the mostly awl-shaped leaves. Spores all of one kind.

LYCOPO'DIUM, L. Club-Moss.

* *Spore-cases in the axils of the ordinary dark-green rigid lanceolate leaves.*

1. **L. sela'go,** L. Stems erect and rigid, forking, forming a level-topped cluster. *Leaves uniform.*—Atl. Prov. and northward.

2. **L. lucid'ulum,** Michx. Stems less rigid than the last, forking. Leaves at first spreading; *then deflexed, arranged in alternate zones* of longer and shorter leaves.

* * *Spore-cases only in the axils of the upper bract-like leaves, forming a spike.*

← *Leaves of the spike not very different from those of the stem.*

3. **L. inunda'tum,** L. A low plant with weak, creeping, sterile stems, and solitary erect fertile stems bearing a

short, thick, leafy spike. Stem-leaves lanceolate, acute, soft, spreading, mostly *entire, those of the spike closely resembling them.*—Sandy bogs.

Var. **Bigelovii,** Tuckerm., has taller stems and leaves more upright and denticulate.—Atl. Prov.

← ← *Leaves of the catkin-like spike quite different from those of the stem.*

4. **L. anno'tinum,** L. Stems creeping, 1-4 feet long. Branches 4-9 inches high, once- or twice-forked. *Spike sessile,* the leaves of it yellowish and scale-like, ovate or heart-shaped, the others spreading or reflexed, rigid, pointed, nearly entire, pale green.—Cold woods.

5. **L. dendroi'deum,** Michx. (GROUND PINE.) Rootstock creeping underground, nearly leafless. Stems much resembling little hemlocks, 6-9 inches high; numerous fan-like spreading branches with shining lanceolate entire leaves. *Spikes sessile,* nearly as in No. 4, 4-10 on each plant. —Moist woods.

6. **L. clava'tum,** L. (CLUB-MOSS.) Stem creeping or running extensively. *Spikes mostly in pairs, raised on a slender peduncle* (4-6 inches long). Leaves linear, awl-shaped, *bristle-tipped.*—Dry woods.

7. **L. complana'tum,** L. Stem creeping extensively. *Branches flattened,* fan-like and spreading, forking above, the branchlets crowded. Leaves awl-shaped, small, in 4 ranks. Spikes 2-4, cylindrical, on *a slender peduncle.*—Dry woods; mostly with evergreens.

ORDER CXV. **SELAGINELLA'CEÆ.**

Small leafy plants, terrestrial or rooted in mud. Stem branching or short and corm-like. Spore-cases solitary, axillary or borne on the upper surface of the leaf at its base and enwrapped in its margins. Spores of two kinds, the large ones (macrospores) developing prothallia which bear archegonia only, and the small ones (microspores) **giving** rise to antheridia only.

Synopsis of the Genera.

1. **Selaginel'la.** Terrestrial. Stem slender. Leaves small. Spore-cases very small, axillary, some containing the orange-coloured powdery microspores, and others with 3 or 4 larger macrospores.
2. **Iso'etes.** Growing in water. Stem corm-like. Leaves elongated and rush-like. Spore-cases large, enwrapped by the spreading bases of the leaves.

1. SELAGINEL'LA, Beauv.

S. rupes'tris, Spring. A little moss-like evergreen, growing on exposed rocks in dense tufts 1-3 inches high. Leaves awl-shaped, with a grooved keel, and tipped with a bristle. Spikes 4-cornered.

2. ISO'ETES. L. QUILLWORT.

I. echinos'pora, Durieu, var. **Braun'ii,** Engelm. (QUILLWORT.) A small aquatic grass-like plant with a corm-like stem, bearing 15-30 slender leaves. The large sporangia axillary, partly enwrapped by the thin edges of the excavated bases of the leaves, beset with small spinules.—Lakes and ponds, chiefly eastward.

ORDER CXVI. SALVINIA'CEÆ.

Small floating moss-like plants, with branching axis, covered with minute 2-lobed imbricated leaves. Sporocarps in pairs beneath the stem, very soft and thin-walled, the smaller ones acorn-shaped, containing a single macrospore, the larger ones globose, containing several microspores.

AZOLLA, Lam.

A. Carolinia'na, Willd. Somewhat deltoid in outline, much branched, of reddish aspect. Floating on quiet waters of Lake Ontario, not common.

GLOSSARY.

A.

Abortive, defective or barren.
Acaulescent, apparently without a stem.
Achene, a dry indehiscent 1-seeded fruit, with the pericarp free from the seed.
Achlamydeous, without calyx and corolla.
Acicular, needle-shaped.
Acuminate, with a long tapering extremity.
Acute, sharp at the end.
Adnate, grown fast to.
Alternate, not opposite.
-androus, at the end of a word, refers to stamens.
Angiospermous, having seeds enclosed in an ovary.
Annual, lasting one year or season only.
Anterior, in front; away from the axis; the same as *lower* and *outer*.
Anther, the pollen-bearing part of the stamen.
Apetalous, without petals.
Appressed, lying close and flat.
Aquatic, growing in water.
Aril, an extra coat or appendage of certain seeds.
Ascending, growing upwards in a slanting direction.
Auricle, an ear-like appendage or lobe.
Auriculate, having auricles.
Awl-shaped, narrowed upward from the base to a sharp rigid point.
Awn, a bristle-like appendage.
Axil, the angle on the upper side of a leaf where it joins the stem.
Axile, central; in the axis.
Axillary, growing from an axil.
Axis, the central line.

B.

Baccate, berry-like.
Barbed, with short sharp points, usually pointing backward.
Beaked, tipped with a distinct beak or point.
Bearded, having long or stiff hairs.
Berry, a pulpy fruit like a currant or gooseberry.
Bi-, a prefix meaning twice, or two.
Biennial, lasting two years or seasons.
Bifid, two-cleft.
Bilabiate, two-lipped.
Bladdery, thin and inflated.
Blade, the broad part of a leaf.
Bract, a modified leaf near a flower or flower-cluster.
Bractlet, a secondary bract.
Bristle, a stiff hair.
Bud, an undeveloped branch or flower.
Bulb, a fleshy underground mass, consisting chiefly of leaf-scales or coats.
Bulblet, a small bulb, especially one borne on the stem.
Bulbous, having the appearance of a bulb.

C.

Caducous, falling off very early.
Calyx, the outer set of flower-leaves, often coloured like a corolla.
Campanulate, bell-shaped.
Canescent, grayish-hoary.
Capitate, like a head.
Capsule, any dry dehiscent syncarpous fruit.
Carinate, having a keel-like ridge or projection.

Carpel, one of the parts which, whether separate or united, make up the pistil.

Cartilaginous, firm and tough.

Caryopsis, a grain or fruit like that of a grass.

Catkin, a slender spike-like cluster of usually imperfect flowers, as in Willow, etc.

Caulescent, having a manifest stem.

Cauline, belonging to the stem.

Cespitose, growing in tufts.

Chaff, a small thin scale or bract, becoming dry and membranous.

Channelled, grooved lengthwise.

Ciliate, with hairs on the edge.

Cinereous, ash-coloured.

Circumcissile, opening by a horizontal circular line.

Clavate, club-shaped, thicker at the top.

Cleistogamous, applied to certain flowers which are fertilized in the bud, without opening.

Cleft, cut to about the middle.

Climbing, ascending by laying hold of neighbouring objects by means of tendrils, etc.

Cohesion, the union of similar parts.

Coma, a tuft of hairs on a seed.

Compressed, flattened.

Conduplicate, folded together lengthwise.

Confluent, blending together into one.

Coniferous, cone-bearing.

Connate, grown together.

Connective, the middle portion of an anther connecting the cells.

Connivent, converging, approaching each other.

Convolute, rolled up lengthwise.

Cordate, indented at the base.

Coriaceous, leathery in texture.

Corm, a solid bulb, like that of Indian-Turnip.

Corolla, the inner of the two sets of the perianth.

Corymb, a flat-topped flower-cluster with pedicels arising at different points on the stem, the outer flowers opening first.

Corymbose, in corymbs; corymb-like.

Cotyledons, the seed-leaves.

Creeping, trailing on the ground and striking root at intervals.

Crenate, with rounded teeth.

Crested, bearing an appendage like a crest.

Crown, an appendage in the throat of a corolla.

Culm, the peculiar stem of a grass or sedge.

Cuneate, wedge-shaped, the broad end upwards.

Cuspidate, tipped with an abrupt but distinctly tapering point.

Cylindraceous, somewhat cylindrical.

Cyme, a flat-topped flower-cluster, with the central flowers opening first.

Cymose, bearing cymes, or cyme-like.

D.

Deciduous, falling off; not persistent.

Decompound, more than once compound or divided.

Decumbent, reclining.

Decurrent, applied to a leaf when the lobes at the base extend down the sides of the stem.

Decussate, with opposite pairs at right angles to each other.

Deflexed, turned abruptly downward.

Dehiscent, splitting open to allow the contents to escape.

Deltoid, triangular, the base downward.

Dentate, toothed, the teeth pointing outward.

Denticulate, minutely dentate.

Depressed, flattened from above.

Di-, twice or two.

Diadelphous, with the filaments cohering in two sets.

Diandrous, having two stamens.

Dichotomous, forking regularly by pairs.
Dicotyledonous, having two cotyledons.
Didynamous, of stamens, when in two pairs of different lengths.
Diffuse, spreading.
Digitate, compound, the members arising from the same point.
Diœcious, of flowers when the staminate and pistillate ones are on separate plants.
Discoid, of composite flowers when ray-florets are absent.
Disk, a development of the receptacle around the base of the pistil; the central part of the head of a composite flower as distinct from the ray.
Dissected, cut up into many segments.
Dissepiment, a partition in an ovary or fruit.
Distinct, not united.
Divergent, spreading apart.
Divided, lobed to the base.
Dorsal, relating to the back of any organ.
Drupe, a fleshy fruit with the seed enclosed in a hard shell, as a plum or cherry.

E.

Elliptical, oval.
Emarginate, slightly notched at the end.
Embryo, the undeveloped plant in the seed.
Endocarp, the innermost layer of the pericarp.
Endogenous, applied to stems whose wood does not grow in rings, but in scattered bundles.
Entire, without indentations of any kind.
Epigynous, growing on the ovary.
Epiphyte, an air-plant, whose roots do not reach the ground.
Equitant, applied to such leaves as those of Iris, which are folded lengthwise, each *astride* of the next one within.
Erect, upright.

Excurrent, applied to stems which can be readily traced through to the top, as in Pine.
Exogenous, applied to stems whose wood grows in layers or rings.
Exserted, thrust out beyond the line of the enveloping organ, as stamens out of a corolla.
Extrorse, facing outward.

F.

Fascicle, a close bundle.
Fertile, applied to flowers having pistils.
Fibrous, thread-like.
Filament, the stalk of the stamen.
Filiform, thread-shaped; long and slender.
Floccose, soft-woolly.
Foliaceous, leaf-like.
-foliate, relating to leaves.
-foliolate, relating to leaflets.
Follicle, a dehiscent fruit of one carpel, splitting down one side only.
Free, not growing fast to any other organ.
Frond, the leaf of a Fern, and some other cryptogams.
Fruit, the ripened ovary, along with any adherent part.
Fugacious, falling away very early.
Fusiform, spindle-shaped, thicker in the middle than at either end.

G.

Galea, a helmet-shaped piece of a perianth, as the upper lip of some labiate corollas.
Gamopetalous, having the petals united together.
Gamophyllous, having the pieces of the perianth united.
Gibbous, bulging outward on one side.
Glabrate, somewhat glabrous; becoming glabrous.
Glabrous, smooth.
Gland, a secreting structure of any kind. Any protuberance having the appearance of such an organ.

Glandular, bearing glands; or gland-like.
Glaucous, whitened with a bloom.
Globose, nearly spherical.
Glumaceous, having glumes; or glume-like.
Glume, a chaffy bract near a grass-flower.
Granular, composed of small grain-like pieces.
Gymnospermous, having seeds which are not enclosed in an ovary.
Gynandrous, of stamens which are borne on the style.

H.

Habit, the general aspect of a plant.
Halberd-shaped, with spreading lobes at the base.
Hastate, the same as halberd-shaped.
Head, a dense cluster of sessile flowers on a very short axis.
Heart-shaped, with an indentation at the base.
Herb, a plant with little or no wood in its stem.
Herbaceous, having the character of an herb; leaf-like in colour and texture.
Hilum, the scar showing where a seed was attached.
Hirsute, pubescent with coarse hairs.
Hispid, with stiff bristly hairs.
Hoary, grayish-white with fine close pubescence.
Hooded, shaped like a hood.
Hypogynous, inserted on the receptacle below the ovary, and free from the latter.

I.

Imbricate, overlapping, as the bracts of the involucre in most composite flowers, having one piece wholly internal and one wholly external, as in the calyx and corolla of many flowers.
Immersed, wholly under water.
Incised, sharply cut.
Included, not at all protruded beyond the surrounding envelope.
Indefinite, many in number.
Indehiscent, not splitting open so as to discharge the contents.
Indigenous, native.
Inferior, lower; outer; of a calyx when below and free from the ovary; of an ovary when attached to the calyx.
Inflorescence, the arrangement of the flowers or flower-clusters.
Inserted, attached to.
Introrse, turned inward.
Involucel, a secondary involucre.
Involucre, a circle of bracts below a flower or flower-cluster.
Involute, rolled inward.
Irregular, with parts differing in size or shape.

K.

Keel, a central longitudinal ridge; the two united petals in the front of a papilionaceous flower.
Kidney-shaped, of a leaf with large rounded lobes at the base.

L.

Labiate, having two lips.
Laciniate, cut into narrow pointed lobes.
Lamella, a thin flat plate.
Lanceolate, rather narrow and tapering from the base to the point.
Lateral, referring to or borne on the side.
Lax, loose.
Leaflet, one of the pieces of the blade of a compound leaf.
Legume, a dehiscent fruit of a single carpel, which opens as a rule by two seams.
Lenticular, shaped like a double-convex lens.
Ligulate, strap-shaped.
Limb, the spreading part of a petal or corolla, etc.
Linear, long and narrow, with both ends alike or nearly so.
Lobe, any segment of an organ.

Lyrate, pinnatifid with the terminal lobe very large and rounded.

M.

Marginal, along or near the edge.
Membranaceous, thin, somewhat transparent.
Midrib, the central or main vein of a leaf.
Monadelphous, with all the filaments grown together.
Monocotyledonous, having only one cotyledon in the embryo.
Monœcious, having staminate and pistillate flowers on different parts of the same plant.
Mucronate, tipped with a minute sharp point.
Multifid, cut into many segments.

N.

Naked, without a covering, as of chaff, hairs, etc.
Nerve, a longitudinal line, as on the calyx of Catnip.
Node, the place on a stem from which a leaf springs.
Numerous, indefinite in number.
Nut, an indehiscent fruit of one cell and one seed with a hard shell, usually the product of a compound pistil.
Nutlet, a body like a little nut.

O.

Obcompressed, compressed from above instead of from side to side.
Obcordate, indented at the apex.
Oblanceolate, narrow and tapering towards the base.
Oblique, unequal-sided, slanting.
Oblong, longer than broad, with rather straight sides, and both ends alike.
Obovate, egg-shaped, with the narrow end down.
Obsolete, not evident.
Obtuse, blunt.
Ocreate, having sheathing stipules.
Operculate, having a lid.
Orbicular, circular in outline.

Oval, somewhat longer than broad, with rounded sides and both ends alike.
Ovary, the lower part of the pistil containing the ovules.
Ovate, egg-shaped, the broad end down.
Ovoid, a solid with an ovate outline.

P.

Palate, an upward projection of the lower lip of a personate corolla, closing the throat.
Palet, one of the inner chaffy scales of a grass-flower.
Palmate, with veins or leaflets radiating from the end of the petiole.
Panicle, an irregularly compound inflorescence of pedicelled flowers.
Panicled, resembling a panicle, borne in a panicle.
Papilionaceous, having a standard, wings, and keel, as in leguminous corollas.
Pappus, the tuft of hairs or bristles answering to the calyx-limb in many composite flowers.
Parasitic, living upon other living plants.
Parietal, borne on the walls.
Parted, very deeply cleft.
Pectinate, cut into comb-like narrow teeth.
Pedate, palmately divided or parted, with the basal segments 2-cleft.
Pedicel, a secondary flower-stalk; the support of a single flower in a cluster.
Peduncle, the support of a solitary flower or of a flower-cluster.
Peltate, shield-shaped; attached by the lower surface.
Pendulous, hanging down.
Perennial, lasting year after year.
Perfect, having both stamens and pistil in the same flower.
Perfoliate, with the stem passing through the leaf.
Perianth, the floral envelopes.
Pericarp, the matured ovary.

Perigynous, inserted on the calyx.
Persistent, remaining after other parts have fallen away.
Personate, applied to a labiate corolla with the throat closed by an upward projection of the lower lip.
Petal, one of the pieces of a corolla.
Petaloid, petal-like.
Petiolate, having a petiole.
Petiole, having a leaf-stalk.
Phanerogamous, having flowers and producing seeds.
Pilose, with soft hairs.
Pinna, a primary division of a frond.
Pinnate, with veins or leaflets proceeding from each side of a mid-rib.
Pinnatifid, deeply pinnately cleft.
Pinnule, a secondary division of a frond; one of the divisions of a pinna.
Pistil, the seed-bearing organ of the flower.
Pistillate, having pistil but no stamens.
Pitted, marked with small pits or depressions.
Placenta, the ridge or projection in the ovary to which the ovules are attached.
Plicate, folded into plaits.
Plumose, feathery; with fine hairs on each side of the axis.
Pod, any dry dehiscent fruit.
Pollen, the grains produced in the anther.
Pollinium (plural *pollinia*), a mass of pollen, as in Milkweeds and Orchids.
Polypetalous, having petals separate from each other.
Pome, an inferior syncarpous fruit with a largely developed adherent calyx, as the apple.
Posterior, the side next the axis.
Procumbent, lying flat on the ground.
Prostrate, the same as procumbent.
Puberulent, minutely pubescent.

Pubescent, covered with fine short hairs.
Punctate, showing transparent dots when held up to the light.
Pungent, acrid to the taste.

R.

Raceme, a flower-cluster of lateral flowers, each on a pedicel of its own.
Racemose, in racemes; resembling a raceme.
Radiate, spreading from a centre; bearing ray-florets.
Radical, proceeding from the base of the stem.
Ray, the branch of an umbel; the circle of marginal florets in a composite flower, as distinguished from the disk.
Receptacle, the enlarged top of the flower-stalk.
Recurved, curved backward.
Reflexed, bent backward or downward.
Regular, with all the parts of the organ of the same size and shape.
Reniform, kidney-shaped.
Repand, with a margin resembling that of an expanded umbrella.
Reticulate, in the form of a network.
Retuse, with a shallow notch at the end.
Revolute, rolled backward from the margin.
Rhachis, the axis of a frond or compound leaf.
Rhizome, an underground stem; a rootstock.
Rib, one of the main veins of a leaf.
Ringent, wide open, gaping.
Rootstock, an underground stem.
Rotate, wheel-shaped.
Rudimentary, but slightly developed.
Rugose, wrinkled.
Runcinate, with teeth on the margin pointing backward.
Runner, a thread-like prostrate branch, proceeding from the base of a stem, and rooting at the extremity.

S.

Saccate, sac-shaped.
Sagittate, arrow-shaped, the basal lobes directed downward.
Salver-shaped, with a border spreading at right angles to the tube.
Samara, a winged fruit, as that of the Ash.
Scabrous, rough.
Scape, a leafless peduncle rising from the ground or near it.
Scarious, thin, dry, and membranaceous.
Scurf, small rusty-looking scales on the epidermis.
Segment, one of the parts of a divided leaf.
Sepal, a division of a calyx.
Septum, a partition.
Serrate, toothed, the teeth directed towards the apex.
Serrulate, finely serrate.
Sessile, without a stalk.
Setaceous, bristle-like.
Setose, beset with bristles.
Sheath, a tubular envelope.
Sheathing, enclosing as with a sheath.
Shrub, a woody perennial smaller than a tree.
Silicle, a short and broad silique.
Silique, the peculiar pod of a Cruciferous flower.
Simple, of one piece.
Sinuate, wavy.
Sinus, the indentation between two lobes.
Sorus, a cluster of sporangia.
Spadix, a spike on a fleshy axis.
Spathe, a bract, subtending or enveloping a spadix.
Spathulate, gradually narrowed downward from a rounded apex.
Spicate, in the form of a spike.
Spike, a cluster of sessile flowers on a more or less elongated axis.
Spikelet, a small or secondary spike.
Spindle-shaped, larger in the middle than at either end.
Spine, a sharp woody outgrowth of the stem.
Spinose, spine-like, or beset with spines.
Sporangium, a spore-case.
Sporocarp, the fruit-case of certain cryptogams.
Spur, a hollow projection.
Squarrose, having spreading tips.
Stamen, a pollen-bearing organ.
Standard, the upper petal of a papilionaceous corolla.
Stellate, star-shaped.
Sterile, not producing seed, without a pistil.
Stigma, the upper end of the pistil, adapted for the reception of pollen.
Stigmatic, stigma-like.
Stipe, the leaf-stalk of a Fern: the stalk supporting a pistil in certain flowers.
Stipular, relating to stipules.
Stolon, a branch which roots.
Stoloniferous, bearing stolons.
Striate, marked with fine longitudinal lines.
Strict, rigid and upright.
Strigose, beset with appressed sharp straight and stiff hairs.
Style, the narrow part of a pistil between the ovary and the stigma.
Stylopodium, a disk-like expansion at the base of a style, as in Umbelliferous plants.
Sub-, a prefix meaning "somewhat."
Subulate, awl-shaped.
Succulent, juicy.
Suffrutescent, slightly shrubby.
Suffruticose, low and woody, diminutively shrubby.
Sulcate, grooved or furrowed.
Superior (calyx), attached to the ovary.
Suture, a seam or line of dehiscence.
Symmetrical, with sets having the same number of parts each, or a multiple of that number.

T.

Tail, any slender prolongation.
Terete, cylindrical.
Terminal, at the apex.
Ternate, in threes.
Tetradynamous, with four long stamens and two short ones.
Throat, the entrance to the tube of a calyx or corolla.
Thyrse, a compact panicle.
Thyrsoid, like a thyrse.
Tomentose, woolly.
Transverse, across.
Tri-, a prefix meaning three or thrice.
Trifoliolate, having three leaflets.
Truncate, as if cut off square across the end.
Tuber, the enlarged end of an underground stem.
Tubercle, a small tuber-like body.
Tuberous, having the appearance of a tuber.
Tunicated, having coats one within another.
Twining, winding spirally about a support.

U.

Umbel, a flower-cluster with radiating pedicels.
Umbellate, umbel-like.
Umbellet, a secondary umbel.
Undulate, wavy.
Urceolate, urn-shaped.
Utricle, an indehiscent 1-seeded fruit, with a thin loose pericarp.

V.

Valvate, with the edges meeting but not overlapping.
Valve, one of the pieces into which a capsule splits.
Veins, the threads of fibro-vascular tissue running through the substance of a leaf.
Ventral, belonging to the inferior or lower side.
Vernation, the folding of a leaf in the bud.
Versatile, attached by the centre (of the anther) and so able to swing about.
Vertical, upright.
Verticillate, whorled.
Villous, with long soft hairs.
Viscid, sticky.

W.

Whorl, a circle of leaves round a stem.
Wing, a thin expansion bordering any organ; one of the side petals of a papilionaceous corolla.
Woolly, with long matted hairs.

AN ALPHABETICAL LIST

OF

COMMON CULTIVATED PLANTS.

Popular Name.	Scientific Name.	Natural Order.
Abutilon.	*Abutilon striatum.*	Malvaceæ.
Aconite.	*Aconitum Napellus.*	Ranunculaceæ.
African Marigold.	*Tagetes erecta.*	Compositæ.
Agapanthus.	*Agapanthus umbellatus.*	Liliaceæ.
Ageratum.	*Ageratum Mexicanum.*	Compositæ.
Almond (Flowering).	*Prunus nana.*	Rosaceæ.
Alyssum (Sweet).	*Alyssum maritimum.*	Cruciferæ.
Apple (Common).	*Pyrus Malus.*	Rosaceæ.
Apple (Siberian Crab).	*Pyrus prunifolia.*	Rosaceæ.
Apricot.	*Prunus Armeniaca.*	Rosaceæ.
Artichoke (True).	*Cynara Scolymus.*	Compositæ.
Artichoke (Jerusalem).	*Helianthus tuberosus.*	Compositæ.
Asparagus.	*Asparagus officinalis.*	Liliaceæ.
Aster (China).	*Callistephus Chinensis.*	Compositæ.
Auricula.	*Primula Auricula.*	Primulaceæ.
Azalea.	*Azalea Indica.*	Ericaceæ.
Balsam.	*Impatiens Balsamina.*	Geraniaceæ.
Banana.	*Musa sapientum.*	Scitamineæ.
Barberry.	*Berberis vulgaris.*	Berberidaceæ.
Barley.	*Hordeum (sev. sp.)*	Gramineæ.
Bean (Windsor).	*Vicia Faba.*	Leguminosæ.

COMMON CULTIVATED PLANTS.

Popular Name.	Scientific Name.	Natural Order.
Bean (Dwarf).	*Phaseolus nanus.*	Leguminosæ.
Bean (Kidney).	*Phaseolus vulgaris.*	Leguminosæ.
Beet.	*Beta vulgaris.*	Chenopodiaceæ.
Begonia.	*Begonia Rex (and many other species).*	Begoniaceæ.
Bleeding Heart.	*Dicentra spectabilis.*	Fumariaceæ.
Bluebottle.	*Centaurea Cyanus.*	Compositæ.
Blue Flag.	*Iris pumila (and other species).*	Iridaceæ.
Borage.	*Borrago officinalis.*	Borraginaceæ.
Boston Ivy.	*Ampelopsis Veitchii.*	Vitaceæ.
Bottle Gourd.	*Lagenaria vulgaris.*	Cucurbitaceæ.
Bouvardia.	*Bouvardia triphylla.*	Rubiaceæ.
Box.	*Buxus sempervirens.*	Euphorbiaceæ.
Broccoli.	*Brassica oleracea, var. Broccoli.*	Cruciferæ.
Buckwheat.	*Fagopyrum esculentum.*	Polygonaceæ.
Cabbage.	*Brassica oleracea.*	Cruciferæ.
Caladium.	*Caladium (sev. sp.)*	Araceæ.
Calceolaria.	*Calceolaria (sev. sp.)*	Scrophularaiceæ.
Calendula.	*Calendula officinalis.*	Compositæ.
Calla Lily.	*Richardia Africana.*	Araceæ.
Camellia.	*Camellia Japonica.*	Camelliaceæ.
Canary-bird Flower.	*Tropæolum peregrinum.*	Geraniaceæ.
Candytuft.	*Iberis umbellata.*	Cruciferæ.
Canna.	*Canna Indica.*	Scitamineæ.
Canterbury Bells.	*Campanula Medium.*	Campanulaceæ.
Capsicum.	*Capsicum annuum.*	Solanaceæ.
Caraway.	*Carum Carui.*	Umbelliferæ.
Carrot.	*Daucus Carota.*	"
Castor-oil Plant.	*Ricinus communis.*	Euphorbiaceæ.
Catalpa.	*Catalpa bignonioides.*	Bignoniaceæ.
Catchfly.	*Silene Armeria.*	Caryophyllaceæ

COMMON CULTIVATED PLANTS.

Popular Name.	Scientific Name.	Natural Order.
Cauliflower.	*Brassica oleracea* (var.)	Cruciferæ.
Celery.	*Apium graveolens* "	Umbelliferæ.
Century Plant.	*Agave Americana.*	Amaryllidaceæ.
Chamomile.	*Anthemis nobilis.*	Compositæ.
Cherry (Red).	*Prunus Cerasus.*	Rosaceæ.
Cherry (English).	" *avium.*	"
China Aster.	*Callistephus Chinensis.*	Compositæ.
Chinese Primrose.	*Primula Sinensis.*	Primulaceæ.
Chives.	*Allium Schœnoprasum.*	Liliaceæ.
Cigar-plant.	*Cuphea platycentra.*	Lythraceæ.
Cineraria.	*Senecio cruentus.*	Compositæ.
Citron.	*Citrullus vulgaris* (var.)	Cucurbitaceæ.
Clarkia.	*Clarkia elegans.*	Onagraceæ.
Clematis.	*Clematis* (sev. sp.)	Ranunculaceæ.
Cobœa.	*Cobœa scandens.*	Polemoniaceæ.
Cockscomb.	*Celosia cristata.*	Amarantaceæ.
Coffee.	*Coffea Arabica.*	Rubiaceæ.
Coleus.	*Coleus Blumei.*	Labiatæ.
Columbine.	*Aquilegia vulgaris.*	Ranunculaceæ.
Convolvulus.	*Ipomœa purpurea.*	Convolvulaceæ.
Corn.	*Zea Mays.*	Gramineæ.
Cornflower.	*Centaurea Cyanus.*	Compositæ.
Cotton.	*Gossipium herbaceum.*	Malvaceæ.
Crocus.	*Crocus vernus.*	Iridaceæ.
Crown Imperial.	*Fritillaria imperialis.*	Liliaceæ.
Cucumber.	*Cucumis sativus.*	Cucurbitaceæ.
Cuphea.	*Cuphea platycentra.*	Lythraceæ.
Currant (Red and White).	*Ribes rubrum.*	Saxifragaceæ.
Currant (Black).	" *nigrum.*	"
Cyclamen.	*Cyclamen Persicum.*	Primulaceæ.
Cypress Vine.	*Quamoclit vulgaris.*	Convolvulaceæ.
Cypripedium.	*Cypripedium* (sev. sp.)	Orchidaceæ.

Popular Name.	Scientific Name.	Natural Order.
Daffodil.	*Narcissus Pseudo-Narcissus.*	Amaryllidaceæ.
Dahlia.	*Dahlia variabilis.*	Compositæ.
Daisy.	*Bellis perennis.*	"
Day-Lily (Common).	*Hemerocallis fulva.*	Liliaceæ.
Deutzia.	*Deutzia gracilis, etc.*	Saxifragaceæ.
Dianthus.	*Dianthus Chinensis (sev. varieties).*	Caryophyllaceæ
Dielytra.	*Dielytra spectabilis.*	Fumariaceæ.
Digitalis.	*Digitalis purpurea.*	Scrophulariaceæ
Dracæna.	*Dracæna (sev. sp.)*	Liliaceæ.
Dusty Miller.	*Centaurea Cineraria (or Candidissima).*	Compositæ.
Dutchman's Pipe.	*Aristolochia Sipho.*	Aristolochiaceæ
Egg Plant.	*Solanum Melongena.*	Solanaceæ.
Elephant's Ear.	*Begonia (many sp.)*	Begoniaceæ.
Endive.	*Cichorium Endivia.*	Compositæ.
Eschscholtzia.	*Eschscholtzia Californica.*	Papaveraceæ.
Evening-Primrose.	*Œnothera biennis(var.)*	Onagraceæ.
Everlasting.	*Gnaphalium (sev. sp.)*	Compositæ.
Feverfew.	*Pyrethrum Parthenium.*	"
Fig.	*Ficus Carica.*	Urticaceæ.
Flax.	*Linum usitatissimum (and other species).*	Linaceæ.
Flower-de-luce.	*Iris Germanica.*	Iridaceæ.
Flower-of-an-hour.	*Hibiscus Trionum.*	Malvaceæ.
Forget-me-not.	*Myosotis palustris.*	Borraginaceæ.
Forsythia.	*Forsythia viridissima.*	Oleaceæ.
Four-o'clock.	*Mirabilis Jalapa.*	Nyctaginaceæ.
Foxglove.	*Digitalis purpurea.*	Scrophulariaceæ
French Marigold.	*Tagetes patula.*	Compositæ.
Fuchsia.	*Fuchsia (many sp.)*	Onagraceæ.

Popular Name.	Scientific Name.	Natural Order.
Gaillardia.	*Gaillardia* (sev. sp.)	Compositæ.
Gardenia.	*Gardenia florida.*	Rubiaceæ.
Garlic.	*Allium sativum.*	Liliaceæ.
Genista.	*Genista tinctoria.*	Leguminosæ.
Geranium.	*Pelargonium (my. sp.)*	Geraniaceæ.
German Ivy.	*Senecio scandens.*	Compositæ.
Gilliflower.	*Matthiola annua.*	Cruciferæ.
Gladiolus.	*Gladiolus* (sev. sp.)	Iridaceæ.
Gloxinia.	*Gloxinia speciosa.*	Gesneraceæ.
Gooseberry.	*Ribes Grossularia.*	Saxifragaceæ.
Grape.	*Vitis vinifera (many var.)*	Vitaceæ.
Grape Hyacinth.	*Muscari botryoides.*	Liliaceæ.
Ground Pink.	*Phlox subulata.*	Polemoniaceæ.
Guelder Rose.	*Viburnum Opulus (var.)*	Caprifoliaceæ
Hawthorn (Eng.)	*Cratægus Oxyacantha.*	Rosaceæ.
Heart's-ease.	*Viola tricolor.*	Violaceæ.
Heliotrope.	*Helitropium Peruvianum.*	Borraginaceæ.
Hellebore (White).	*Veratrum viride.*	Liliaceæ.
Hemp.	*Cannabis sativa.*	Urticaceæ.
Henbane.	*Hyoscyamus niger.*	Solanaceæ.
Holly.	*Ilex Aquifolium.*	Aquifoliaceæ.
Hollyhock.	*Althæa rosea.*	Malvaceæ.
Honesty.	*Lunaria biennis.*	Cruciferæ.
Honeysuckle.	*Lonicera* (sev. sp.)	Caprifoliaceæ.
Hop.	*Humulus Lupulus.*	Urticaceæ.
Horehound.	*Marrubium vulgare.*	Labiatæ.
Horse-Chestnut.	*Æsculus Hippocastanum.*	Sapindaceæ.
Horse-radish.	*Nasturtium Armoracia.*	Cruciferæ.
Houseleek.	*Sempervivum tectorum.*	Crassulaceæ.
Hyacinth.	*Hyacinthus orientalis.*	Liliaceæ.
Hydrangea.	*Hydrangea Hortensia.*	Saxifragaceæ.

Popular Name.	Scientific Name.	Natural Order.
Ice Plant.	*Mesembryanthemum crystallinum.*	Mesembryanthemeæ.
Indian Corn.	*Zea Mays.*	Gramineæ.
Indian Cress.	*Tropæolum majus.*	Geraniaceæ.
Indian Mallow.	*Abutilon Avicennæ.*	Malvaceæ.
Indian Shot.	*Canna (sev. sp.)*	Scitamineæ.
India-Rubber Tree.	*Ficus elastica.*	Urticaceæ.
Iris.	*Iris (sev. sp.)*	Iridaceæ.
Ivy.	*Hedera Helix.*	Araliaceæ.
Jerusalem Artichoke.	*Helianthus tuberosus.*	Compositæ.
Jessamine.	*Jasminum (sev. sp.)*	Oleaceæ.
Jonquil.	*Narcissus Jonquilla.*	Amaryllidaceæ.
Kale.	*Brassica oleracea.*	Cruciferæ.
Kidney Bean.	*Phaseolus vulgaris.*	Leguminosæ.
Kohlrabi.	*Brassica oleracea (var.)*	Cruciferæ.
Laburnum.	*Laburnum vulgare.*	Leguminosæ.
Ladies' Eardrops.	*Fuchsia (many sp.)*	Onagraceæ.
Lamb's Quarters.	*Chenopodium album.*	Chenopodiaceæ.
Lantana.	*Lantana (sev. sp.)*	Verbenaceæ.
Larkspur.	*Delphinium (sev. sp.)*	Ranunculaceæ.
Lavender.	*Lavandula vera.*	Labiatæ.
Leek.	*Allium Porrum.*	Liliaceæ.
Lemon-scented Verbena.	*Lippia citriodora.*	Verbenaceæ.
Lentil.	*Lens esculenta.*	Leguminosæ.
Lettuce.	*Lactuca sativa.*	Compositæ.
Lilac (Common)	*Syringa vulgaris.*	Oleaceæ.
Lily.	*Lilium (many sp.)*	Liliaceæ.
Lily-of-the-Valley.	*Convallaria majalis.*	Liliaceæ.
Linden (Europe).	*Tilia Europæa.*	Tiliaceæ.
Live-for-Ever.	*Sedum Telephium.*	Crassulaceæ.
Lobelia (blue).	*Lobelia Erinus.*	Lobeliaceæ.

Popular Name.	Scientific Name.	Natural Order.
Locust-tree.	*Robinia (sev. sp.)*	Leguminosæ.
Love-in-a-mist.	*Nigella Damascena.*	Ranunculaceæ.
Love-lies-bleeding.	*Amarantus melancholicus.*	Amarantaceæ.
Lychnis.	*Lychnis (sep. sp.)*	Caryophyllaceæ.
Lycopodium.	*Selaginella (sev. sp.)*	Lycopodiaceæ.
Madder.	*Rubia tinctoria.*	Rubiaceæ.
Marigold.	*Calendula officinalis.*	Compositæ.
Marvel-of-Peru.	*Mirabilis Jalapa.*	Nyctaginaceæ.
Matrimony-Vine.	*Lycium vulgare.*	Solanaceæ.
Maurandia.	*Maurandia (sev. sp.)*	Scrophulariaceæ.
Melilotus (white).	*Melilotus alba.*	Leguminosæ.
Melon (Musk).	*Cucumis Melo.*	Cucurbitaceæ.
" (Water).	*Citrullus vulgaris.*	"
Mignonette.	*Reseda adorata.*	Resedaceæ.
Mimosa.	*Mimosa pudica.*	Leguminosæ,
Mock-Orange.	*Philadelphus Coronarius.*	Saxifragaceæ.
Molucca Balm.	*Moluccella lævis.*	Labiatæ.
Monkshood.	*Aconitum Napellus.*	Ranunculaceæ.
Morning Glory.	*Ipomœa purpurea.*	Convolvulaceæ.
Moss Pink.	*Phlox subulata.*	Polemoniaceæ.
Mountain Ash.	*Pyrus Americana.*	Rosaceæ.
Mourning Bride.	*Scabiosa atropurpurea.*	Dipsaceæ.
Mulberry.	*Morus alba.*	Urticaceæ.
Muskmelon.	*Cucumis Melo.*	Cucurbitaceæ.
Musk-plant.	*Mimulus moschatus.*	Scrophulariaceæ.
Narcissus (Polyanthus).	*Narcissus Tazetta.*	Amaryllidaceæ.
Nasturtium.	*Tropæolum majus.*	Geraniaceæ.
Nemophila.	*Nemophila (sev. sp.)*	Hydrophyllaceæ
Nolana.	*Nolana atriplicifolia.*	Solanaceæ.
Oat.	*Avena sativa.*	Gramineæ.

COMMON CULTIVATED PLANTS.

POPULAR NAME.	SCIENTIFIC NAME.	NATURAL ORDER.
Oleander.	*Nereum Oleander.*	Apocynaceæ.
Onion.	*Allium Cepa.*	Liliaceæ.
Orange (Sweet).	*Citrus Aurantium.*	Rutaceæ.
Orange-Gourd.	*Cucurbita ovifera.*	Cucurbitaceæ.
Oxalis.	*Oxalis (sev. sp.)*	Oxalidaceæ.
Oyster-Plant.	*Tragopogon porrifolius.*	Compositæ.
Pœony (Common).	*Pæonia officinalis.*	Ranunculaceæ.
Pansy.	*Viola tricolor.*	Violaceæ.
Parsley.	*Petroselinum sativum.*	Umbelliferæ.
Parsnip.	*Pastinaca sativa.*	"
Passion-Flower (Common).	*Passiflora cærulea.*	Passifloraceæ.
Pea.	*Pisum sativum.*	Leguminosæ.
Peach.	*Prunus Persica.*	Rosaceæ.
Peanut.	*Arachis hypogœa.*	Leguminosæ.
Pear.	*Pyrus communis.*	Rosaceæ.
Peppermint.	*Mentha piperita.*	Labiatæ.
Perilla.	*Perilla Nankinensis.*	Labiatæ.
Periwinkle (Common).	*Vinca Minor.*	Apocynaceæ.
Petunia.	*Petunia (sev. sp.)*	Solanaceæ.
Phlox.	*Phlox (sev. sp.)*	Polemoniaceæ.
Pie-Plant.	*Rheum Rhaponticum.*	Polygonaceæ.
Pimpernel.	*Anagallis arvensis.*	Primulaceæ.
Pine-Apple.	*Ananassa sativa.*	Bromeliaceæ.
Pink.	*Dianthus (many sp.)*	Caryophyllaceæ.
Plum.	*Prunus Domestica (many var.)*	Rosaceæ.
Plumbago (Blue).	*Plumbago Capensis.*	Plumbaginaceæ
Poinsettia.	*Euphorbia pulcherrima.*	Euphorbiaceæ.
Polyanthus.	*Primula officinalis (sev. var.)*	Primulaceæ.
Pomegranate.	*Punica Granatum.*	Lythraceæ.
Poppy.	*Papaver somniferum.*	Papveraceæ.

Popular Name.	Scientific Name.	Natural Order.
Portulaca.	*Portulaca grandiflora.*	Portulacaceæ.
Potato.	*Solanum tuberosum.*	Solanaceæ.
Primrose (Chinese).	*Primula Sinensis.*	Primulaceæ.
Princes' Feather.	*Polygonum orientale.*	Polygonaceæ.
Privet (Common).	*Ligustrum vulgare.*	Oleaceæ.
Pumpkin.	*Cucurbita Pepo.*	Cucurbitaceæ.
Radish.	*Raphanus sativus.*	Cruciferæ.
Raspberry (Red or Yellow).	*Rubus Idœus.*	Rosaceæ.
Red Pepper.	*Capsicum annuum.*	Solanceæ.
Red-hot-poker.	*Tritoma Uvaria.*	Liliaceæ.
Rhododendron.	*Rhododendron Dauricum.*	Ericaceæ.
Rhubarb.	*Rheum Rhaponticum.*	Polygonaceæ.
Rice.	*Oryza sativa.*	Gramineæ.
Rocket	*Hesperis matronalis.*	Cruciferæ.
Rose.	*Rosa (many sp.)*	Rosaceæ.
Rose-Mallow.	*Hibiscus Syriacus.*	Malvaceæ.
Rosemary.	*Rosmarinus officinalis*	Labiatæ.
Rowan-Tree.	*Pyrus Americana.*	Rosaceæ.
Rutabaga.	*Brassica campestris.*	Cruciferæ.
Rye.	*Secale cereale.*	Gramineæ.
Sage (Common).	*Salvia officinalis.*	Labiatæ.
Salsify.	*Tragopogon porrifolius.*	Compositæ.
Salvia.	*Salvia (many sp.)*	Labiatæ.
Savory (Summer).	*Satureia hortensis.*	"
Scabiosa.	*Scabiosa atropurpurea.*	Dipsaceæ.
Scarlet Lychnis.	*Lychnis Chalcedonica.*	Caryophyllaceæ
Scarlet-Runner.	*Phaseolus multiflorus.*	Leguminosæ.
Sensitive-plant.	*Mimosa pudica.*	"
Shell-Flower.	*Moluccella lœvis.*	Labiatæ.
Smilax.	*Myrsiphyllum asparagoides.*	Liliaceæ.

Popular Name.	Scientific Name.	Natural Order.
Snapdragon.	*Antirrhinum majus.*	Scrophulariaceæ
Snowball-Tree.	*Viburnum Opulus.*	Caprifoliaceæ.
Snowberry.	*Symphoricarpus racemosus.*	" "
Snowdrop.	*Galanthus nivalis.*	Amaryllidaceæ.
Spearmint.	*Mentha viridis.*	Labiatæ.
Spiderwort.	*Tradescantia zebrina.*	Commelinaceæ.
Spinach.	*Spinacia oleracea.*	Chenopodiaceæ.
Spiræa.	*Spiræa (sev. sp.)*	Rosaceæ.
Spurge.	*Euphorbia (sev. sp.)*	Euphorbiaceæ.
Squash.	*Cucurbita maxima.*	Cucurbitaceæ.
Star-of-Bethlehem.	*Ornithogallum umbellatum.*	Liliaceæ.
Stephanotis.	*Stephanotis floribunda*	Asclepiadaceæ.
Stock (Ten-Weeks).	*Matthiola annua.*	Cruciferæ.
Stone-crop (Mossy.)	*Sedum acre.*	Crassulaceæ.
Strawberry.	*Fragaria (sev. sp.)*	Rosaceæ.
Strawberry Geranium.	*Saxifraga sarmentosa*	Saxifragaceæ.
Strawberry Tomato.	*Physalis Alkekengi.*	Solanaceæ.
Sugar Cane.	*Saccharum officinarum.*	Gramineæ.
Sweet Basil.	*Ocimum Basilicum.*	Labiatæ.
Sweet-Brier.	*Rosa rubiginosa.*	Rosaceæ.
Sweet Clover.	*Melilotus alba.*	Leguminosæ.
Sweet-William.	*Dianthus barbatus.*	Caryophyllaceæ.
Syringa.	*Philadelphus coronarius.*	Saxifragaceæ.
Tansy.	*Tanacetum vulgare.*	Compositæ.
Tea Plant.	*Thea viridis.*	Camelliaceæ.
Thimbleberry.	*Rubus occidentalis.*	Rosaceæ.
Thyme.	*Thymus Serpyllum.*	Labiatæ.
Tiger-Flower.	*Tigridia pavonia (sev. var.)*	Iridaceæ.

POPULAR NAME.	SCIENTIFIC NAME.	NATURAL ORDER.
Tobacco (Common).	*Nicotiana Tabacum.*	Solanaceæ.
Tomato.	*Licopersicum esculentum.*	"
Tree of Heaven.	*Ailanthus glandulosus.*	Simarubaceæ.
Trumpet Creeper.	*Tecoma (sev. sp.)*	Bignoniaceæ.
Tulip.	*Tulipa Gesneriana (sev. var.)*	Liliaceæ.
Turnip.	*Brassica Napus.*	Cruciferæ.
Vegetable Marrow.	*Cucurbita verrucosa.*	Cucurbitaceæ.
Verbena.	*Verbena (sev. sp.)*	Verbenaceæ.
Veronica.	*Veronica (sev. sp.)*	Scrophulariaceæ.
Vetch (Common).	*Vicia sativa.*	Leguminosæ.
Violet (Sweet).	*Viola odorata.*	Violaceæ.
Virginia Creeper.	*Ampelopsis quinquefolia.*	Vitaceæ.
Wallflower.	*Cheiranthus Cheiri.*	Cruciferæ.
Watermelon.	*Citrullus vulgaris.*	Cucurbitaceæ.
Wax-Plant.	*Hoya carnosa.*	Asclepiadaceæ.
Wheat.	*Triticum vulgare.*	Gramineæ.
Windsor Bean.	*Vicia Faba.*	Leguminosæ.
Wistaria.	*Wistaria Sinensis.*	"
Wolfsbane.	*Aconitum Napellus.*	Ranunculaceæ
Woodbine.	*Lonicera (sev. sp.)*	Caprifoliaceæ.
Wormwood (Common).	*Artemisia Absinthium*	Compositæ.
Yucca.	*Yucca aloifolia.*	Liliaceæ.
Zinnia.	*Zinnia elegans.*	Compositæ.

INDEX.

The names of the Orders, Classes, and Divisions are in large capitals; those of the Sub-orders in small capitals. The names of Genera, as well as popular names and synonyms, are in ordinary type.

	PAGE.		PAGE.
Abies	216	Ambrosia	111
ABIETINEÆ	214	Amelanchier	74
Abutilon	39	American Brooklime	153
Acalypha	197	American Colombo	177
Acanthaceæ	159	American Cowslip	146
Acanthus Family	159	American Laurel	142
Acer	47	Amorpha	57
Acerates	181	Ampelopsis	45
Achillea	131	Amphicarpæa	59
Acnida	188	AMYGDALEÆ	62
Acorus	218	Anacharis	226
Actæa	9	ANACARDIACEÆ	43
Actinomeris	129	Anagallis	148
Adam-and-Eve	234	Androsace	146
Adder's-Mouth	233	Andromeda	142
Adder's-Tongue	267	Anemone	3
Adiantum	259	ANGIOSPERMS	1
Adlumia	15	ANONACEÆ	10
Æthusa	90	Antennaria	114
Agrimonia	65	Anthemis	127
Agrimony	65	APETALOUS EXOGENS	182
Alchemilla	65	Aphyllon	150
Alder	209	Apios	59
Alisma	225	Aplectrum	234
ALISMACEÆ	224	Aplopappus	116
Alkanet	170	APOCYNACEÆ	179
Allium	243	Apocynum	179
Alnus	209	Apple	73
Alum-root	78	Apple of Peru	176
Alyssum	23	AQUIFOLIACEÆ	145
Amaranth	187	Aquilegia	8
Amaranth Family	187	Arabis	20
Amarantus	187	ARACEÆ	217
AMARANTACEÆ	187	Aralia	92
AMARYLLIDACEÆ	236	ARALIACEÆ	92
Amaryllis Family	236	Arbor Vitæ	216

292 INDEX.

	Page.
Archangelica	90
Archemora	89
Arctium	110
Arctostaphylos	141
Arenaria	34
Arethusa	233
Arisæma	218
ARISTOLOCHIACEÆ	182
Arnica	120
Aromatic Wintergreen	141
Arrow-Arum	218
Arrow-Grass	225
Arrow-Head	225
Arrow-Wood	97
Artemisia	112
ARTOCARPEÆ	198
Arum Family	217
Asarum	182
ASCLEPIADACEÆ	179
Asclepias	180
Ash	181
Ash-leaved Maple	48
Asimina	10
Asparagus	242
Aspen	212
Aspidium	262
Asplenium	260
Aster	121
Astragalus	54
Atriplex	186
Avens	66
Azalea	142
Azolla	271
Baked-apple Berry	70
Ballota	167
Balm	163
Balm of Gilead	212
Balsam Family	42
BALSAMINACEÆ	42
Baneberry	9
Baptisia	61
Barberry	11
Barbarea	19
Barberry Family	10

	Page.
Barren Strawberry	66
Bartsia	158
Basil	164, 165
Basswood	39
Bastard Toad-flax	195
Bayberry	203
Beach Pea	59
Bearberry	141
Beard-Tongue	155
Beaver-Poison	91
Beech	207
Beech-Drops	150
Beech-Fern	261
Bedstraw	99
Beggar's Lice	169
Beggar-ticks	130
Bellflower	136
Bellis	127
Bellwort	240
BERBERIDACEÆ	10
Berberis	11
Bergamot	165
Betula	208
BETULACEÆ	207
Bidens	130
Bindweed	174
Birch	208
Birch Family	207
Birthwort Family	182
Bishop's Cap	77
Bitter-Cress	19
Bitter-Nut	203
Bittersweet	175
Black Alder	145
Blackberry	70, 71
Black Bindweed	191
Black Grass	245
Black Horehound	167
Black Mustard	22
Black Snake-root	9, 89
Bladder Campion	33
Bladder Fern	264
Bladder-Kalmia	39
Bladder-Nut	47
Bladder-pod	19

INDEX.

	PAGE.
Bladderwort	149
Bladderwort Family	149
Blazing-Star	114
Blite	185
Blitum	185
Blood-root	14
Blue Ash	182
Blue Beech	207
Blueberry	139
Bluebottle	110
Blue Cohosh	11
Blue Flag	235
Bluets	101
Blue-eyed Grass	236
Blue Lettuce	134
Blue-weed	168
Bœhmeria	201
Boneset	115
Borage Family	167
BORRAGINACEÆ	167
Botrychium	266
Bouncing Bet	32
Bowman's Root	65
Box Elder	48
Bracted Bindweed	173
Bracken	260
Brake	260
Bramble	70
Brasenia	12
Brasssica	22
Bristly Sarsaparilla	92
Brooklime	153
Brook-weed	148
Broom Crowberry	213
Broom-rape Family	150
Brunella	166
Buckbean	178
Buckthorn	45
Buckthorn Family	45
Buckwheat	193
Buckwheat Family	188
Buda	36
Bugbane	9
Bugseed	186
Bugle-weed	162

	PAGE.
Bugloss	169
Bunch-berry	93
Bupleurum	91
Burdock	110
Bur-Marigold	130
Burnet	65
Burning-Bush	46
Bur-reed	220
Bush-Clover	61
Bush-Honeysuckle	97
Butter-and-Eggs	154
Buttercup	6
Butterfly-weed	181
Butternut	202
Butter-weed	125
Butterwort	150
Button-bush	100
Buttonwood	201
Cacalia	115
CACTACEÆ	86
Cactus Family	86
Cakile	25
Calamintha	165
Calaminth	165
Calamus	218
Calla	218
Callitriche	80
Calluna	142
Calopogon	233
Caltha	8
Calypso	233
Calystegia	173
Camelina	24
Campanula	136
CAMPANULACEÆ	136
Campanula Family	136
Campion	33
Camptosorus	261
Cancer-root	150
CANNABINEÆ	199
Cannabis	201
Caper Family	25
CAPPARIDACEÆ	25
CAPRIFOLIACEÆ	95
Capsella	24

INDEX

	PAGE.		PAGE.
Caraway	91	Chickweed-Winterberry	146
Carduus	110	Chimaphila	144
Carex	251	Chiogenes	141
Cardamine	19	Choke-berry	73
Cardinal Flower	135	Choke-Cherry	64
Carpet-weed	86	Chrysanthemum	126
Carpinus	207	Chrysopsis	116
Carrion Flower	237	Chrysosplenium	78
Carrot	89	Cichorium	132
Carum	91	Cichory	132
Carya	202	Cicuta	91
CARYOPHYLLACEÆ	32	Cimicifuga	9
Cashew Family	43	Cinnamon Fern	266
Cassandra	142	Cinque-foil	67
Castanea	206	Circæa	81
Castilleia	157	Cirsium	109
Catbrier	237	CISTACEÆ	29
Catchfly	33	Claytonia	37
Catmint	165	Clearweed	201
Catnip	165	Cleavers	99
Cat-tail Family	219	Clematis	3
Cat-tail Flag	220	Cleome	25
Caulophyllum	11	Cliff-brake	260
Ceanothus	45	Climbing-Bittersweet	46
Cedar	216	Clintonia	240
Celandine	14	Clotbur	111
CELASTRACEÆ	46	Cloud-berry	70
Celastrus	46	Clover	52
Celtis	200	Club-Moss	269, 270
Centaurea	110	Club-Moss Family	269
Cephalanthus	100	Cnicus	109
Cerastium	35	Cockle	33
CERATOPHYLLACEÆ	213	Cocklebur	111
Ceratophyllum	213	Cockspur Thorn	73
Chain-Fern	261	Cohosh	11
Chamærhodos	69	Collinsia	155
Chamomile	127	Collinsonia	164
Charlock	22	Collomia	173
Chelidonium	14	Coltsfoot	121
Chelone	155	Columbine	8
CHENOPODIACEÆ	184	Comandra	195
Chenopodium	185	Comfrey	169
Cherry	64	COMPOSITÆ	102
Chestnut	206	Composite Family	102
Chickweed	34	Comptonia	204

	PAGE.
ione-Flower	127
CONIFERÆ	214
Conioselinum	90
Conium	92
Conopholis	150
CONVOLVULACEÆ	173
Convolvulus	174
Convolvulus Family	173
Coptis	8
Corallorhiza	234
Coral-root	234
Corema	213
Coreopsis	129
Corispermum	186
CORNACEÆ	93
Corn-Cockle	33
Cornel	93
Corn-Spurrey	37
Cornus	93
Corpse-Plant	144
Corydalis	16
Corylus	207
Costmary	127
Cottonwood	2.2
Cowbane	89
Cow-herb	32
Cow-Parsnip	89
Cowslip	146
Cow-Wheat	158
Crab-Apple	73
Cranberry	139
Cranberry-tree	98
Cranesbill	41
CRASSULACEÆ	78
Cratægus	72
Creeping-Snowberry	141
Crepis	133
Cress Family	16
Crowberry Family	213
Crowberry	213
Crowfoot	6
Crowfoot Family	2
CRUCIFERÆ	16
CRYPTOGAMS	252
Cryptotænia	91

	PAGE.
Cuckoo-flower	19
CUCURBITACEÆ	85
Cudweed	113
Cup-plant	131
CUPRESSINEÆ	215
CUPULIFERÆ	204
Currant	75
Cuscuta	174
Custard-Apple Family	10
Cycloloma	186
Cynoglossum	169
Cynthia	136
CYPERACEÆ	248
Cyperus	249
Cypripedium	234
Cystopteris	264
Daisy	127
Daisy Fleabane	126
Dalibarda	70
Dandelion	134
Daphne	194
Datura	176
Daucus	89
Day Lily	243
Dead-Nettle	167
Deer-berry	141
Deer-Grass	84
Delphinium	9
Dentaria	19
Desmodium	59
Dewberry	71
Dianthera	159
Dicentra	15
Dicksonia	265
DICOTYLEDONS	1
Diervilla	97
Dioscorea	236
DIOSCOREACEÆ	236
Diplopappus	124
Diplotaxis	23
DIPSACEÆ	101
Dipsacus	102
Dirca	194
Disporum	241

INDEX.

	PAGE.
Ditch-stone Crop	78
Dock	192
Dockmackie	98
Dodder	174
Dodecatheon	146
Dogbane	179
Dogbane Family	179
Dog's-tooth Violet	242
Dogwood	93
Dogwood Family	93
Downy Arrow-wood	97
Draba	23
Dracocephalum	165
Dragon-head	165
Drosera	30
DROSERACEÆ	30
Duckweed	219
Duckweed Family	218
Dutchman's Breeches	15
Dwarf Dandelion	131
Echinocystis	85
Echinospermum	169
Echium	168
Eel-Grass	224, 226
ELÆAGNACEÆ	194
Elæagnus	194
Elder	97
Elecampane	116
Eleocharis	250
Ellisia	172
Elm	199
Elm Family	198
Elodea	226
Elodes	31
EMPETRACEÆ	213
Empetrum	213
ENDOGENS	217
Enchanter's Nightshade	81
Epigæa	141
Epilobium	81
Epipactus	232
Epiphegus	150
EQUISETACEÆ	267
Equisetum	267

	PAGE.
Erechtites	113
ERICACEÆ	137
ERICINEÆ	138
Erigenia	92
Erigeron	125
Eriogonum	189
Eriocaulon	247
ERIOCAULONACEÆ	247
Eriophorum	250
Erodium	42
Erythronium	242
Erysimum	21
Euonymus	46
Eupatorium	115
Euphorbia	196
EUPHORBIACEÆ	195
Euphrasia	157
Evening Primrose	83
Evening Primrose Family	81
Everlasting	113, 114
Everlasting Pea	59
EXOGENS	1
Eyebright	157
Fagopyrum	193
Fagus	207
Fall Dandelion	132
False Asphodel	240
False Dragon-head	166
False Flax	24
False Gromwell	170
False Hellebore	241
False Indigo	57, 61
False Lettuce	134
False Loosestrife	83
False Mallow	39
False Mermaid	42
False Mitre-Wort	77
False Nettle	201
False Pennyroyal	163
False Pimpernel	156
False Solomon's Seal	241
False Spikenard	241
Fern Family	257
FERNS	252

	PAGE.		PAGE.
Fetid Horehound	167	Gentian	177
Fever-bush	193	Gentiana	177
Feverfew	127	GENTIANACEÆ	176
Fever-wort	97	Gentian Family	176
FICOIDEÆ	86	GERIANACEÆ	41
Figwort	155	Geranium	41
Figwort Family	151	Geranium Family	41
Filbert	207	Gerardia	156
FILICES	257	Germander	162
Fir	216	Geum	66
Fire-Pink	33	Giant-Hyssop	165
Fireweed	113	Gillenia	65
Five-Finger (Cinque-Foil)	67	Gilia	173
Flax	40	Ginseng	92
Flax Dodder	174	Ginseng Family	92
Flax Family	40	Glaux	147
Fleabane	125	Gleditschia	62
Floating-Heart	179	Glycyrrhiza	57
Flœrkea	42	Gnaphalium	113
Flower-de-Luce	235	Goat's Beard	135
Flowering Fern	265	Golden Aster	116
FLOWERING PLANTS	1	Golden Ragwort	115
FLOWERLESS PLANTS	252	Golden-Rod	117
Fool's Parsley	90	Golden Saxifrage	78
Forget-me-not	171	Gold-Thread	8
Fragaria	69	Goodyera	232
Franseria	111	Gooseberry	75
Frasera	177	Goosefoot	185
Fraxinus	181	Goosefoot Family	184
French Weed	24	Goose-Grass	99
Frog's-bit Family	226	Gourd Family	85
Frostweed	29	GRAMINEÆ	251
Fumaria	16	Grape	45
FUMARIACEÆ	15	Grass Family	251
Fumitory	15, 16	Grass of Parnassus	76
Fumitory Family	15	Grass-wrack	224
		Gratiola	156
Gaillardia	129	Great Angelica	90
Galeopsis	166	Green Ash	182
Galium	99	Green-brier	237
GAMOPETALOUS EXOGENS	95	Green Dragon	218
Garlic	243	Green Milkweed	181
Gaultheria	141	Green Violet	28
Gaura	83	Grindelia	126
Gaylussacia	139	Gromwell	170

298 INDEX.

	PAGE.		PAGE.
Ground Cherry	175	Heracleum	89
Ground Hemlock	216	Herb-Robert	41
Ground Ivy	165	Hesperis	22
Ground Laurel	141	Heteranthera	247
Ground-nut	59	Heuchera	78
Ground-Pine	270	Hibiscus	39
Groundsel	115	Hickory	202
Gutierrezia	126	Hieracium	132
Gymnocladus	62	Hippuris	80
GYMNOSPERMS	214	Hoary Pea	57
		Hoary Puccoon	170
		Hobble-bush	98
Habenaria	229	Hog Pea-nut	59
Hackberry	200	Hogweed	111
Halenia	177	Holly	145
HALORAGEÆ	79	Holly Family	145
HAMAMELACEÆ	79	Honey-Locust	62
Hamamelis	79	Honeysuckle	96
Harbinger-of-spring	92	Honeysuckle Family	95
Harebell	136	Honewort	91
Hart's-Tongue	261	Hop	201
Hawkweed	132	Hop-Hornbeam	207
Hawthorn	72	Hop-tree	43
Hazel-nut	207	Horehound	166
Heal-all	166	Hornbeam	207
Heather	142	Horned Pondweed	224
Heath Family	137	Hornwort	213
Hedeoma	164	Hornwort Family	213
Hedge Bindweed	173	Horse-Balm	164
Hedge-Hyssop	156	Horse-Mint	164
Hedge-Mustard	22	Horseradish	18
Hedge-Nettle	167	Horsetail	267
Hedysarum	61	Horsetail Family	267
Helenium	126	Horse-weed	125
Helianthemum	29	Hosackia	53
Helianthus	128	Hound's Tongue	169
Heliopsis	130	Houstonia	101
Hemerocallis	243	Huckleberry	139
Hemlock	216	Hudsonia	29
Hemlock-Parsley	90	Humulus	201
Hemlock-Spruce	216	Huntsman's Cup	13
Hemp	201	Hydrastis	9
Hemp Family	199	HYDROCHARIDACEÆ	226
Hemp-Nettle	166	Hydrocotyle	89
Henbane	176	HYDROPHYLLACEÆ	175
Hepatica	5		

	PAGE.		PAGE.
Hydrophyllum	172	Kalmia	142
Hyoscyamus	176	Kentucky Coffee-tree	62
HYPERICACEÆ	30	Knapweed	110
Hypericum	30	Knotgrass	189
Hypopitys	144	Knotweed	189
Hypoxys	236	Knotwort Family	183
Hyssop	164	Krigia	131
Hyssopus	164		
		LABIATÆ	160
Ice-Plant Family	86	Labrador Tea	142
Ilex	115	Lactuca	134
ILLECEBRACEÆ	183	Lady's Mantle	65
Ilysanthes	156	Lady's Slipper	234
Impatiens	43	Lady's Thumb	190
Indian Cucumber-root	240	Lady's Smock	19
Indian Hemp	179	Ladies' Tresses	231
Indian Mallow	39	Lake-Cress	18
Indian Physic	65	Lambkill	142
Indian Pipe	144	Lamb's Quarters	185
Indian Plantain	115	Lamium	167
Indian Tobacco	135	Lampsana	131
Indian Turnip	218	Laportea	200
Innocence	101	Lappa	110
Inula	116	Larch	216
IRIDACEÆ	235	Larix	216
Iris	235	Larkspur	9
Iris Family	235	Lathyrus	59
Iron-weed	114	LAURACEÆ	193
Iron-wood	207	Laurel Family	193
Isanthus	163	Laurestinus	97
Isoetes	271	Lead-Plant	57
Iva	115	Leadwort Family	144
		Leaf-Cup	131
Jeffersonia	12	Leather-leaf	142
Jerusalem Artichoke	129	Leatherwood	194
Jerusalem Oak	185	Lechea	29
Jewel-Weed	43	Ledum	142
Joe-Pye Weed	115	Leek	243
Juglans	202	LEGUMINOSÆ	49
JUGLANDACEÆ	202	Lemna	219
JUNCACEÆ	243	LEMNACEÆ	218
Juncus	244	LENTIBULARIACEÆ	149
June-berry	74	Leontodon	132
Juniper	216	Leonurus	167
Juniperus	216	Lepidium	24

	PAGE.		PAGE.
Lespedeza	61	Lycopodium	269
Lettuce	134	Lycopus	163
Leucanthemum	126	Lycopsis	169
Liatris	114	Lygodesmia	133
LIGULIFLORÆ	108	Lysimachia	147
Ligusticum	90	LYTHRACEÆ	84
LILIACEÆ	237	Lythrum	84
Lilium	242		
Lily	242	Madder Family	98
Lily Family	237	MAGNOLIACEÆ	9
Limnanthemum	179	Magnolia Family	9
LINACEÆ	40	Maidenhair	259
Linaria	154	Maianthemum	241
Linden Family	39	Mallow	38
Lindera	193	Mallow Family	38
Linnæa	96	Malva	38
Linum	40	MALVACEÆ	38
Liparis	233	Malvastrum	39
Liquorice	57	Mamillaria	86
Liriodendron	10	Mandrake	12
Listera	233	Maple	47
Lithospermum	170	Mare's-Tail	80
Live-for-ever	79	Marrubium	166
Liver-leaf	5	Marsh-Cress	18
Lizard's-tail	183	Marsh-Elder	115
Lobelia	135	Marsh-Marigold	8
LOBELIACEÆ	135	Marsh Rosemary	144
Lobelia Family	135	Marsh St. John's-wort	31
Locust-tree	54	Maruta	127
Lonicera	93	Matricaria	127
Loosestrife	84, 147	Matrimony-Vine	176
Loosestrife Family	84	May-Apple	12
Lophanthus	165	Mayflower	141
Lopseed	159	Mayweed	127
Lousewort	158	Meadow-Beauty	84
Lovage	90	Meadow-Parsnip	90
Lucerne	53	Meadow-Rue	5
Ludwigia	83	Meadow-Sweet	64
Lungwort	170	Medeola	240
Lupine	52	Medicago	53
Lupinus	52	Medick	53
Luzula	244	Melampyrum	158
Lychnis	33	Melastoma Family	84
Lycium	176	MELASTOMACEÆ	84
LYCOPODIACEÆ	269	Melilot	53
		Melilotus	53

INDEX.

	PAGE.
Melissa	166
MENISPERMACEÆ	10
Menispermum	10
Mentha	163
Menyanthes	178
Mermaid-weed	80
Mertensia	170
Mexican Tea	185
Mezereum Family	194
Microstylis	233
Milfoil	131
Milk-Vetch	54
Milkweed	180
Milkweed Family	179
Milkwort	48
Milkwort Family	48
Mimulus	156
Mint	163
Mint Family	160
Mitchella	100
Mitella	77
Mitrewort	77
Mocassin Flower	234
Mock Pennyroyal	164
Mollugo	86
Monarda	164
Moneses	143
Moneywort	147
Monkey-Flower	156
MONOCOTYLEDONS	217
Monolepis	186
Monotropa	144
MONOTROPEÆ	139
Montelia	188
Montia	88
Moonseed	10
Moonseed Family	10
Moonwort	266
Moosewood	194
Morus	200
Moss-Campion	33
Moss Pink	173
Mossy Stone-crop	78
Motherwort	167
Mountain Ash	74
Mountain Holly	145
Mountain Maple	47
Fountain Mint	164
Mouse-ear Chickweed	35
Mousetail	6
Mugwort	112
Mulberry	200
Mulgedium	134
Mullein	153
Musenium	91
Musk-Mallow	38
Musk Thistle	110
Mustard	22
Myosotis	171
Myosurus	6
Myrica	203
MYRICACEÆ	203
Myriophyllum	80
Nabalus	133
NIADACEÆ	221
Naias	224
Naked Broom-rape	150
Nasturtium	18
Neckweed	154
Negundo	48
Nemopanthes	145
Nepeta	165
Nesæa	85
Nettle	200
Nettle Family	198
Nettle-tree	200
New Jersey Tea	45
Nicandra	176
Nicotiana	176
Nightshade	175
Nightshade Family	174
Nine-Bark	64
Nipple-wort	131
Nuphar	13
Nymphæa	13
NYMPHÆACEÆ	12
Nyssa	94
Oak	205
Oak Family	204

	PAGE		PAGE
Œnothera	83	Pear	73
OLEACEÆ	181	Pearl-wort	36
Oleaster Family	194	Pedicularis	158
Olive Family	181	Pellæa	260
ONAGRACEÆ	81	Pellitory	201
Onion	243	Peltandra	218
Onoclea	264	Pennycress	24
Onopordon	110	Pennyroyal	164
Onosmodium	170	Penthorum	78
OPHIOGLOSSACEÆ	259	Pentstemon	155
Ophioglossum	267	Pepper Family	182
Opuntia	86	Pepperidge	94
Orache	186	Peppergrass	24
Orange-root	9	Peppermint	163
ORCHIDACEÆ	226	Pepper-root	19
Orchis	229	Petalostemon	58
Orchis Family	226	Petasites	121
OROBANCHACEÆ	150	Peucedanum	89
Orpine	78	PHANEROGAMS	1
Orpine Family	78	Phegopteris	261
Orthocarpus	157	Phlox	172
Osmorrhiza	91	Phryma	159
OSMUNDACEÆ	259	Physalis	175
Osmunda	265	Physostegia	166
Ostrich Fern	264	Phytolacca	183
Ostrya	207	PHYTOLACCACEÆ	183
Oswego Tea	164	Picea	215
OXALIDACEÆ	42	Pickerel-weed	247
Oxalis	42	Pickerel-weed Family	247
Ox-Eye	130	Pignut	203
Ox-eye Daisy	126	Pigweed	185
Oxytropis	56	Pilea	201
Painted-Cup	157	Pimpernel	138
Papaver	14	Pine	215
PAPAVERACEÆ	14	Pine-drops	144
Papaw	10	Pine Family	214
Parietaria	201	Pine-sap	144
Parnassia	76	Pinguicula	150
Paronychia	183	Pink Family	32
Parsley Family	87	Pinus	215
Parsnip	89	Pinweed	29
Partridge-berry	100	PIPERACEÆ	182
Pasque-Flower	3	Pipewort	247
Pastinaca	89	Pipewort Family	247
Pasture Sage-brush	113	Pipsissewa	144

INDEX. 303

	PAGE.
Pitcher-Plant Family	13
Plane-tree	201
Plane-tree Family	201
PLANTAGINACEÆ	148
Plantago	148
Plantain	148
Plantain Family	148
PLATANACEÆ	201
Platanus	201
Pleurisy-root	181
Plum	64
PLUMBAGINACEÆ	144
Plumeless Thistle	110
Podophyllum	12
Pogonia	233
Poison Elder	44
Poison Hemlock	92
Poison Ivy	44
Poison Oak	44
Poke Milkweed	180
Pokeweed	183
Pokeweed Family	183
Polanisia	25
POLEMONIACEÆ	172
Polemonium Family	172
Polygala	48
POLYGALACEÆ	48
POLYGONACEÆ	188
Polygonatum	242
Polygonum	180
Polymnia	131
POLYPETALOUS EXOGENS.	1
POLYPODIACEÆ	257
Polypodium	259
Polypody	259
POMEÆ	63
Pondweed	221
Pondweed Family	221
Pontederia	247
PONTEDERIACEÆ	247
Poplar	212
Poppy	14
Poppy Family	14
Populus	212
Portulaca	37

	PAGE.
PORTULACACEÆ	37
Potamogeton	221
Potentilla	67
Poterium	65
Prairie Clover	58
Prairie Dock	131
Prairie Rocket	21
Prenanthes	133
Prickly Ash	43
Prickly Pear	86
Primrose	146
Primrose Family	145
Primula	146
PRIMULACEÆ	145
Prince's Pine	144
Prosartes	241
Proserpinaca	80
Prunus	64
P oralea	57
Ptelea	43
Pteris	260
Pterospora	144
Puccoon	170
Pulse Family	49
Purslane	37
Purslane Family	37
Putty Root	234
Pycnanthemum	164
Pyrola	143
PYROLEÆ	139
Pyrus	73
Quercitron	206
Quercus	205
Quillwort	271
Radish	25
Ragweed	111
Ragwort	116
RANUNCULACEÆ	2
Ranunculus	6
Raphanus	25
Raspberry	70
Rattlesnake-Plantain	232
Rattlesnake-root	133
Rattlesnake-weed	132

304 INDEX.

	PAGE.		PAGE.
Rein-Orchis	229	Samphire	186
RHAMNACEÆ	45	Sandalwood Family	195
Rhamnus	45	Sand-Spurrey	36
Rhexia	84	Sandwort	34
Rhinanthus	158	Sanguinaria	14
Rhododendron	142	Sanicle	89
Rhus	44	Sanicula	89
Ribes	75	SANTALACEÆ	195
Rib-grass	148	Sea Blite	187
Rich-weed	201	SAPINDACEÆ	47
Robinia	54	Saponaria	32
Robin's-Plaintain	125	Sarracenia	13
Rock-Cress	20	SARRACENIACEÆ	13
Rocket	22	Saskatoon-berry	74
Rock-Rose	29	Sassafras	193
Rock-Rose Family	29	Satureia	164
Rosa	71	Saururus	183
Rose	71	Savory	164
ROSACEÆ	62	Saxifraga	77
Rose Family	62	SAXIFRAGACEÆ	75
Rose Bay	142	Saxifrage	77
Rose Mallow	39	Saxifrage Family	75
Rose-root	79	Scheuchzeria	225
Rosin-Plant	131	Schollera	247
RUBIACEÆ	98	Scirpus	250
Rubus	70	Scolopendrium	261
Rudbeckia	127	Scotch Thistle	110
Rue Family	43	Scouring Rush	267, 269
Rumex	192	Scrophularia	155
Rush	244	SCROPHULARIACEÆ	151
Rush Family	243	Scutellaria	166
RUTACEÆ	43	Sea Milkwort	147
		Sea Rocket	25
Sage-brush	113	Sedge Family	248
Sagina	36	Sedum	78
Sagittaria	225	Seed-box	84
St. John's-wort	30	Selaginella	271
St. John's-wort Family	30	SELAGINELLACEÆ	270
SALICACEÆ	209	Self-Heal	166
Salicornia	186	Seneca Snakeroot	49
Salix	209	Senecio	115
Salsify	135	Sensitive Fern	264
Salsola	187	Service-berry	74
SALVINIACEÆ	271	Shad-bush	74
Sambucus	97	Sheep-berry	97
Samolus	118	Sheep-Laurel	142

	PAGE		PAGE
Shepherdia	195	Spindle-tree	46
Shepherd's Purse	24	Spiny Clotbur	111
Sherardia	100	Spiræa	64
Shield's Fern	262	Spiranthes	231
Shin-leaf	143	Spleenwort	260
Shrubby Trefoil	43	Spotted Cowbane	91
Sickle-pod	20	Spring-Beauty	37
Sicyos	85	Spring Cress	19
Side-saddle Flower	13	Spruce	215
Silene	33	Spurge	196
Silphium	131	Spurge Family	195
Silver-weed	69	Spurred Gentian	177
Sisymbrium	22	Spurrey	37
Sisyrinchium	236	Squaw-Huckleberry	141
Sium	91	Squaw-root	150
Skullcap	166	Squaw-weed	115
Skunk Cabbage	218	Squirrel-Corn	15
Smart-weed	190, 191	Stachys	167
SMILACEÆ	237	Staff-tree	46
Smilacina	241	Staff-tree Family	46
Smilax	237	Staphylea	47
Smilax Family	237	Star-Cucumber	85
Sneeze-weed	126	Star-Flower	146
Sneeze-wort	131	Star-Grass	236
Snowberry	96	Star-Thistle	110
Soapberry Family	47	Starwort	34, 121
Soapwort	32	Statice	144
SOLANACEÆ	174	Steironema	147
Solanum	175	Stellaria	34
Solea	28	Stickseed	169
Solidago	117	Stitchwort	34
Solomon's Seal	242	Stone-crop	78
Sonchus	134	Stone-root	164
Sorrel	192	Stork's-bill	42
Sour-gum Tree	94	Stramonium	176
Sow Thistle	134	Strawberry	69
Sparganium	220	Strawberry Blite	185
Spearmint	163	Streptopus	241
Spearwort	6	Strophostyles	59
Specularia	137	Struthiopteris	264
Speedwell	153	Suæda	187
Spergula	37	Succory	132
Spergularia	36	Sugar-berry	200
Spice-bush	193	Sumach	41
Spikenard	92	Summer Savory	164

	PAGE
Sundew	30
Sundew Family	30
Sunflower	128
Swamp Dock	192
Swamp Loosestrife	85
Sweet Brier	72
Sweet Cicely	91
Sweet Clover	53
Sweet Coltsfoot	121
Sweet Fern	204
Sweet Flag	218
Sweet-Gale	203
Sweet-Gale Family	203
Sycamore	201
Symphoricarpus	96
Symphytum	169
Symplocarpus	218
Tamarac	216
Tanacetum	111
Tansy	111
Tansy Mustard	22
Tape-Grass	226
Taraxacum	134
Tare	58
TAXINEÆ	215
Taxus	216
Tea-berry	141
Tear-thumb	191
Teasel	102
Teasel Family	101
Tephrosia	57
Teucrium	162
Thalictrum	5
Thaspium	90
Thermopsis	61
Thistle	109
Thorn	73
Thorn-Apple	176
Thoroughwax	91
Thoroughwort	115
Thlaspi	24
Three-seeded Mercury	197
Thuja	216
THYMELEACEÆ	194
Tiarella	97
Tick-Trefoil	59
Tickseed	129
Tilia	39
TILIACEÆ	39
Toad-Flax	154
Tobacco	176
Tofieldia	240
Toothache-tree	43
Toothwort	19
Touch-me-not	43
Tower Mustard	20
Trailing Arbutus	141
Tragopogon	135
Treacle-Mustard	21
Trefoil	52
Trientalis	146
Trifolium	52
Triglochin	225
Trillium	240
Triosteum	97
Troximon	133
Trumpet-weed	115
Tsuga	216
TUBULIFLORÆ	102
Tulip-tree	10
Turtle-head	155
Tussilago	121
Twablade	233
Twin-flower	96
Twin-leaf	12
Twisted Stalk	241
Typha	220
TYPHACEÆ	219
ULMACEÆ	198
Ulmus	199
UMBELLIFERÆ	87
Urtica	200
URTICACEÆ	198
URTICEÆ	199
Utricularia	149
Uvularia	240
VACCINEÆ	137
Vaccinium	139
Valerian	101

	PAGE.		PAGE.
Valeriana	101	Water-Milfoil Family	79
VALERIANACEÆ	101	Water-Parsnip	91
Valerian Family	101	Water-Pennywort	89
Vallisneria	226	Water-Pepper	191
Velvet-leaf	39	Water-Pimpernel	148
Venus's Looking-glass	137	Water Plantain	225
Veratrum	241	Water-Plantain Family	224
Verbascum	153	Water-Purslane	83
Verbena	159	Water-Shield	12
VERBENACEÆ	158	Water Star-Grass	247
Vernonia	114	Water-Starwort	80
Veronica	153	Water-weed	226
Vervain	159	Water-Willow	159
Vervain Family	158	Wax-Myrtle	203
Vesicaria	19	Wax-work	46
Vetch	58	White Lettuce	133
Viburnum	97	White Snakeroot	115
Vicia	58	White-weed	126
Vine Family	44	Whitewood	39
Viola	26	Whitlow Grass	23
VIOLACEÆ	26	Whitlow-wort	183
Violet	26	Wild Allspice	193
Violet Family	26	Wild Balsam-Apple	85
Viper's Bugloss	168	Wild Bean	59
Virginia Creeper	45	Wild Bergamot	165
Virgin's Bower	3	Wild Chamomile	127
VITACEÆ	44	Wild Comfrey	169
Vitis	45	Wild Elder	92
		Wild Ginger	182
Wake-Robin	240	Wild Indigo	61
Waldsteinia	66	Wild Liquorice	99
Walking-leaf	261	Wild Radish	25
Walnut	202	Wild Sarsaparilla	92
Walnut Family	202	Willow	209
Water-Arum	218	Willow Family	209
Water-Beech	207	Willow-herb	81
Water-Cress	18	Wind Flower	4
Water-Hemlock	91	Winterberry	145
Water-Horehound	163	Winter Cress	19
Waterleaf	172	Wintergreen	141, 143
Waterleaf Family	171	Witch-Hazel	79
Water-Lily	13	Witch-Hazel Family	79
Water-Lily Family	12	Withe-rod	97
Water-Marigold	130	Wolf-berry	96
Water-Milfoil	80	Wolffia	219

INDEX.

	PAGE.		PAGE.
Wood-Betony	158	Yam	236.
Woodbine	96	Yam Family	236
Wood-Fern	262	Yarrow	131
Wood-Nettle	200	Yellow Adder's-tongue	242
Wood-Rush	244	Yellow-eyed Grass	247
Wood-Sage	162	Yellow-eyed Grass Family	247
Woodsia	265	Yellow Pond-Lily	12
Wood-Sorrel	42	Yellow Puccoon	9
Wood-Sorrel Family	42	Yellow-Rattle	158
Woodwardia	261	Yellow Rocket	19
Worm-seed Mustard	21	Yew	216
Wormwood	112		
Xanthium	111	Zannichellia	224
Xanthoxylum	43	Zizia	90
XYRIDACEÆ	247	Zostera	224
Xyris	247	Zygadenus	240

www.ingramcontent.com/pod-product-compliance
Lightning Source LLC
Chambersburg PA
CBHW030728230426
43667CB00007B/638